MARS' MAGNETISM AND ITS INTERACTION WITH THE SOLAR WIND

Space Sciences Series of ISSI

Volume 18

The International Space Science Institute is organized as a foundation under Swiss law. It is funded through recurrent contributions from the European Space Agency, the Swiss Confederation, the Swiss National Science Foundation, and the University of Bern. For more information, see the homepage at http://www.issi.unibe.ch/.

The titles published in this series are listed at the end of this volume.

MARS' MAGNETISM AND ITS INTERACTION WITH THE SOLAR WIND

Edited by

DANIEL WINTERHALTER
Jet Propulsion Laboratory, California Institute of Technology, U.S.A.

MARIO ACUÑA
NASA/Goddard Space Flight Center, U.S.A.

ALEXANDER ZAKHAROV
Space Research Institute, Moscow, Russia

Reprinted from *Space Science Reviews*, Volume 111, Nos. 1–2, 2004

KLUWER ACADEMIC PUBLISHERS
DORDRECHT / BOSTON / LONDON

A.C.I.P. Catalogue record for this book is available from the Library of Congress

ISBN 978-1-4020-2048-3

Published by Kluwer Academic Publishers,
P.O. Box 990, 3300 AZ Dordrecht, The Netherlands

Sold and distributed in North, Central and South America
by Kluwer Academic Publishers,
101 Philip Drive, Norwell, MA 02061, U.S.A.

In all other countries, sold and distributed
by Kluwer Academic Publishers,
P.O. Box 322, 3300 AH Dordrecht, The Netherlands

Printed on acid-free paper

TABLE OF CONTENTS

Introduction and Foreword

The present volume stems from a workshop on the "Interaction of the Solar Wind with Mars", which was held at the International Space Science Institute (ISSI) October 22–26, 2001, in Bern, Switzerland. At the workshop, 36 scientists from 8 countries, most of them team members of either NASA's Mars Global Surveyor mission or the then – USSR Phobos mission, participated in an exciting and engaging dialog concerning the particle and field environment of Mars, and the origin of Mars' newly discovered remnant magnetic field.

The workshop concentrated on the Solar Wind/Mars interaction, with the goal to integrate the results from the Phobos 2 mission with those from Mars Global Surveyor (MGS). From both missions much information about the Mars/solar wind interaction had been disseminated, and each had produced descriptions and vocabularies mirroring both the capabilities and the limitations of the respective mission. In the workshop, complementary information from the two missions was emphasized. As expected, the integration of the results led to a much-improved understanding of the interaction process.

The interaction of the solar wind with Mars is, to first order, determined by whether or not the planet has an intrinsic magnetic field. From the very beginning of Mars exploration, this question has been hotly debated. When Mariner 4 took the first measurements near Mars in 1965, a bow shock was observed by the magnetometer. Its position was interpreted by some to indicate the presence of a small planetary field, while others argued against it.

The next seven missions to Mars space, including the Viking mission, added only marginally to the understanding, primarily do to the lack of appropriate instrumentation on these spacecraft, and unsuitable trajectories. Then the Phobos mission was launched in 1988, carrying a nearly complete suite of field and particle instruments on board. Upon arrival in 1989, Phobos 2 entered a highly elliptical, 77 h period orbit, with periapsis at 850 km altitude. After four orbits the spacecraft was maneuvered into an 8 h period circular orbit of 6203 km altitude (2.8 Rm). Unfortunately, contact with Phobos 2 was lost then (Phobos 1 was lost during cruise), ending much too prematurely an exiting and ambitious mission. Nevertheless, during its brief time in orbit, Phobos 2 measurements showed that a complex and probably unique plasma environment existed around Mars. However, it still could not be ascertained whether or not a planetary magnetic field was present.

The Mars Global Surveyor (MGS) mission finally answered this question. After the loss of Mars Observer during its preparation for the Mars orbit insertion maneuver (1993), MGS was build and launched on November 7, 1996. The primary objective of the MGS mission was, and still is, to conduct a global mapping mission of Mars by performing an extended orbital study of the planet's surface, atmosphere, and gravitational and magnetic fields.

For field-and-particle instrumentation, MGS carries only two magnetometers and an Electron Reflectometer (MAG/ER), and thus is blind to ion, plasma wave, and energetic particle events. However, it did gather the magnetic and electron data in great detail and with very good statistics from an optimal orbit. Phobos 2, on the other hand, had many field and particle instruments, but took measurements at large distances and for a relatively short period of time.

MGS' planned mapping orbit was a low altitude (about 350 km), near-circular, near-polar orbit that is Sun-synchronous with the day-side equatorial crossing at 2:00 PM local mean solar time. To deliver the spacecraft into this orbit, aerobraking was used over a period of about 18 month. This was longer than originally planned because one of the solar panels was damaged during launch. However, the extended aerobraking turned out to be a great bonus to the MAG/ER investigation, because it carried the spacecraft down to 100 km altitudes on each orbit, and often lower, well below the ionospheric peak. It was thus possible for the magnetometer to peek below the hiding cover of the ionosphere and check directly for any planetary magnetic fields.

Due to these advantageous orbits, MGS could now show conclusively that no dipole-like, dynamo generated planetary field exist on Mars at present. However, large remnant magnetic fields were detected in localized regions on the planet, indicating that a dynamo field did exist in the past, probably until some 4 billion years ago [see Connerney et al., this issue, and reference therein]. The facts that Mars could sustain a fully developed dynamo field during its first few hundred million years, and that the dynamo stopped as early as 4 billion years ago, have strong implications as to how Mars formed from pre-planetary matter, and how it has evolved thermally since.

The absence of a large-scale planetary dipole field sets stringent conditions on the present solar wind/Mars interaction process - in fact it now defines how we look at the near Mars plasma environment. We knew before that the interaction was not lunar-like, because Mars has an ionosphere. But compared to Venus' dense ionosphere it is translucent, too thin to routinely stand-off the solar wind. We now know that the interaction is not magnetospheric, i.e., Earth-like, either.

The remnant magnetic field, while larger than any seen on Earth, does not appear to significantly influence the Mars solar wind interaction process on a global scale. Only locally, the remnant fields add a dynamic character to the interaction as they rotate into view of the solar wind every Mars rotation period, adding periodically magnetic pressure to help standoff the solar wind.

But on the global scale the interaction is between the exosphere of Mars, consisting of heavy ions of planetary origin, and the solar wind protons. The interaction produces a thin, permanent boundary between the two populations. This was, for the first time, brought into sharp focus at the workshop, where it was realized that the many boundaries and pauses that were observed between the bow shock and the ionosphere by different Phobos and MGS instruments are actually one and the same. In order to reduce the confusion caused by different terminology, the workshop attendees adapted, more or less arbitrarily, the name "Magnetic Pile-Up Boundary" (MPB) for the boundary (see Nagy et al., this issue). The analog for the MPB at Earth is the magnetopause, the similarity being that both boundaries divide the regions dominated by the planet from that dominated by the solar wind. But it should be stressed that the interaction process is entirely different at the two kinds of boundaries. Work in progress suggests that MPB-like signatures are also present in Venus and comet data, suggesting that the interaction mechanism detailed in Nagy et al. (this issue) may be one generally at work around unmagnetized or weakly magnetized bodies with atmospheres.

The present issue is divided into two parts. The first part consists of three detailed review papers, the first of which discusses the remnant field observations and their implications. The second paper concerns itself with the observations and theories of the major boundaries and regions between the upper atmosphere and the bow shock, particularly the ionosphere, the Magnetic Pile-Up Region (MPR), the MPB, and the magnetosheath. The third paper discusses the bow shock structure and the prominent upstream waves.

Part 2 consists of contributed papers of shorter length, which discuss certain details and analyses that may not have been appropriate for a review paper. In addition, the contributed papers may accommodate points of view more controversial in nature.

The organizer and all the participants of the workshop are grateful to have received substantial support from the International Space Science Institute to conduct the workshop and to prepare this volume. All the staff and colleagues at ISSI were superbly helpful and engaging, and particularly Professor Johannes Geiss, the founder and then-executive director of ISSI, is profoundly thanked. ISSI and its people provided the environment that allowed the workshop participants to solve a long outstanding problem of Mars exploration.

Substantial support of equal measure was also received from the Mars Global Surveyor Project, a National Aeronautics and Space Administration mission. This effort could not have been done without the encouragement and support by Thomas E. Thorpe, the Science Office Manager and now Project manager of MGS, who understands the need for scientific discourse during exploration missions such as MGS. Many thanks to him.

Further, the editors would like to give thanks to the reviewers of the papers in this issue, particularly Dave Stevenson (Caltech, Pasadena), Oleg Vaisberg (Space Research Institute, Moscow), and three others.

For the co-editors Mario Acuña and Alexander Zakharov,

Daniel Winterhalter
Jet Propulsion Laboratory
Pasadena, California

Part I: Review Articles

MARS CRUSTAL MAGNETISM

J.E.P. CONNERNEY[1], M.H. ACUÑA[1], N.F. NESS[2], T. SPOHN[3] and G. SCHUBERT[4]
[1]*NASA Goddard Space Flight Center, Greenbelt, MD, U.S.A.*
[2]*University of Delaware, Wilmington, DE, U.S.A.*
[3]*Westfalische Wilhelms-Universitat, Munster, Germany*
[4]*University of California, Los Angeles, CA, U.S.A.*

Accepted in final form 26 July 2003

Abstract. Mars lacks a detectable magnetic field of global scale, but boasts a rich spectrum of magnetic fields at smaller spatial scales attributed to the spatial variation of remanent magnetism in the crust. On average the Mars crust is 10 times more intensely magnetized than that of the Earth. It appears likely that the Mars crust acquired its remanence in the first few hundred million years of evolution when an active dynamo sustained an intense global field. An early dynamo era, ending in the Noachian, or earliest period of Mars chronology, would likely be driven by thermal convection in an early, hot, fluid core. If crustal remanence was acquired later in Mars history, a dynamo driven by chemical convection associated with the solidification of an inner core is likely. Thermal evolution models cannot yet distinguish between these two possibilities. The magnetic record contains a wealth of information on the thermal evolution of Mars and the Mars dynamo, but we have just begun to decipher its message.

1. Introduction

The Mars Global Surveyor magnetic field investigation provided the first unambiguous detection of the magnetic field associated with Mars (Acuña *et al.*, 1998). Mars presently has no significant global-scale magnetic field of internal origin, as would be expected of a source in the deep interior, e.g., an active dynamo. Remarkably, the magnetic field of Mars is crustal in origin, resulting from the variation in remanent magnetization of the crust. The most intense magnetic fields appear in association with the heavily-cratered southern highlands (Acuña *et al.*, 1999), although extended regions surrounding the giant impact basins Argyre and Hellas appear weakly magnetized or non-magnetic. Likewise, the magnetic field of crustal origin over much of the northern lowlands is relatively weak, but significant sources have been mapped at high northern latitudes (Acuña *et al.*, 1999) along the 330 deg West longitude meridian.

A similar hemispheric dichotomy is reflected in Mars geology. Terrain north of the dichotomy boundary is apparently lightly cratered and presumably younger, topographically smooth and flat, and low in elevation (Smith *et al.*, 1999). Terrain south of the boundary is heavily cratered and presumably older, more rough at all spatial scales and at high elevation. Relative ages across the dichotomy boundary have become less certain with the discovery of faint topographical features called 'quasi-circular depressions' in both hemispheres (Frey *et al.*, 1999). If these are counted as vestigial remains of ancient craters, the underlying crust in the northern lowlands might be a great deal older than suggested by surface appearance. The northern lowlands are thought to have experienced sedimentary or volcanic resurfacing in the Late Hesperian and Amazonian ages. The origin of the crustal dichotomy remains a mystery, one that is fundamental to Mars evolution. It has been attributed to a large impact (Wilhelms and Squyres, 1984) or impacts (Frey and Schultz, 1988) and alternatively to tectonic processes (Wise *et al.*, 1979; Sleep, 1994). The similar apparent dichotomy in crustal magnetization gives some hope that the magnetic record, correctly deciphered, can help solve one of the most important outstanding questions of Mars evolution. How did the crust form?

1.1. Prior Observations

There have been in excess of 30 attempted missions to planet Mars, beginning with an unnamed Soviet probe launched in 1960 (Table I). Only 1 in 4 achieved any measure of success, a grim reminder of the difficulty of planetary exploration, particularly in the early days of space exploration. In the last decade, the demonstrated rate of success has improved to about 50%.

The first successful mission to Mars was that of the USA spacecraft Mariner 4, which passed within 4 Mars radii on July 15, 1965. The magnetic field was indistinguishable from that of the undisturbed interplanetary environment, with no indication of a Mars magnetosphere, magnetopause or bow shock (Smith *et al.*, 1965). Thus Smith *et al.* set an upper limit on a present-day magnetic dipole moment of 3×10^{-4} Me (Me is the Earth's dipole moment, 8×10^{25} G cm^3), equivalent to an equatorial field of less than 100 nT.

For the next 32 years, despite numerous opportunities, no space probe instrumented to measure magnetic fields would pass close enough to the planet's surface to establish the presence of an intrinsic magnetic field. Soviet spacecraft were all instrumented to measure magnetic fields, and while Mars 2, 3 and 5, and Phobos 2 returned useful magnetic field data, they did not pass close enough to unambiguously detect an intrinsic field. The first US Mars missions were instrumented with magnetic sensors, but like the early Soviet missions, they either failed or remained too distant. Beginning with the launch of Mariner 6 in 1969, and excepting Mars Observer (MO) and its replacement, Mars Global Surveyor (MGS), spacecraft and landers sent by the USA to Mars did not carry magnetometers. The very successful Viking orbiters, with periapses at an altitude of 300 km, would surely have detected the crustal magnetic field in 1976 had they carried magnetometers.

The lack of a definitive measurement in the thirty years leading up to MGS, and the great interest in the question of Mars magnetism, spawned an enormous body of literature and much debate on the subject. By all accounts, the intrinsic magnetic field of Mars had to be quite small, so as to not produce a well developed and easily identified magnetosphere like that of Earth. But was a modest intrinsic

TABLE I
Chronology of Mars mission attempts[a]

Attempt	Launch date	Name	Country	Magnetic field investi-gation?	Mission outcome
1	October 14 1960	None	USSR	Y	Failed to reach Earth orbit
2	October 24 1962	None	USSR	Y	Failed to leave Earth Orbit
3	November 1 1962	Mars 1	USSR	Y	Flyby June 19, 1963; telemetry failed
4	November 4 1962	None	USSR	Y	Failed to leave Earth Orbit
5	November 5 1964	Mariner 3	USA	Y	Shroud separation failure
6	November 28 1964	Mariner 4	USA	Y	Successful flyby; imagery and magnetic measurements (distant)
7	November 30 1964	Zond 2	USSR	Y	Contact lost before Mars encounter
8	July 18 1965	Zond 3	USSR	Y	No Mars encounter
9	February 24 1969	Mariner 6	USA		Successful flyby July 31
10	March 27 1969	Mariner 7	USA		Successful flyby August 5
11	March 27 1969	None	USSR	Y	Launch Failure
12	April 14 1969	None	USSR	Y	Launch failure
13	May 8 1971	Mariner 8	USA		Launch failure
14	May 10 1971	Kosmos 419	USSR	?	Failed to leave Earth orbit
15	May 19 1971	Mars 2	USSR	Y	Orbited Mars; descent module crashed
16	May 28 1971	Mars 3	USSR	Y	Orbited Mars; descent module landed, telemetry failure
17	May 30 1971	Mariner 9	USA		MOI November 13; successful mission
18	July 21 1973	Mars 4	USSR	Y	Failed to orbit Mars
19	July 25 1973	Mars 5	USSR	Y	Orbited Mars; partially successful mission
20	August 5 1973	Mars 6	USSR	Y	Flyby; descent module landed, little data
21	August 9 1973	Mars 7	USSR	Y	Flyby; descent module missed Mars
22	August 20 1975	Viking 1	USA		MOI June 19, 1976; landed July 20
23	September 5 1975	Viking 2	USA		MOI August 7 1976; landed September 3
24	July 7 1988	Phobos 1	USSR	Y	Contact lost before Mars encounter
25	July 12 1988	Phobos 2	USSR	Y	MOI January 29, 1989; lost telemetry March 27
26	September 25 1992	Mars Observer	USA	Y	Failed to achieve Mars orbit
27	November 7 1996	Mars Global Surveyor	USA	Y	MOI September 12, 1997; successful mission (currently in extended mission)
28	December 4 1996	Mars Pathfinder	USA		Landed July 4, 1997; successful mission
29	July 4 1998	NOZOMI	JAPAN	Y	Mission replan for 2004 arrival at Mars
30	December 11 1998	Mars Climate Orbiter	USA		Failed to achieve Mars orbit
31	January 3 1999	Mars Polar Lander	USA		No contact subsequent to landing attempt; No contact with 2 microprobes deployed
32	April 7 2001	Mars Odyssey	USA		MOI October 24, 2001; successful mission

[a]Entries through 1988 from Snyder and Moroz (1992).

field – almost exclusively assumed to be a planetary dipole – necessary to explain observations at Mars? The observations considered were largely those from the Soviet Mars 2, 3, and 5 probes, reviewed by Russell (1979) and Ness (1979), but also included ionospheric electron density profiles from numerous radio occultations and the Viking landers (reviewed by Slavin and Holzer, 1982). Estimates of a putative Mars dipole ranged from 0.8 to 2.55×10^{22} G-cm^3 (equivalent equatorial surface field of 20 to 65 nT), but the evidence was weak at best and widely criticized (e.g., Russell, 1978a, b; Ness, 1979).

The more recent Phobos 2 mission provided the closest observations (to 800 km altitude) and numerous more distant tail passages but provided 'no conclusive evidence for a magnetic field of internal origin' according to Riedler *et al.* (1989). However, others saw evidence of a planetary field in the same data (Dolginov and Zhuzgov, 1991; Slavin *et al.*, 1991; Verigin *et al.*, 1991; Mohlmann *et al.*, 1991) and the controversy did not end (e.g., Russell *et al.*, 1995; Gringauz *et al.*, 1993) until MGS entered Mars orbit in September 1997.

1.2. Mars Global Surveyor Observations

The Mars Global Surveyor spacecraft and Mission Plan were designed to recover a subset of the science objectives of the Mars Observer Mission, which ended with Mars Observer's unsuccessful orbit insertion maneuver. The Mars Observer Mission was designed to provide a direct entry into a high inclination mapping orbit at 400 km altitude. Mars Global Surveyor, in contrast, was designed to enter into a highly elliptical initial orbit, and achieve a circular (polar) mapping orbit at 400 km altitude by aerobraking in the Mars atmosphere (Albee *et al.*, 1998). This approach uses less propellant and makes more efficient use of launch mass.

Aerobraking works by using the planet's atmosphere to slow the spacecraft, reducing apoapsis by repeated passages through the atmosphere. The desired drag is obtained by adjusting the spacecraft periapsis with trim maneuvers to raise or lower the altitude of the spacecraft at closest approach, targeting the appropriate atmospheric density. As a result of a spacecraft anomaly experienced early in the mission, MGS used many more drag passes (891 in all) than was originally intended to slow the spacecraft. The magnetometer investigation was an unintended beneficiary of the revised MGS mission plan, which allowed a far greater sampling of the magnetic field well below the nominal 400 km mapping phase altitude.

This rather unique set of circumstances resulted in a mission that can best be described in three distinct mission modes, separated by orbital characteristics (Table II). The aerobraking phases (AB1 and AB2) occurred with the spacecraft in elliptical orbit about Mars and these mission phases provided in-situ observations along most of the orbit, extending down to altitudes of as little as 100 km. During each aerobraking pass MGS acquired measurements of the vector magnetic field along the orbit track at varying altitude above the surface. These passes are distrib-

Table II.
Mars global surveyor mission phases

Phase	Begin date Year mm day (DOY)	End date Year mm day (DOY)	Mapping cycles
	Pre-mapping		
ab1	1997 09 13 (256)	1998 03 25 (084)	
spo1	1998 03 26 (085)	1998 05 27 (147)	
spo2	1998 05 28 (148)	1998 09 22 (265)	
ab2	1998 09 23 (266)	1998 03 08 (067)	
	Mapping		
map1	1999 03 09 (068)	1999 06 01 (152)	1–3
map2	1999 06 02 (153)	1999 08 24 (236)	4–6
map3	1999 08 25 (237)	1999 11 16 (320)	7–9
map4	1999 11 17 (321)	2000 02 08 (039)	10–12
map5	2000 02 09 (040)	2000 05 02 (123)	13–15
map6	2000 05 03 (124)	2000 07 25 (207)	16–18
map7	2000 07 26 (208)	2000 10 17 (291)	19–21
map8	2000 10 18 (292)	2001 01 31 (031)	22–24 (25)[a]
	Extended Mapping		
ext1	2001 02 01 (032)	2001 04 03 (093)	(25)[a] 26–27
ext2	2002 04 04 (094)	2001 06 26 (177)	28–30
ext3	2001 06 27 (178)	2001 09 18 (261)	31–33
ext4	2001 09 19 (262)	2001 12 11 (345)	34–36
ext5	2001 12 12 (346)	2002 03 05 (064)	37–39
ext6	2002 03 06 (065)	2002 05 28 (148)	40–42
ext7	2002 05 29 (149)	2002 08 20 (232)	43–45
ext8	2002 08 21 (233)	2002 11 12 (316)	46–48
ext9	2002 11 13 (317)	2003 02 04 (035)	49–51
ext10	2003 02 05 (036)	2003 04 29 (119)	52–54
ext11	2003 04 30 (120)	2003 07 22 (203)	55–57

[a]map8 was extended through the end of primary mission to include part (3/4) of the next mapping cycle (25)

uted more or less randomly over the surface of Mars and provide a sparse sampling of the field nearest the surface.

The pre-mapping mission phase included 'Science Phasing Orbits' (SPO1 and SPO2), during aerobraking operations, in which the spacecraft was 'parked' at a fixed periapsis awaiting the next mission phase. The SPO orbits provided relatively dense sampling of the vector field above the north pole of Mars at altitudes between 170 and 200 km. Mapping phase observations began in March, 1999 and continued for a full Mars year (687 days), through the end of January 2001, the end of the MGS Primary Mission. Since then the mission has operated, and continues to operate, in extended mission, nearing the end of the second Mars year in mapping operations. The mapping phases provided abundantly oversampled coverage of the magnetic field at a nominal altitude of 400 km (370 km–438 km) and fixed (2 am –2 pm) local time.

The mapping phases, and extended mapping phases, listed in Table II nominally consist of 3 mapping cycles each (with the exception of map8 and ext1). A mapping cycle is 28 days long and provides coverage of the entire planet by repeated application of a 7 day near-repeat cycle (Albee *et al.*, 2001). The 7 day cycle consists of 88 orbits, each crossing the equator 28.6 deg westward of the previous orbit. At the end of a 7 day repeat cycle, the spacecraft ground track is displaced eastward by a nominal 242 km from the initial ground track. Subsequent cycles 'fill in the gaps', providing under ideal conditions ground tracks crossing the equator every 3 km (1 deg longitude = 59.2 km) at the end of 1 Mars year.

From the perspective of mapping a potential field, this is a particularly generous spatial coverage and a very desirable pattern of mapping. The density of sampling greatly exceeds that necessary to follow variations in the magnetic field of crustal origin with position (at orbit altitude). The spatial resolution needed to map a potential field scales with the altitude of observation; for example, a survey at 400 km altitude measures variations of comparable scale (e.g., Blakely, 1995) and need not be mapped at 3 km resolution. Just how densely the field needs to be sampled depends also on the quality or signal-to-noise (S/N) of the measurement, higher quality measurements allowing for increased spatial resolution. So there is ample opportunity to average the field over many nearby samples to achieve improved signal fidelity, reducing 'noise' from external sources. The mapping scheme is well suited for this, providing widely separated orbit tracks and a relatively long interval of time between nearby samples. For example, it takes 7 days to return to within 242 km of a particular location on the equator. This greatly exceeds the correlation length of variations ('noise') due to external sources, so to a first approximation physically adjacent (in longitude) samples may be treated as if they are statistically independent.

1.2.1. *MGS Aerobraking Observations*

An early MGS aerobraking pass is shown in Figure 1, where the spacecraft trajectory and the vector magnetic field both appear in projection on the sky as viewed from the Sun. The magnetic field is illustrated by a vector originating at the spacecraft position at 3-s intervals along the trajectory. This example, from the 6th orbit of MGS on day 264 of 1997, was the first pass over the crust of Mars to unambiguously establish the nature of the magnetic field of Mars. From this image one can see that the magnetic field of Mars is not global in scale, as it would necessarily be if it were generated in the deep interior, for example by a dynamo. Instead, for this and many other examples one finds that the spatial scale of the observed field is comparable to the altitude of observation. This is what one expects of mag-

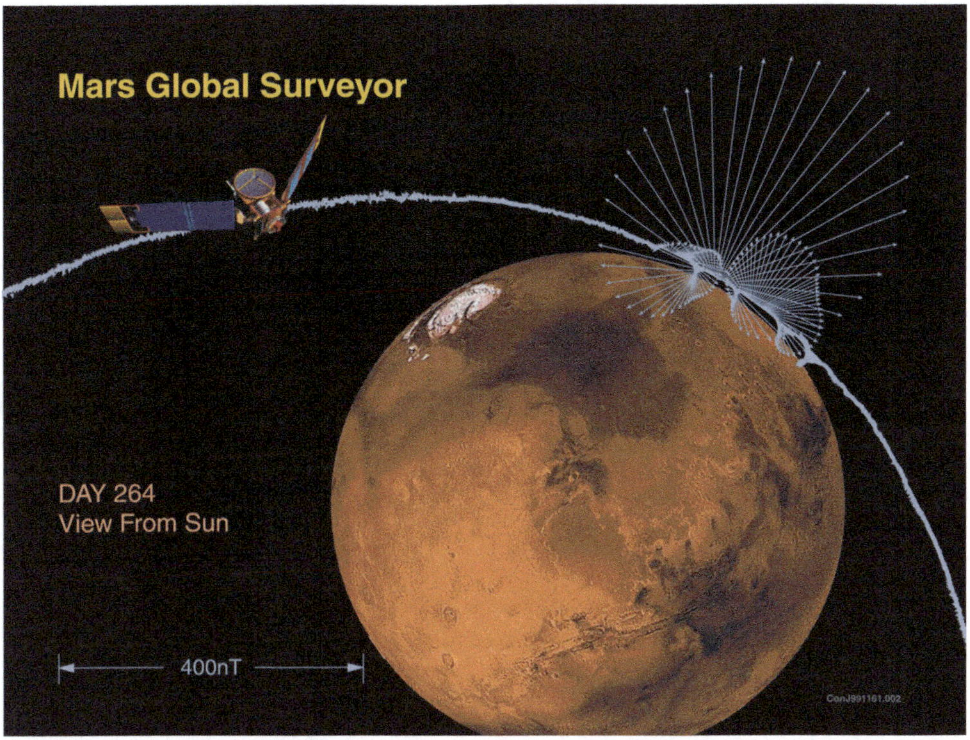

Figure 1. Projection of the MGS spacecraft trajectory and observed magnetic field onto a plane perpendicular to the Mars–Sun line and the Mars orbit plane for periapsis pass 6, 1997 day 264. The magnetic field is illustrated at 3 s intervals by a scaled vector projection of **B** originating from the spacecraft position at such times.

netic sources located near the surface of the planet, i.e., in the crust of Mars. The magnetic field of Mars is dominated by intensely magnetized sources distributed, non-randomly, in the Mars crust (Acuña *et al.*, 1998, 1999).

The MGS MAG-ER investigation obtained nearly continuous data coverage throughout most of the pre-mapping phase of the mission, with particular emphasis on acquisition of data near periapsis, as this presented a unique opportunity to sample the field at low altitudes (Albee *et al.*, 2001). Magnetometer data were acquired during 127 of 197 periapsis passes occurring in ab1; 60 of 130 passes in spo1; 211 of 244 passes in spo2; and 896 of 1111 passes in ab2. During each pass the field was sampled at low altitudes along a limited ground track centered on the latitude and longitude of periapsis and extending approximately ± 25 deg north and south of closest approach. The latitude of periapsis, about 30 deg N at the beginning of ab1, progressed slowly northward toward and over the pole through spo2. During ab2, the latitude of periapsis progressed southward, reaching a maximum of 87 deg S by the end of aerobraking. These occurred more or less randomly in longitude, so by the end of pre-mapping almost the entire surface

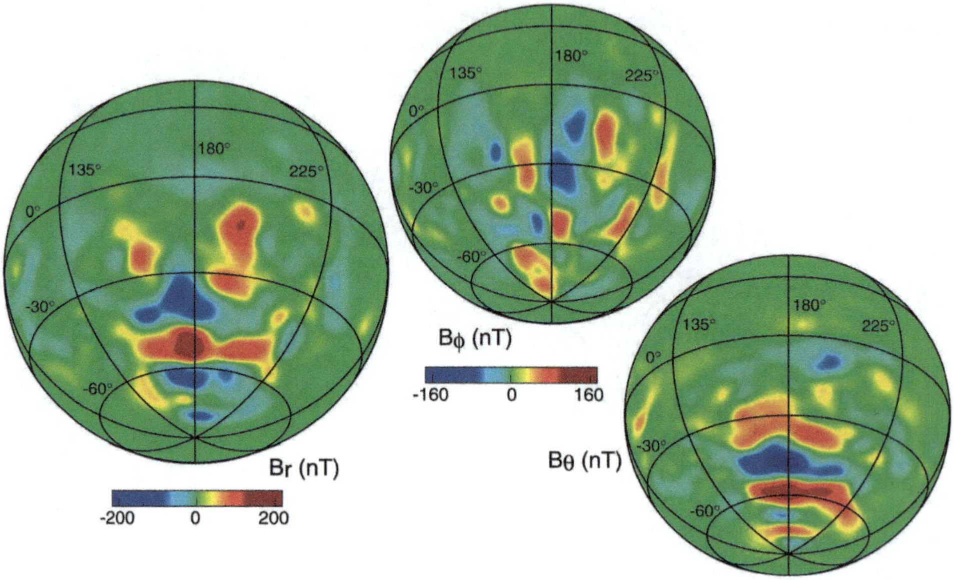

Figure 2. Orthographic projections of the three components of the magnetic field (B_r, B_θ, B_ϕ) at a nominal 400 km mapping orbit altitude, viewed from 30 deg S and 180 deg East longitude (after Connerney *et al.*, 2001).

of Mars had been sampled, albeit sparsely, at low altitude. Thus it was possible for Acuña *et al.* (1999) to map the global distribution of magnetic sources in the crust of Mars, revealing a remarkable association with the dichotomy boundary and southern highlands (see Figure 2 of Acuña *et al.*, 1999).

1.2.2. *MGS Mapping Observations*

The Mars Global Surveyor mapping phase, beginning in March of 1999, provided measurements of the vector magnetic field at nearly constant altitude (370–438 km) and fixed (2 am–2 pm) local time. These observations approximate the ideal in mapping a potential field, in that they offer a measure of the field on a closed surface about the source. A 'mapping cycle' of 28 days duration in orbit provides nearly uniform coverage of the planet with track-to-track separation of about 1 deg (59 km separation) at the equator.

At the 400 km nominal mapping altitude, magnetic fields generated by the interaction of Mars' atmosphere with the solar wind can at times be appreciable. The remainder of this volume is devoted to the study of this interaction. However, our focus in this chapter concerns the mapping of the crustal field itself and minimizing the influence of external sources on the results. To a very good first approximation, external fields draped over a conducting obstacle will align with the conducting surface, appearing largely in the horizontal component of the field. Magnetic fields generated by the solar wind interaction are time variable, reflecting variations in the solar wind, and greater in magnitude near the sub-solar point. With data from many mapping cycles, allowing measurement redundancy and statistical reduction of external noise, the crustal magnetic field has been mapped with extraordinary signal fidelity (Acuña *et al.*, 1999, 2001; Connerney *et al.*, 2001).

Figure 2 shows a series of orthographic projections of the three components of the vector field observed in mapping orbit (Connerney *et al.*, 2001) from a vantage point located at 30 S and 180 E, above the intensely magnetized southern highlands. The map was constructed by collecting night-time observations in 1 by 1 deg lat/lon bins (59 km at the equator) and using a median value for each component as an approximation to the vector crustal field. The field due to the solar wind interaction is both quieter and weaker in magnitude over the darkened hemisphere, so elimination of dayside observations helps reduce the effects of external fields. Use of the median value for an estimate of the crustal field further reduces the effects of a time-variable external field, but it is important to note that a residual bias of a few nT remains. This residual contains some part of the global scale response of Mars' atmosphere to the solar wind (Crider *et al.*, this volume) and can also be identified in the low degree and order terms of a spherical harmonic expansion (Arkani-Hamed, 2002a). It is also important to recognize that these maps are essentially images of the observations, as opposed to images obtained from a model fit to observations, and that estimates in adjacent longitude bins are largely statistically independent. The high signal fidelity (of order 100) evident in the images is truly representative of the data.

It is interesting to compare the global maps of the crustal magnetic field of Mars (Connerney *et al.*, 2001) with those of the Earth derived from MAGSAT observations at comparable altitude (Langel *et al.*, 1982), and the more recent maps derived from Orsted (Olsen, 2002) and Champ (Maus *et al.*, 2002) at somewhat greater altitude. In the case of the Earth, the much larger internal field must first be removed, with some uncertainty, and observations need be detrended or otherwise filtered along track to remove residual internal fields as well as external fields. What remains is a crustal field at least an order of magnitude weaker than that of Mars. The Earth's magnetic anomalies must be estimated in the presence of a large inducing field, so it is not possible to distinguish between induced and remanent magnetism, except, perhaps in a statistical sense (Maus and Haak, in press). Free of such ambiguities, Mars is an ideal laboratory for the study of crustal magnetic fields and their origin.

2. Interpretation of the Magnetic Field

Before we turn to a more detailed discussion of how to interpret the crustal magnetic field of Mars, it is useful to summarize what can reasonably be inferred from the observations themselves. The lack of a present-day global field means that Mars does not now have an active dynamo in its deep interior. Since parts of the crust are intensely magnetized, it is hard to escape the conclusion that Mars once had an active dynamo (Acuña *et al.*, 1999), whether or not remanence was acquired by thermoremanent magnetization (TRM), as seems likely, or some other process (e.g., chemical remanent magnetization, or CRM). This observation requires that Mars had a convecting, electrically-conducting core within which a dynamo operated for a geologically significant interval of time. Most models of Mars evolution would have a molten iron core forming early, after or during hot accretion some 4.5–4.6 billion years (Gyr) before the present (Schubert *et al.*, 1992; Stevenson, 2001; Spohn *et al.*, 2001; see, however, Senshu *et al.*, 2002).

It also appears likely that Mars' dynamo ceased operation relatively early in the planets evolutionary history, as suggested by Acuña *et al.* (1999), although this inference is not indisputable. Acuña *et al.* noted that expansive regions above the large impact basins (Hellas, Argyre, Isidis) lacked measurable crustal fields, as did much of the northern lowlands. For the moment let us accept the notion that this observation implies that the associated crust is not magnetized. Had these structures formed in the presence of a large ambient field, they would likely have acquired intense remanent magnetization as the rock cooled below the Curie temperature. This process would leave identifiable magnetic signatures above these features, where none is observed. Either they formed after cessation of the dynamo, or the magnetic imprint originally acquired was subsequently erased. This would require reworking the entire crust in the vicinity of these basins (e.g., reheating above the Curie temperature or demagnetization by shock) without leaving a visible trace. The formation of these large impact basins is believed to have occurred about 4 Gyr ago, leading to the conclusion that the dynamo had ceased operation after just a few hundred million years.

An alternative proposal by Schubert and colleagues (Schubert *et al.*, 2000) would have the large impact basins form prior to the onset of the dynamo (see also Stevenson, 2001), reversing the chronology proposed by Acuña *et al.* (1999). This proposal was motivated, in part, by the lack of detection of a distinct magnetic signature, or 'fringing field', associated with the large impact basins (which depends on the scale length of crustal magnetization, discussed in detail in Section 2.3). Schubert *et al.* propose that the magnetization of the southern highlands arose mainly from localized heating and cooling events that post-date the era of large impacts and basin formation. They observe that early onset and cessation of the dynamo is difficult to reconcile with the notion of a dynamo driven by solidification of an inner core (as is thought to be the case for Earth). There is as yet no unambiguous observational constraint on the timing of the dynamo, but for the fact that it has existed in the past and has ceased to operate since. Whatever the cause, the dichotomy in magnetization of the Mars crust is perhaps the most significant clue in its evolution.

It is also difficult to escape the conclusion that the crust of Mars is intensely magnetized, and that the magnetization displays a coherence over hundreds of km, persisting for billions of years. This requires an iron rich crust with a magnetic mineralogy that can acquire and preserve, over aeons, a large remanent field. Together with inferences on the composition of the Mars mantle from studies of the SNC meteorites (Wadhwa, 2000; McSween *et al.*, 2001), this implies an increased oxidation state relative to mantle-derived rock, consistent with assimilation of an aqueous component at crustal depths (Connerney *et al.*, 2001).

These observations provide very important constraints for Mars evolution, but it is clear that there is much more to be learned if the magnetic record in the Martian crust can be deciphered. In the following sections we consider models of the magnetic field, both global and regional, and what might be learned from them.

2.1. Basic Concepts

Interpretation of the magnetic field is an exercise in the field of potential theory, the study of harmonic functions that are solutions of Laplace's equation. They often arise in studies of gravity, magnetic and electric fields, and problems of heat and fluid flow, for example (Kellogg, 1929; Blakely, 1995). The principle of superposition holds for harmonic functions, so given a distribution of sources (e.g., magnetic dipoles) one can compute the field throughout space by summing up that due to individual sources or by integration over a volume. Theorems of uniqueness on harmonic functions ensure that a potential is uniquely determined within a closed regular region by its values on the boundary of that region, or, with some restriction and to within an additive constant, by the values of its normal derivative on the boundary (e.g., Kellogg, 1929). Alternatively, if the potential and its normal derivative are known on any arbitrary surface, it is uniquely determined in the three dimensional neighborhood of any point of that surface. So, for example, the properties of such functions can be used to continue the field from one surface to another, as is often done in geophysical surveys, apart from practical concerns.

However, the distribution of sources is not determined by knowledge of the potential, however complete. A potential can be due to various source distributions. As a simple example, consider the magnetic field of a dipole at the origin and that of a uniformly magnetized sphere of radius a. Observed in a region exterior to the sphere, both source distributions produce the same (dipole) magnetic field and are therefore indistinguishable. In general, given a source some distance from the point of observation, there is always an alternate source distribution nearer the point of observation that can approximate to any desired accuracy the field of the more distant source. So, for example, when fitting an observed field with a dipole source at some depth in the crust, it is likely that the true source is both closer to the observer (not as deep in the crust) and spatially distributed.

One may also point to sources that produce no magnetic field, and by superposition show that any model is non-unique. The magnetic field external to a uniformly-magnetized infinite plate is zero (e.g., Blakely, 1995). This might be a good approximation if the observer is close to a uniformly-magnetized source, compared to its extent. Such a plate with a hole in it produces the same field as an isolated, uniformly magnetized (in the opposite sense) disc sized to fit the hole. This works as well with a spherical cavity; a dipole field observed above the crust might equally well indicate a hole in such a plate or an isolated, uniformly magnetized sphere of the same size.

Consider a spherical shell (e.g., planetary crust) in which the intensity of magnetization is constant and the direction of magnetization is everywhere radial. Symmetry and Gauss' theorem dictate that the magnetic field produced by such a source is zero. In fact, there are infinitely many ways to magnetize a spherical shell (and therefore a solid sphere) in such a way as to produce no magnetic field exterior to the shell (Runcorn, 1975a, b). Runcorn's theorem states that if a shell of any thickness acquires a magnetization, the intensity of which is proportional and parallel to a magnetizing field of internal origin, which later disappears, its external field is zero (aside from practical concerns; see, e.g., Goldstein, 1975; Leweling and Spohn, 1997). Runcorn applied this theorem to argue in favor of a lunar magnetic field of internal origin. However, one can also posit a crust with many 'Runcorn spheres' of all sizes scattered about, undetectable by external measurements of the magnetic field.

Measurements of the vector magnetic field, exterior to a source region, cannot, therefore, uniquely determine the properties of the source. Additional information is required if we are to learn about the intensity and direction of magnetization, or the distribution of magnetization, within the source region (i.e., crust, or interior of a planet). The validity of a magnetization model and the utility of the model depend critically on the validity of the additional information, or constraints, imposed on the solution.

Spherical harmonic analysis is by far the most popular method used in the interpretation of potential fields of global scale, dating back to its introduction by Gauss in 1838. It is particularly useful to describe the field produced by a source distributed near the origin, and has been applied to the Earth's field (Langel, 1987) as well as all of the magnetized planets (Connerney, 1993). It is far less efficient when applied to a source distribution that is not compact, such as crustal magnetization. On a global scale, crustal sources have been modeled by fitting a large number of equally spaced dipoles in a shell approximating the crust (Mayhew and Galliher, 1982). They assumed that the Earth's crustal field is largely due to induction, the direction of each dipole being fixed by that of the main field. This constraint is not applicable in the case of Mars, where no global field exists today.

On a local scale, observations can be fit to any model for which the field can be calculated, beginning with simple dipoles, thin sheets, uniformly magnetized volumes, etc. In geophysical surveys, one often finds the direction of magnetization constrained, as appropriate for induced fields, in modeling the source. Constraints provided by local geology, gravity, seismic data, or borehole measurements and samples are often introduced to constrain the distribution of magnetization (e.g., Telford *et al.*, 1976) in applied geophysics.

2.2. Global Models

The first global scale model of the magnetic field of Mars was produced by Purucker and colleagues (Purucker *et al.*, 2000) using Mars Global Surveyor aerobraking phase observations. They used an equivalent source technique, representing the crustal field by superposition of many dipoles, approximately equally spaced (about 111 km apart), on a reference sphere of radius 3394 km ($\sim 1\ R_m$). The magnetization direction of all 11,550 dipoles was assigned (radial) as were the locations; only the magnitude of each was varied to fit the observations. The model was used to construct a nearly complete altitude-normalized map of the radial magnetic field at a reference altitude of 200 km. The model should be regarded as a tool for interpolating the field between orbit tracks in a manner that satisfies Laplace's equation (one of many). The model will approximate the field in the immediate vicinity of orbit tracks. At low altitudes, the calculated field would merely reveal the (arbitrary) locations of the many dipoles used to represent the field. Special care is needed to remove source dipoles that are not well constrained by the observations (too distant from the trajectory), and to identify individual passes with large external field variations, to avoid bias in the source dipoles fitted.

Spherical harmonic models of the field have been presented by Arkani-Hamed (2001a, 2002a), Cain *et al.* (2003), and Voorhies *et al.* (2002). Arkani-Hamed's model is an expansion to degree and order 50, most recently improved by addition of mapping orbit observations, while both Cain *et al.* (2003) and Voorhies *et al.* (2002) use expansions to degree and order 90. The higher the degree and order of the expansion, the greater the spatial resolution afforded by the model, in general, although one will find much discussion regarding the appropriate choice of $n_{\max}$ for a specific dataset. Arkani-Hamed (2002) represents the spatial resolution afforded by a 50-deg expansion as comparable to a wavelength of 420 km on the surface and argues that this is the maximum resolution afforded by the data. However, Cain *et al.* (2003) show several examples of low-altitude aerobraking data that are reasonably well fit by their 90-deg spherical harmonic model and poorly represented by an expansion limited to 50-deg.

Like the many-dipoles equivalent source models, the spherical harmonic models provide a useful tool for interpolating the field in the general vicinity of the orbit tracks along which the original data were acquired. Upward continuation of the field with a spherical harmonic expansion can be expected to work well, within a region devoid of local currents, providing that the external field is also taken into account. At the 400 km altitude of the mapping mission, MGS spends about half its time above the ionopause (Mitchell *et al.*, 2001), and local currents due to the interaction of the atmosphere and crustal field with the solar wind are already significant.

In contrast, downward continuation of a spherical harmonic model works well only for compact sources, i.e., fields due to sources deep in the interior. The (un-modeled) magnetic field associated with small spatial scale variations in crustal

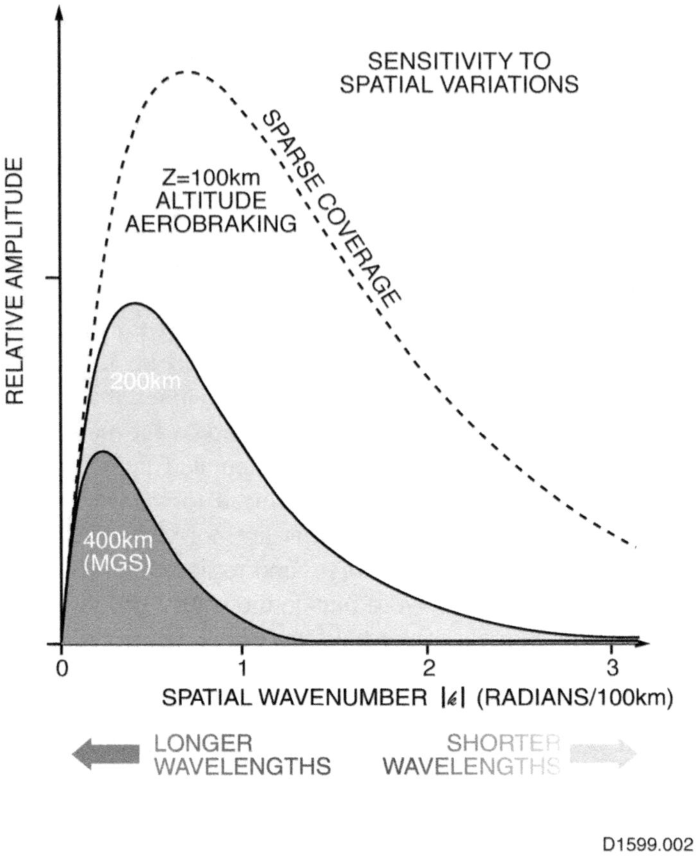

Figure 3. Sensitivity to spatial variations in magnetization depends on the altitude of observation. We compare the relative amplitude of crustal signals as a function of spatial wavenumber (wavelength) for three different mapping altitudes. Peak sensitivity shifts to shorter wavelengths as altitude decreases (after Blakely, 1995).

magnetization grows exponentially larger as one continues downward. Such fields will quickly overtake those associated with large spatial scales. Indeed, the altitude of observation essentially selects a range of observable spatial wavelengths, comparable to the survey altitude; signals associated with smaller spatial scales, *and* longer spatial scales, are greatly attenuated (Figure 3). Downward continuation of a truncated spherical harmonic will do little more than slightly alter the field as observed at altitude, providing little additional insight. Any information about small spatial scales – those that will be far more prominent near the surface – has already been removed from the data!

Figure 3 is a good illustration of the power of altitude in a geophysical survey, or in our application, the global mapping of a planet. The altitude of the mapping orbit determines the wavelength of spatial variations that the survey will be most sensitive to; lower altitudes favor higher spatial resolution. The curve corresponding to 100 km altitude (reached in aerobraking) is rendered with a dashed line to distinguish it from those (at 200 km and 400 km) that correspond to practical circular orbit altitudes. The atmospheric drag at 100 km altitude on Mars is much too large for completion of a global survey at that altitude. A collection of aerobraking passes provides a wealth of information but these data cannot be continued without additional information, in contrast to high-fidelity measurements on a surface.

The spatial resolution afforded by a complete survey at a particular altitude is easily deduced from Figure 3. The commonly-held view is that the spatial resolution afforded by a survey is comparable to the survey altitude. This is about right in the case of the Earth, where fields of crustal origin must be measured, with considerable difficulty, in the presence of the much larger background field, resulting in limited signal-to-noise maps. In this case, for S/N ratios of a few, the spatial resolution obtained is comparable to that of the peak amplitude, at about 1 radian per h where h is the mapping altitude. However, if you measure the field with much higher signal-to-noise, you gain information at spatial scales above and below the peak. The highest spatial resolution afforded by the survey is obtained by following the curve in Figure 3 to the right until the amplitude is reduced by a factor comparable to the S/N of the map. For the MGS mapping of Mars at 400 km altitude (Connerney *et al.*, 2001) the vector field has been mapped with extraordinary signal fidelity (S/N $\sim$ 100), yielding information to spatial scales of about 1 radians/100 km.

The global representations will be most useful in studies of the solar wind interaction with Mars, where details of the field below mapping altitude are less important. The global representations, it should be noted, are very inefficient when applied to crustal sources, which by nature are local. The magnetization acquired and preserved in the crust is strongly dependent on the rock composition and mineralogical form of the iron, its magnetic microstructure, magnetizing field intensity, and the subsequent chemical, thermal, and mechanical history of the sample. These are largely localized phenomena with small spatial scales, compared with satellite mapping altitudes. An isolated source is more easily and accurately described by a simple dipole than with a set of orthogonal basis functions (spherical harmonics) spanning the globe. For example, a spherical harmonic expansion to $n_{\max} = 90$ requires $(n_{\max} + 1)^2 - 1 = 8280$ coefficients to fit such a source where 6 free parameters would do better.

For this reason, we prefer regional analyses and source modeling techniques, described in the next section, to address crustal magnetization. However, spherical harmonic analyses can provide a useful representation for some applications and an interesting comparison with Earth's crust. Figure 4, from Voorhies *et al.* (2002), shows a comparison of the 'magnetic spectra' of Earth and Mars, i.e., the mean-square amplitude of the field due to harmonics of degree n as a function of n. The Earth's spectrum for $n < 13$ or so is identified with the core field, produced by the dynamo below 0.55 the planet's radius. The higher degree terms are identified with crustal sources (Langel and Estes, 1982) and are reasonably well fit by a

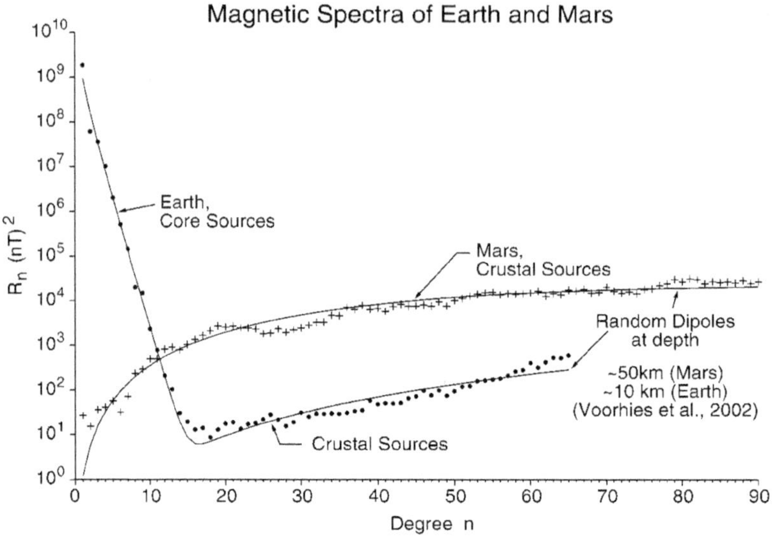

Figure 4. Mean square amplitude of the magnetic field on the surface of a sphere (radius a) from spherical harmonics of degree n for Earth ($a = 6371$ km) and Mars ($a = 3394$ km). The fitted curve for the Earth is the sum of spectra from core sources (dynamo) and that of a crustal source shell of random dipoles at shallow depth. The fitted curve for Mars is that of a source shell of random dipoles in the crust only (after Voorhies *et al.*, 2002).

distribution of random dipoles near the surface (top 18 km of crust). In contrast, the entire Mars spectrum is well described by a random distribution of dipoles at a depth of about 50 km (Voorhies *et al.*, 2002) beneath the surface. Connerney *et al.* (1999) claimed that the Mars crust was an order of magnitude more intensely magnetized than Earth's, on average. Voorhies *et al.* (2002) show more precisely that the Mars crust is a factor of 9.6 ($\pm$ 3) more intensely magnetized over all spatial scales for which a comparison is possible.

2.3. Source models

Perhaps the most fruitful way one might address crustal magnetization is through the use of source models, whereby the field produced by a volume of magnetized material is compared with that observed. Of course, one must be cognizant of the non-uniqueness inherent in potential field modeling, and critical of any assumptions made to constrain the range of solutions. However, source modeling allows for the introduction of additional constraints based on other knowledge, where appropriate. In general, magnetic survey data can be enhanced with knowledge of local geology, sample measurements, gravity, borehole measurements and samples, and so on. This is why *forward modeling* is so popular in geophysical surveys. It is relatively easy to incorporate such constraints into the model, e.g., constraining the volume and location of magnetized material, or the direction of magnetization (induced fields), or intensity of magnetization (if one has samples).

In the case of Mars, we currently have few if any constraints applicable on a scale appropriate to the observations. Most of the Mars data is acquired at about 400 km altitude, supplemented by aerobraking data down to about 100 km altitude. We are thus limited to spatial scales *roughly* comparable to 100 km (Figure 3) and require constraints of comparable scale. It is difficult to imagine a source of such dimensions with uniform magnetization, for example, or uniform direction of magnetization. However, one can envision the wholesale demagnetization of the crust in the vicinity of large impact craters and basins, many of which have appropriate spatial scales (e.g., Acuña *et al.*, 1999).

It may be worthwhile to digress briefly to address questions of scale in more detail. Consider the regional demagnetization of the crust in the vicinity of the major impact basins (e.g., Hellas, Argyre, Isidis), used by Acuña *et al.* (1999) to establish constraints on the history of the dynamo. They assumed that crustal remanence was erased by a thermal event or shock demagnetization, or both, associated with the large impact events. The magnetic field mapped over these regions is very weak or non-existent, compared to that in neighboring regions where variations in the field are large over a range of spatial scales appropriate to the mapping altitude. This is the expected result if the crust was characterized throughout by variations in magnetization over a broad range of spatial scales, subsequently erased in the vicinity of the large impact basins (diameter 1000–2000 km). This interpretation requires that the spatial scale of the largest basins exceed that of the spatial variations in crustal magnetization.

The opposite extreme in spatial scales would lead to a very different conclusion. Suppose that the crust were uniformly magnetized throughout, so that the spatial scale of variations in crustal magnetization greatly exceeded that of the largest impact basins. In this case, demagnetization of a circular disc associated with the impact event would produce a 'hole' in the otherwise uniformly magnetized infinite plate. Mapped at an altitude significantly less than the dimensions of the disc, the magnetic field would simply reveal the edge of the disc, tending to zero towards the center of the disc and away from the rim (Ochadlick, 1991). This kind of signature is not evident in the MGS observations (Acuña *et al.*, 1999; Arkani-Hamed, 2001a). The lack of such a signature led Schubert *et al.* (2000) to argue against the possibility of an early Mars dynamo. The disparate interpretations result directly from different assumptions regarding the relative spatial scales.

Figure 5 shows examples of uniformly magnetized sources and the fields observed along a traverse above them. The fields produced by the sphere and long cylinder at a depth d are similar, but the cylinder can be distinguished by the relative amplitudes of horizontal and vertical fields, and by the less rapid decay of the field with distance from the source. The depth of the source in both cases can be estimated from the peak-to-peak separation of the field (either component, depending on magnetization direction) in the ideal case. The thin plate example shows a traverse just above the plate (depth $\ll$ width) to illustrate the edge effect visible in passing over a uniformly magnetized source at an altitude that is much

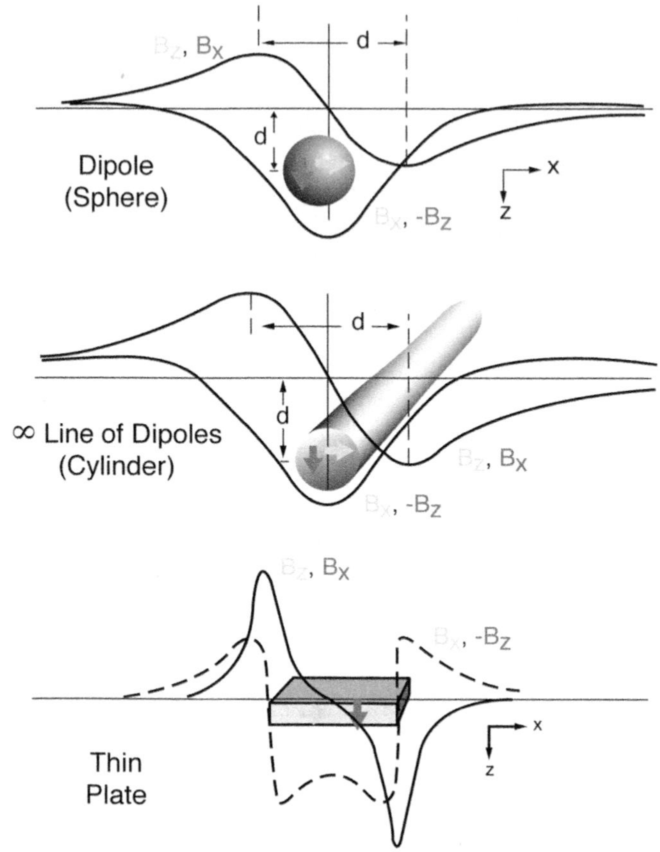

Figure 5. Magnetic field above some simple sources (red labels for $+z$ direction of magnetization, blue for $+x$) (a) dipole source, equivalent to a uniformly magnetized sphere, at depth d (b) infinite line of dipoles, or uniformly magnetized cylinder, at depth d (c) uniformly magnetized thin plate, at depth 0.1d.

less than the width of the plate. In exploration geophysics, the horizontal gradient of the field is often used to outline separate sources. A traverse above the same source at depth comparable to or greater than the width would look much like those above.

In source modeling, one attempts to learn something about the source by fitting one or more such simple sources to the observed vector field. It is very common in exploration geophysics to approximate more complex sources by summation over many uniformly magnetized thin plates. Connerney *et al.* (1999) fit a source model to the vector data acquired by MGS at low altitude (to 100 km) over the intensely magnetized southern highlands. The magnetic field observed over this region is organized in east-west-trending lineations, some extending over 2000 km in length. A remarkably good fit to the vector data acquired during several aerobraking passes over Terra Cimmeria and Terra Sirenum was obtained using a

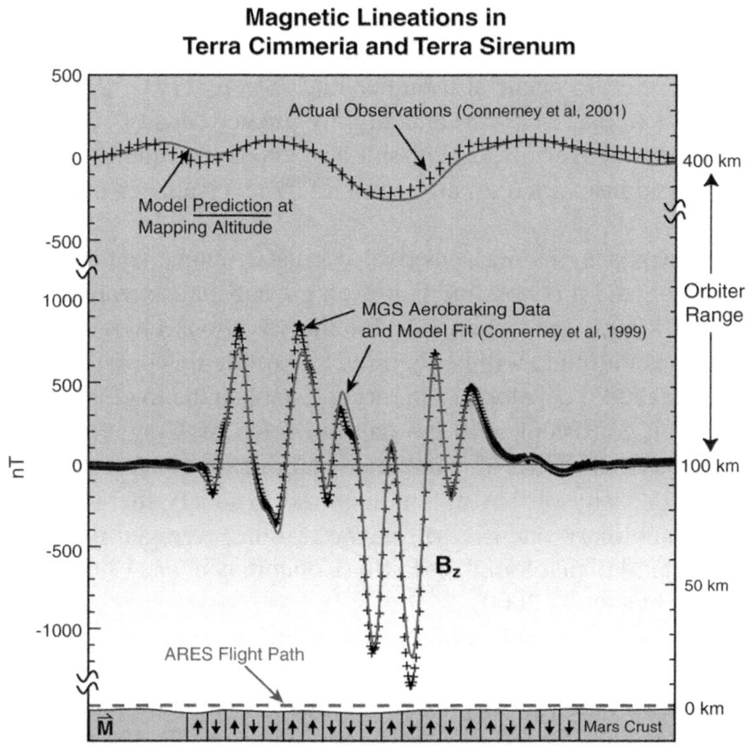

Figure 6. MGS observations (crosses) at approximately 100 km (aerobraking) and 400 km (mapping orbit) altitudes compared with that computed from the 20-plate model of Connerney *et al.* (1999). The model, fit to the aerobraking observations only, does a very good job fitting the high altitude observations obtained two years later during mapping phase. Model magnetizations indicated below are schematic only. This comparison illustrates well the rapid increase in magnitude of variations of smaller spatial scale observed at lower altitude.

multiple thin plate model. The model consisted of 20 uniformly magnetized thin (30 km) plates, 200 km in width, extending to infinity in the east-west direction (Figure 6). Plate magnetizations, obtained by least squares fitting the underparameterized linear system, varied from approximately $+20$ A m^{-1} to -20 A m^{-1}, in a pattern reminiscent of the alternating magnetic polarities observed over mid ocean ridges on Earth (Vine and Matthews, 1963).

Connerney *et al.* (1999) concluded that crustal magnetizations on Mars were at least an order of magnitude greater than those observed on Earth, and noted that the observations could be fit with a pattern of magnetization like that found on Earth in association with crustal spreading in the presence of a reversing dynamo (considered in more detail by Nimmo and Stevenson, 2000). The characteristic horizontal scale length of such features on Earth, determined by the crustal spreading rate and the characteristic time interval between dynamo reversals, is of order 10 km, a spatial scale well beyond the reach of orbiting spacecraft. Variations of order 100 km width might result on Mars if the rate of spreading was greater and/or if dynamo reversals were less frequent, relative to Earth. Crustal spreading on Mars is expected to occur at a higher rate (Sleep, 1994), if at all, because of lower gravity (0.38 that of Earth) and slightly greater density of the mantle. Little is known of dynamo reversals, except on Earth where the time between reversals of the magnetic field has varied greatly, with a few reversals per million years being typical.

The MGS observations are sensitive to crustal magnetizations averaged over ~ 100 km scales, and it is possible that even greater magnetization contrasts, over smaller spatial scales, are responsible for the variations observed by MGS. Observations at lower altitude will be required to explore this possibility. At the other extreme, Parker (2003) developed a theory to establish the lowest possible intensity of magnetization consistent with the data modeled by Connerney *et al.*, (1999). Interpolating his results to a comparable crustal thickness, Parker finds a minimum magnetization intensity of 9 A m^{-1} is required to satisfy the MGS observations. This is an extraordinary intensity of magnetization (averaged over huge volumes of crust), compared to that found on Earth (Connerney *et al.*, 1999; Arkani-Hamed, 2001a; Kletetschka *et al.*, 2000).

Sprenke and Baker (2000) modeled the same observations under a variety of assumptions, and found that a crustal spreading model fit the observations well, particularly if a variation in magnetization intensity along the east-west direction was allowed. They too found magnetization intensities of 10s of A m^{-1}, assuming a 30 km thick crust. Their model allowed normal and reversed magnetization directions only, fixed in the direction of an assumed aerocentric dipole. They also examined models with only 1 polarity (magnetic shell concept), creating a similar field by separation of many similarly magnetized slabs. Given the insensitivity of all such models to addition or subtraction of a uniformly magnetized layer, the models differ little in practical terms; the distinction is in the mechanism of origin.

How the Mars highlands crust acquired remanence of this magnitude, and with the intriguing east-west trending (largely two dimensional) geometry, remains a mystery. Among the possibilities considered by Connerney *et al.* (1999), crustal spreading in the presence of a reversing dynamo, like sea floor spreading on Earth, is perhaps the most controversial and far reaching. This is a mechanism that is in process on Earth, and it has the potential of satisfying the need for huge volumes of intensely magnetized crust. Alternatively, the lineations may have formed as a series of volcanic or intrusive events with the appropriate geometry (Connerney *et al.*, 1999), forming giant dike swarms (Nimmo, 2000) like the fragmented dike swarms observed on Earth (Ernst *et al.*, 2001). The latter are abundant within the most ancient, Precambrian terranes on Earth, and coeval dikes can be traced from one continent to another after the continents are reconstructed by closure of the Atlantic Ocean. The formation of a series of suitable dikes on Mars is plausible (Nimmo, 2000), but even with very intense magnetization (15 A m^{-1}) they would necessarily be great in number and orders of magnitude wider than the 10 meter width typical of terrestrial dikes (Ernst *et al.*, 2001). Relatively few large dikes have been identified on Mars, compared with Earth and Venus; those that are visible in Terra Sirenum and Terra Cimmeria are not oriented in the direction of the magnetic features.

Fairen *et al.* (2002) proposed that the magnetic lineations on Mars were formed at a convergent plate margin through collision and accretion of teranes. In their model, the strongly magnetized southern highlands are 'a jigsaw of continental and oceanic pieces' related to a past era of plate tectonics, formed by a process of crustal convergence rather than crustal generation.

While much attention thus far has focused on the southern highlands, source models have also been developed for other regions. Hood and Zakharian (2001) have done detailed modeling of the north polar sources mapped during the SPO mission phase (Acuña *et al.*, 1998), concluding that they formed prior to the end of heavy bombardment. They also computed putative paleopoles, assuming that the sources are uniformly magnetized discs of approximately 400 km diameter. Their paleopoles lie between about 30 and 60 deg N along the 225 deg E meridian. Arkani-Hamed (2001b) isolated 10 sources, computing paleopole positions for each by fitting uniformly magnetized bodies (in the shape of an ellipsoid of revolution) of similar dimension to each. He finds a cluster of such pole positions centered on 25 deg N and 230 deg E. In contrast, Sprenke and Baker's (2000) preferred paleopole lies near the equator at the prime meridian.

All of the authors discussed above acknowledge that it is impossible to uniquely determine the direction of magnetization in the crust from the measurements available. It would seem unlikely that sources with such dimensions (~ 100 km $\times$ 30 km) are uniformly magnetized; indeed, cooling such a large body by thermal diffusion could require tens of million years. It is also likely that the actual causative bodies are non-uniformly magnetized and that they assume volumes poorly approximated by simple geometrical bodies. The intensity of magnetization, in particular, typically varies by a large factor within the same geologic formation. In general, the direction of magnetization inferred for a particular source covaries with the location, geometry, and size of the source; absent *a priori* knowledge or geophysical constraints, such determinations need be viewed with caution.

3. Implications for Mars Interior and Evolutionary History

The observation of strong remanent crustal magnetization provides useful constraints on the interior structure and evolution of Mars. It is generally accepted that terrestrial planetary magnetic fields are produced by regenerative dynamo action in a fluid iron-rich core or outer core shell. The observation of strong crustal remanence thus provides further evidence for the existence of an iron-rich core. Models of Mars' interior structure (e.g., Goettel, 1981; Sohl and Spohn, 1997) have cores with radii between 40% and 50% of the planetary radius. There is little doubt that the core formed early (Schubert *et al.*, 1992; Stevenson, 2001; Spohn *et al.*, 2001). Pb and H/W isotope data (Chen and Wasserburg, 1986; Lee and Halliday, 1997; Kleine *et al.*, 2002) suggest that the core formed contemporaneous with accretion or within a few 10 million years thereafter.

The Earth's dynamo is thought to be driven by vigorous thermal and compositional convection associated with the freezing of the solid inner core. Numerical calculations (e.g., Glatzmeier and Roberts, 1997; Roberts and Glatzmeier, 2000) have shown that both thermal and compositional convection can drive a dynamo. Thermal convection in the liquid part of the core is driven by a sufficiently large super-adiabatic temperature gradient in the core. Because the core fluid has a comparatively low viscosity (approximately 1 Pa s), only a small super-adiabatic temperature difference across the core is required. This arises naturally when mantle convection removes heat from the core at a sufficiently large rate. This heat flow must be larger than the heat flow supplied by thermal conduction along the core adiabat. The latter therefore serves as a lower limit, below which thermally driven convection in the core is not supported. The critical value of core heat flow depends on the adiabatic temperature gradient in the core and the thermal conductivity. Nimmo and Stevenson (2000) estimate the critical heat flow for the Martian core to be between 5 and 20 mW m^{-2}.

In the present Earth, latent heat release at the inner core-outer core boundary probably keeps the core super adiabatic (Stevenson *et al.*, 1983; Buffet *et al.*, 1996). Estimates of the growth rate of the Earth's inner core suggest that the inner core is substantially younger (1–2 Gyrs of age) than the oldest magnetized continental crust rocks (> 3.5 Gyrs). If the inner core is indeed this young, then the Earth's early dynamo was driven by cooling and thermal convection alone without the latent heat and chemical buoyancy supplied by the inner core. Alternatively, the Earth's inner core could be substantially older if the core contains radiogenic heat sources. By analogy, either a purely thermally driven dynamo or a dynamo driven by latent heat release and chemical buoyancy might have operated in the Martian core.

A comparison of two thermal history models of Mars appears in Figure 7. This figure shows the heat flow from the core as a function of time calculated for two different evolutionary models of Mars (Breuer and Spohn, 2003). The first model type assumes that Mars had an early epoch of plate tectonics for about 500 Myr after accretion and core formation (see also Nimmo and Stevenson, 2000). That epoch ended with a transition to the present tectonic style, which is characterized by an immobile or stagnant lid, under which convection continues in the mantle. Heat transfer under these conditions, including strongly temperature dependent mantle viscosity, has become known as stagnant lid convection (e.g., Solomatov, 1995; Grasset and Parmentier, 1998). The second model type assumes that stagnant lid convection prevailed throughout the entire evolution of Mars. Two curves are drawn for each model type, differing in the assumed initial temperature difference between the core and the mantle evaluated at the core mantle boundary.

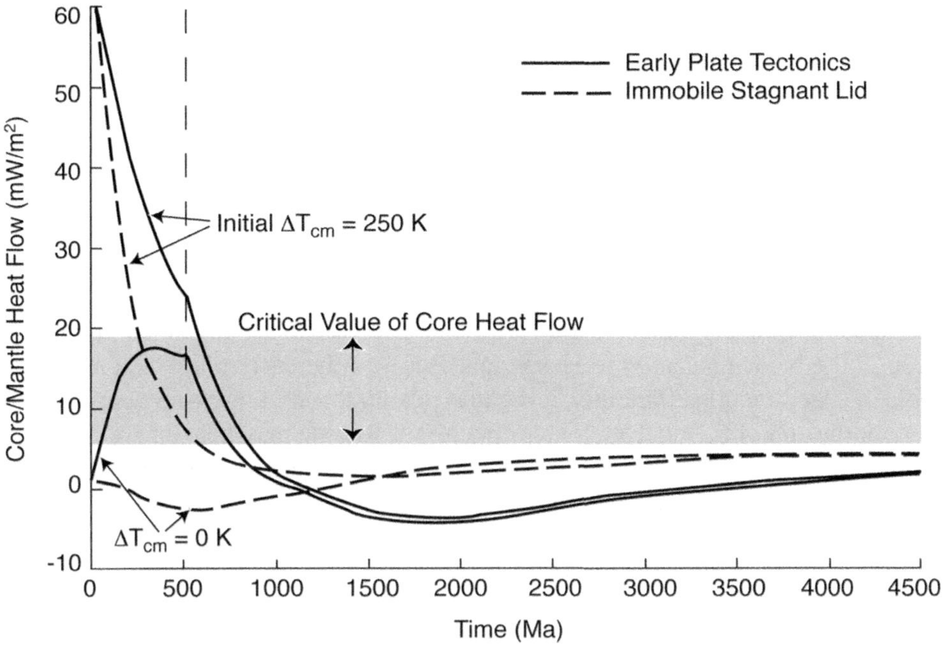

Figure 7. Core/mantle heat flow as a function of time for models of Mars evolution. Solid line: models with an early epoch of plate tectonics followed after 500 Myr by the formation of an immobile stagnant lid or lithosphere above the (convecting) mantle. Dashed line: models with an immobile stagnant lid throughout Mars history. Two curves are drawn for each model, differing in the assumed initial temperature difference between the core and the mantle, evaluated at the core-mantle boundary.

The heat flow for all of the models shown in Figure 7 is below the critical value required to sustain thermal convection and thermally driven dynamo action at the present time. Model calculations by Stevenson *et al.* (1983), Schubert and Spohn (1990), and Spohn (1991), using constant mantle viscosity parameterizations, led to much the same conclusion. The models of Breuer and Spohn (2003) predict thermal convection early in Martian history, and are consistent with dynamo action for periods of hundreds of million years, depending on model details such as initial mantle and core temperatures, mantle and core rheology, and the style of heat transfer. These models are consistent with cessation of the dynamo by the time of formation of the Hellas and Isidis basins, providing thermal convection is sufficient to drive the dynamo.

The immobile lid model requires an initial temperature difference between the core and the mantle of a few hundred degrees Kelvin to drive dynamo action, whereas a model with an early era of plate tectonics does not. The stagnant lid model is relatively inefficient at cooling the deep interior of the planet. The necessary temperature difference might arise from superheating the core during core formation. Stevenson (1990) and Nimmo and Stevenson (2000) have argued that this superheating may be difficult to obtain because the iron in the initial iron silicate mixture might be finely distributed and in thermal equilibrium with the silicates. But Stevenson (2000) has more recently argued for early core superheat to help explain early molten cores.

Convection in the plate tectonics regime brings cold, near surface material to the base of the mantle and cools the core more effectively than stagnant lid convection. As a result, it is more effective in producing core thermal convection and possibly dynamo action. Breuer and Spohn (2003), however, argue that the model with an early plate tectonics episode of mantle convection is difficult to reconcile with the crust formation history of Mars. Geologic evidence (e.g., Frey *et al.*, 2002; Head *et al.*, 2002) suggests that crustal formation and volcanic activity continued well through the Noachian into the Hesperian. Because the early plate tectonics model cools the deep mantle efficiently, it has less potential for widespread volcanic activity after the end of the plate tectonics epoch. The mantle may heat up after cessation of plate tectonics, depending on the heat source inventory and mantle rheology, and may resume volcanic activity. According to Breuer and Spohn (2003), the temperature would peak in the Amazonian, too late to explain the observed crustal formation. Core convection may also be driven by the presence of radiogenic heat sources in the core, as thought possible for the Earth (e.g., Zindler and Hart, 1986; Breuer and Spohn, 1993). Recent studies provide experimental evidence both in support of (Gessmann and Wood, 2002) and contrary to (Chabot and Drake, 1999) this hypothesis.

Chemical convection in the core can drive a dynamo more effectively than thermal convection (e.g., Braginsky, 1964; Stevenson *et al.*, 1983). While the conversion of thermal energy to magnetic field energy is subject to a Carnot efficiency factor, a chemically driven dynamo is not subject to a similar restriction in efficiency. Chemical convection can occur when a solid inner core freezes in a liquid core containing some light, alloying element. Sulfur is usually taken as the most likely candidate because of its cosmochemical abundance and its chemical affinity for iron. If the core has a non-eutectic composition, inner core growth will release buoyant fluid at the surface of the inner core and drive compositional convection even if the core is stably stratified against thermal convection. However, Schubert and Spohn's (1990) thermal evolution models suggest that the present Martian core is likely to be entirely liquid if it contains about 15 weight-% sulfur, as is consistent with the chemistry of SNC meteorites. This would preclude the possibility of compositionally driven convection. A stagnant lid model predicts even higher temperatures in the lower mantle and core and, therefore, strengthens this conclusion.

A present day entirely liquid core also results from the model with an early phase of plate tectonics (Breuer and Spohn, 2003). A stably stratified liquid core at present, but with an early phase of thermal convection and dynamo action, provides a simple explanation for the present absence of a magnetic field at Mars and its early crustal magnetization. However, it is not necessary that the present Martian core be entirely liquid if the compositional dynamo can be otherwise inhibited.

Jault (1996) has suggested that dynamo action may be strongly impeded if an inner solid core grows larger than about 0.35 of the core size.

Evolutionary models both with and without the formation of a solid inner core have been described in recent literature. Formation of an inner core depends on the assumed mantle rheology, liquidus curve (e.g., effect of sulfur on the melting point depression, and how that is calculated), and mode of heat transfer (e.g., plate tectonics or stagnant lid convection). Schubert and Spohn's (1990) model used a relatively stiff rheology (2×10^{21} Pa S at 1600 K) along with Stevenson *et al.*'s (1983) approximation to the liquidus to conclude that a core with 15% sulfur would remain liquid. Longhi *et al.* (1992) used a more sophisticated method due to Williams and Jeanloz (1990) to calculate the liquidus, finding a smaller melting point depression that will favor formation of an inner core. However, the stagnant lid model, with higher temperatures in the mantle and core, results in a liquid core if a similarly stiff rheology is assumed (Breuer and Spohn, 2003). Hauck and Phillips (2002) assumed a relatively weak mantle rheology (10^{19} Pa S at 1600 K) and the Stevenson *et al.* (1983) approximation to the liquidus, finding a molten core at present in their model of Mars evolution. However, had they used the Longhi *et al.* (1992) liquidus, it is likely that a solid inner core would have grown. A present day liquid core also results from models with an early era of plate tectonics, 15% sulfur, a stiff mantle rheology, and the Longhi *et al.* (1992) liquidus (Breuer and Spohn, 2003). It is also possible to have early plate tectonics, form an inner core, and not have a dynamo. In these models the core heats up after the end of the plate tectonics era to a maximum temperature from which it begins to cool. The inner core will keep its radius as long as the core temperature remains below the melting temperature of iron and above the temperature it had at the end of plate tectonics. Inner core growth resumes, along with a chemical dynamo, only when the core temperature falls below its value at the end of the plate tectonics epoch (see also Stevenson, 2001).

Schubert *et al.* (2000) have argued that the remanent magnetization of the crust was acquired much later in Martian history. (See also Stevenson (2001) and Zuber (2001) for a discussion of the timing of the dynamo.) Late magnetization of the crust requires the operation of a strong dynamo after formation of the giant basins, as well as a late addition to the southern highlands crust during the same time. This proposal addresses the observation that not all of the southern highlands crust appears intensely magnetized at this time. Schubert's proposal provides for this by postulating that only the region now characterized by intense magnetization experienced the crustal growth event during the time the dynamo was operative. How this occurred, leaving no visible trace, and why it was confined to specific parts of the southern highlands, remains a mystery.

Schubert's 'late dynamo' proposal allows sufficient time for core solidification to develop as an energy source to power the dynamo. The lack of an operating dynamo today is then attributed to a cessation of inner core growth, e.g., the present core might be either completely frozen or stably stratified and not continuing to freeze. A recent analysis of MGS radio tracking data yielded an estimate for the tidal Love number k(2) that strongly suggests a fluid or partly fluid core (Yoder *et al.*, 2003) for present day Mars. The Love number is not sensitive enough to the presence and size of a putative inner core, in light of other uncertainties, to distinguish between the two possibilities. The Yoder *et al.* estimate is inconsistent with an earlier analysis (Smith *et al.*, 2001) which found a value of the Love number consistent with a completely solid core. While it is not yet entirely clear, observationally, if all or part of the core remains fluid, the thermodynamic models and arguments described above lead to this expectation. There is also the possibility that the dynamo may not be operative even though the core is freezing and convecting, as suggested by Jault (1996). If sulfur is the light alloying element, then a completely frozen core must be colder than the eutectic temperature of Fe-FeS. The eutectic temperature at Martian core/mantle boundary pressures is between 1400 and 1500 K. This low a temperature is difficult to reconcile with models of the thermal evolution of Mars. It is also difficult to see how the inner core could stop growing as long as the planet cools. If Mars has a partially solidified core at present it must be concluded that solidification of the core is simply not sufficient to sustain dynamo action. Dynamo action is not sufficiently well understood to claim that the presence of an inner solid core requires there to be a dynamo.

Phase transitions in the Martian mantle and their interactions with the mantle flow may provide an alternative explanation for late onset of a purely thermal dynamo that could stop operating similar to the early dynamo discussed above. Spohn *et al.* (1998) have discussed how a putative perovskite layer above the core may affect the thermal history of the core. Such a layer is possible if the Martian core is small and iron-rich so that the pressure in the mantle exceeds the spinel-perovskite transition pressure in the lower mantle (Bertka and Fei; 1997 and Sohl and Spohn; 1997). The perovskite layer could be about 100 to 200 km thick. Since the transition is temperature dependent and has a negative Clapeyron slope it is more likely to occur in early Mars when the mantle temperature was hotter than today. According to 2-D and 3-D convection calculations by Breuer *et al.* (1997), the perovskite layer reduces the heat flow from the core because it forms an additional (convective) layer between the overlying mantle and the core. Therefore, an early perovskite layer would tend to work against an early dynamo. Core superheating during core formation and core heat sources might counter the reduction in core heat flow by such a layer. A perovskite layer could thus be responsible for a late onset of the dynamo, or re-birth of a dynamo, at a later time.

The crystallization ages of the Martian meteorites, except for ALH84001, which is 4.4 billion years old, suggest that they were formed less than about 1 billion years ago. As discussed in Spohn *et al.* (1998), the perovskite layer might thin and eventually disappear as the planet cools. When the perovskite layer becomes so thin that it no longer provides a barrier to the heat flow from the hot core, the layer will become unstable and the core cooling rate will suddenly increase. As a consequence, a thermally driven or chemically driven dynamo could be started or reactivated. A thermally driven dynamo, as model calculations suggest, could then operate for a few 100 million years before it (again) becomes extinct. A chemically driven dynamo should operate longer, but it would have to cease functioning before the present day.

Models of core evolution are consistent with both an early magnetic field and a late onset magnetic field. These models cannot predict the strength of the magnetic field, but an early dynamo would most likely be driven by relatively inefficient thermal convection while a late onset dynamo would most likely be driven by more efficient chemical compositional convection and latent heat due to inner core freezing. If the core were superheated during its formation, then an early thermally driven dynamo would not require plate tectonics to operate. In this case, a model with a single, immobile plate on top of a convecting mantle can explain early dynamo action, crustal growth, and volcanic activity extending into the Hesperian and beyond, albeit at an ever decreasing rate.

If the magnetization of the crust *postdates* the early Noachian then models with a growing inner core provide an efficient dynamo mechanism but need to explain why and how the dynamo stopped operating. The weak magnetization of the Martian meteorites may suggest a late onset or reactivation of a dynamo that would stop operating before the present. It is also entirely possible that the Martian meteorites were magnetized late in Martian history in the magnetic field of the earlier magnetized crust. More observational data are required to better constrain the history of the magnetic field and the tectonics of Mars.

4. Summary and Conclusions

The discovery of intense crustal magnetization on Mars is one of the most significant surprises of the Mars Global Surveyor era (Stevenson, 2001), one that bodes well for future Mars exploration. Interpretation of the MGS results is still in its infancy, and there is much healthy debate on the interpretation of the observations. As MGS continues to gather data in its mapping orbit, improved global magnetic field maps will become available, and with increasing S/N comes increasing spatial resolution. Some questions may be resolved as the global map comes into better focus, but many will remain, awaiting further observations.

The MGS aerobraking phase offered a tantalizing glimpse of what might be accomplished with a global magnetic field map at low altitude. An orbiter might practically acquire such a map at altitudes of < 150 km, if designed with low drag and/or if fueled to maintain orbit against atmospheric friction. The current MEPAG (Mars Exploration Payload Analysis Group) document identifies such a mission as a next logical step in the exploration of Mars.

The history of Earth Science in the 1960's – and the widespread acceptance of the unifying theory of plate tectonics – was facilitated by the decoding of magnetic records acquired a few km above mid ocean ridges (e.g., Vine and Matthews, 1963).

How exciting it is to imagine what secrets await us at Mars! Access to those secrets will require a magnetic survey with km scale spatial resolution, performed by a low-altitude aerial vehicle or a sufficiently mobile landed vehicle. This too, is a high priority for Mars exploration in the next decade.

Acknowledgements

We thank J. Geiss of the International Space Science Institute in Bern, Switzerland, for hosting the workshop, supported by ISSI, NASA, and the Mars Global Surveyor Project at JPL. We thank our colleagues for helpful discussions, review of the manuscript, and so on. We thank D. Winterhalter of the JPL for organizing the workshop and the workshop proceedings. This research was funded in part by the MGS Project.

References

Acuña, M. H. *et al.*: 1992, 'Mars Observer Magnetic Fields Investigation', *J. Geophys. Res.* **97**, 7799–7814.

Acuña, M. H. *et al.*: 1998, 'Magnetic Field and Plasma Observations at Mars: Initial Results of the Mars Global Surveyor Mission', *Science* **279**, 1676–1680.

Acuña, M. H. *et al.*: 1999, 'Global Distribution of Crustal Magnetism Discovered by the Mars Global Surveyor MAG/ER Experiment', *Science* **284**, 790–793.

Acuña, M. H. *et al.*: 2001, 'The magnetic Field of Mars: Summary of Results from the Aerobraking and Mapping Orbits', *J. Geophys. Res.* **106**, 23403–23417.

Albee, A. L., Palluconi, F. D. and Arvidson, R. E.: 1998, 'Mars Global Surveyor Mission: Overview and Status', *Science* **279**, 1671–1672.

Albee, A. L., Arvidson, R. E., Palluconi, F. D. and Thorpe, T.: 2001, 'Overview of the Mars Global Surveyor Mission', *J. Geophys. Res.* **106**, 23291–23316.

Arkani-Hamed, J.: 2001a, 'A 50-degree Spherical Harmonic Model of the Magnetic Field of Mars', *J. Geophys. Res.* **106**(E10), 23197–23208.

Arkani-Hamed, J.: 2001b, 'Paleomagnetic Pole Positions and Pole Reversals on Mars', *Geophys. Res. Lett.* **28**(17), 3409–3412.

Arkani-Hamed, J.: 2002a, 'An Improved 50-degree Spherical Harmonic Model of the Magnetic Field of Mars Derived from Both High-altitude and Low-altitude Data', *J. Geophys. Res.* **107**(E5), 10.1029/2001 JE001496.

Arkani-Hamed, J.: 2002b, 'Magnetization of the Mars Crust', *J. Geophys. Res.* **107**(E10), 5083, doi: 10.1029/2001 JE001835.

Bertka, C. M. and Fei, Y.: 1997, 'Mineralogy of the Martian Interior Up to Core-mantle Boundary Pressures', *J. Geophys. Res.* **102**, 5251–5264.

Blakely, R. J.: 1995, *Potential Theory in Gravity and Magnetic Applications*, Cambridge University Press, Cambridge, 437 pp.

Breuer, D. and Spohn, T.: 1993, 'Cooling of the Earth, Urey Ratios, and the Problem of Potassium in the Core', *Geophys. Res. Lett.* **20**, 1655–1658.

Breuer, D., Yuen, D. A. and Spohn, T.: 1997, 'Phase Transitions in the Martian Mantle: Implications for Partially Layered Convection', *Earth Planetary Sci. Lett.* **148**, 457–469.

Breuer, D. and Spohn. T.: 2003, 'Early Plate Tectonics vs. Single Plate Tectonics on Mars: Evidence from Magnetic Field History and Crust Evolution', *J. Geophys. Res.* submitted.

Braginsky, S. I,: 1964, 'Magnetohydrodynamics of the Earth's Core', *Geomag. Aeron.* **4**, 698–712.

Buffett, B. A., Huppert, H. E., Lister, J. R. and Woods, A. W.: 1996, 'On the Thermal Evolution of the Earth's Core', *J. Geophys. Res.* **101**, 7989–8006.

Cain, J. C., Ferguson, B. and Mozzoni, D.: 2002, 'An $n = 90$ Internal Potential Function of the Magnetic Field of the Martian Crustal Magnetic Field', *J. Geophys. Res.* **107**(E10), doi: 10.1029/2001 JE0011487, in press..

Chabot, N. L. and Drake, M. J.: 1999, 'Potassium Solubility in Metal: The Effects of Composition at 15 kbar and 1900 deg C on Partitioning Between Iron Alloys and Silicate Melts', *Earth Planetary Sci. Lett.* **172**, 323–335.

Chen, J. H. and Wasserburg, G. J.: 1986, Formation Ages and Evolution of Shergotty and its Parent Planet from U-Th-Pb Systematics', *Geochim. Cosmochim. Acta* **50**, 955–968.

Christensen, P. R. *et al.*: 2000, 'Detection of Crystalline Hematite Mineralization on Mars by the Thermal Emission Spectrometer: Evidence for Near-surface Water', *J. Geophys. Res.* **105**, 9623–9642.

Cisowski, S. M.: 1986, 'Magnetic Studies on Shergotty and Other SNC Meteorites', *Geochimica and Cosmochimica Acta.* **50**, 1043–1048.

Connerney, J. E. P.: 1993, 'Magnetic Fields of the Outer Planets', *J. Geophys. Res.* **98** (E10), 18,659–18679.

Connerney, J. E. P. *et al.*: 1999, 'Magnetic Lineations in the Ancient Crust of Mars', *Science* **284**, 794–798.

Connerney, J. E. P. *et al.*: 2001, 'The Global Magnetic Field of Mars and Implications for Crustal Evolution', *Geophys. Res. Lett.* **28**, 4015–4018.

Davaille, A. and Jaupart, C.: 1993, 'Transient High-Rayleigh-number Thermal Convection with Large Viscosity Variations', *J. Fluid Mech.* **253**, 141–166.

Dolginov, Sh. Sh. and Zhuzgov, L. N.: 1991, 'The Magnetic Field and Magnetosphere of the Planet Mars', *Planetary Space Sci.* **39**, 1493–1510.

Ernst, R. E., Grosfils, E.B. and Mege, D.: 2001, 'Giant Dike Swarms: Earth, Venus, and Mars', *Ann. Rev. Earth Planetary Sci.* **29**, 489–534.

Fairen, A. G., Ruiz, J. and Anguita, F.: 2002, 'An Origin for the Linear Magnetic Anomalies on Mars Through Accretion of Terranes: Implications for Dynamo Timing', *Icarus* **160**, 220–223.

Frey, H. and Schultz, R. A.: 1988, 'Large Impact Basins and the Mega-impact Origin for the Crustal Dichotomy on Mars', *Geophys. Res. Lett.* **15**, 229–232.

Frey, H. V., Roark, J. H., Shockey, K. M., Frey, E. L. and Sakimoto, S. E. H.: 2002, 'Ancient Lowlands on Mars', *Geophys. Res. Lett.* **29**, 10.1029/2001 GLO13832.

Gessmann, C. K. and Wood, B. J.: 2002, 'Potassium in the Earth's Core?', *Earth Planetary Sci. Lett.* **200**, 63–78.

Glatzmeier, G. A. and Roberts, P. H. 'Simulating the Geodynamo', *Contemp. Phys.* **38**, 269–288.

Goettel, K. A.: 1981, 'Density of the Mantle of Mars', *Geophys. Res. Lett.* **8**, 497–500.

Goldstein, M. L.: 1975, 'Lunar Magnetism', *Nature* **258**, 175.

Grasset, O. and Parmentier, E. M.: 1998, 'Thermal Convection in a Volumetrically Heated, Infinite Prandtl Number Fluid With Strongly Temperature-Dependent Viscosity: Implications for Planetary Thermal Evolution', *J. Geophys. Res.* **103**, 18171–18181.

Gringauz, K. I., Verigin, M., Luhmann, J., Russell, C. T. and Mihalov, J. D.: 1993, 'On the Compressibility of the Magnetic Tails of Mars and Venus', in *Plasma Environments of Non-magnetic Planets*, Pergammon, New York, pp. 265–270.

Hauck, S. A. and Phillips, R. J.: 2002, 'Thermal and Crustal Evolution of Mars', *J. Geophys. Res.* **107**, 10.029/2001 JE001801.

Head, J. W., Kreslavsky, M. A. and Pratt, S.: 2002, 'Northern Lowlands of Mars: Evidence for Widespread Volcanic Flooding and Tectonic Deformation in the Hesperian Period', *J. Geophys. Res.* **107**, 10.1029/2000 JE001445.

Hood, L. L. and Zakharian, A.: 2001, 'Mapping and Modeling of Magnetic Anomalies in the Northern Polar Region of Mars', *J. Geophys. Res.* **106** (E7), 14601–14619.

Jault, D.: 1996, 'Magnetic Field Generation Impeded by Inner Cores of Planets', *Cr. Acad. Sci. II A* **323**, 451–458.

Kellog, O. D.: 1929, *Foundations of Potential Theory*, Frederick Ungar Publishers, New York.

Kleine, T., Münker, C., Metzger and Palme, H.: 2002, 'Rapid Accretion and Core Formation on Asteroids and the Terrestrial Planets from Hf-W Chronometry', *Nature* **418**, 952–955.

Kletetschka, G., Wasilewski, P. J. and Taylor, P. T.: 2000, 'Hematite vs. Magnetite as the Signature for Planetary Magnetic Anomalies?', *Earth Planetary Sci. Lett.* **176**, 469–479.

Langel, R. A.: 1987, in J. A. Jacobs (ed.), 'The Main Field', *Geomagnetism* Academic Press, London, pp. 249–512.

Langel, R. A., Schnetzler, C. C., Phillips, J. D. and Horner, R. J.: 1982, 'Initial Vector Magnetic Anomaly Map from MAGSAT', *Geophys. Res. Lett.* **9**, 273–276.

Langel, R. A. and Estes, H.: 1982, 'A Geomagnetic Field Spectrum', *Geophys. Res. Lett.* **9**, 250–253.

Lee, D. C. and Halliday, A. N.: 1997, 'Core Formation on Mars and Differentiated Asteroids', *Nature* **388**, 854–857.

Leweling, M. and Spohn, T.: 1997, 'Mars: a Magnetic Field Due to Thermoremanence?', *Planetary Space Sci.* **45**, 1389–1400.

Longhi, J., Knittle, E., Holloway, J. R. and Wanke, H.: 1992, in Kieffer, Hugh H., Jakosky, Bruce, M., Snyder, Conway W., Matthews, Mildred S., (Eds), 'The Bulk Composition, Mineralogy, and Internal Structure of Mars', *Mars*, University of Arizona Press, Tucson, AZ, pp. 184–208.

Maus, S., Rother, M., Holme, R., Luhr, H., Olsen, N. and Haak, V.: 2002, 'First Scalar Magnetic Anomaly Map from CHAMP Satellite Data Indicates Weak Lithospheric Field', *Geophys. Res. Lett.* **29** (14), 10.1029/2001 GL013685.

Maus, S. and Haak, V.: 2002, 'Is the Long Wavelength Crustal Magnetic Field Dominated by Induced or Remanent Magnetization?', *Geophys. Res. Lett.* in press.

Mayhew, M. A. and Galliher, S. C.: 1982, 'An Equivalent Source Magnetization Model for the United States Derived from MAGSAT Data', *Geophys. Res. Lett.* **9**, 311–313.

McSween, H. Y., Jr. *et al.*: 2001, 'Geochemical Evidence Magmatic Water Within Mars from Pyroxenes in the Shergotty Meteorite', *Nature*, **409**, 487–490.

Mitchell, D. L. *et al.*: 2001, 'Probing Mars' Crustal Magnetic Field and Ionosphere with the MGS Electron Reflectometer', *J. Geophys. Res.* **106**(E10), 23419–23427.

Mohlmann, D. *et al.*: 1991, 'The Question of an Internal Martian Magnetic Field', *Planetary Space Sci.* **39**, 83.

Ness, N. F.: 1979, 'The Magnetic Fields of Mercury, Mars and Moon', *Ann. Rev. Earth Planetary Sci.* **7**, 248–288.

Nimmo, F.: 2000, 'Dike Intrusion as a Possible Cause of Linear Martian Magnetic Anomalies', *Geology* **28**, 391–394.

Nimmo, F. and Stevenson, D.: 2000, 'Influence of Early Plate Tectonics on the Thermal Evolution and Magnetic Field of Mars', *J. Geophys. Res.* **105**, 11969–11979.

Ochadlick, A. R., Jr.: 1991, 'Magnetic Exploration of Ocean Crust for Craters of Impact Origin', *Geophysics* **56**(6), 1153–1157.

Olsen, N.: 2002, 'A Model of the Geomagnetic Field and its Secular Variation for Epoch 2000', *Geophys. J. Int.* **149**, 454–462.

Parker, R. L.: 2003, 'Ideal Bodies for Mars Magnetics', *J. Geophys. Res.* **108**(E1), 5006, doi: 10.1029/2001 JE001760.

Purucker, M. *et al.*: 2000, 'An Altitude-normalized Magnetic Map of Mars and Its Interpretation', *Geophys. Res. Lett.* **27**, 2449–2452.

Riedler, W. *et al.*: 1989, 'Magnetic Fields Near Mars: First Results', *Nature* **341**, 604–607.

Roberts, P. H. and Glatzmeier, F. A.: 2000, 'Geodynamo Theory and Simulation', *Rev. Modern Physics* **72**, 1081–1123.

Runcorn, S. K.: 1975, 'An Ancient Lunar Magnetic Dipole Field', *Nature* **253**, 701–703.

Runcorn, S. K.: 1975, 'On the Interpretation of Lunar Magnetism', *Phys. Earth planetary Int.* **10**, 327–335.

Russell, C. T.: 1978a, 'The Magnetc Field of Mars: Mars 3 Evidence Re-examined', *Geophys. Res. Lett.* **5**, 81–84.

Russell, C. T.: 1978b, 'The Magnetic Field of Mars: Mars 5 Evidence Re-examined', *Geophys. Res. Lett.* **5**, 85–88.

Russell, C. T.: 1979, in C. F. Kennel (ed.), 'The Interaction of the Solar Wind With Mars, Venus and Mercury', *Solar System Plasma Physics*, North-Holland, Amsterdam, pp. 208–252.

Russell, C. T. *et al.*: 1995, 'A Simple Test of the Induced Nature of the Martian Tail', *Planetary Space Sci.* **43**(7), 875–879.

Schubert, G. and Spohn. T.: 1990, 'Thermal History of Mars and the Sulfur Content of Its Core', *J. Geophys. Res.* **95**, 14095–14104.

Schubert, G, Solomon, S. C., Turcotte, D. L., Drake, M. J. and Sleep,: 1992, in Kieffer, Hugh, H., Jakosky, Bruce M., Snyder, Conway, W., Matthews, Mildred S. (ed.), 'Origin and Thermal Evolution of Mars, *Mars* University of Arizona Press, Tucson, AZ, pp. 147–183.

Schubert, G., Russell, C. T. and Moore, W. B.: 2000, 'Geophysics - Timing of the Martian dynamo', *Nature* **408**, 666–667.

Senshu, H., Kuramoto, K. and Matsui, T.: 2002, 'Thermal Evolution of a Growing Mars', *J. Geophys. Res.* **107**(E12), 5118.

Slavin, J. A., Schwingenschuh, K., Riedler, W. and Yeroshenko, Y.: 1991, 'The Solar Wind Interaction With Mars: Mariner 4, Mars 2, Mars 3, Mars 5, and Phobos 2 Observations of Bow Shock Position and Shape', *J. Geophys. Res.* **96**, 11235–11241.

Slavin, J. A. and Holzer, R. E.: 1994, 'The Solar Wind Interaction With Mars Revisited', *J. Geophys. Res.* **87**, 10285–10296.

Sleep, N. H.: 1994, 'Martian Plate Tectonics', *J. Geophys. Res.* **99**, 5639–5655.

Smith, D. E. *et al.*: 1999, 'The Global Topography of Mars and Implications for Surface Evolution', *Science* **284**, 1495–1503.

Smith, D. E., Zuber, M. T. and Neumann, G. A.: 2001, 'Seasonal Variations of Snow Depth on Mars', *Science* **294**(5549), 2141–2146.

Smith, E. J., Davis, L. Jr., Coleman, P. J. and Jones, D. E.: 1965, 'Magnetic Field Measurements Near Mars', *Science* **149**, 1241–1242.

Sohl, F. and Spohn, T.: 1997, 'The Structure of Mars: Implications from SNC-Meteorites', *J. Geophys. Res.* **102**, 1613–1635.

Solomatov, V. S.: 1995, 'Scaling of Temperature- and Stress-Dependent Viscosity', *Phys. Fluids* **7**, 266–274.

Spohn, T.: 1991, 'Mantle Differentiation and Thermal Evolution of Mars, Mercury, and Venus'. *Icarus* **90**(2), 222–236.

Spohn T., Acuña, M. A., Breuer, D., Golombek, M., Greeley, R., Halliday, A., Hauber, E., Jaumann, R. and Sohl, F.: 2001, 'Geophysical Constraints on the Evolution of Mars', *Space Sci. Rev.* **96**, 231–262.

Spohn, T., Sohl, F. and Breuer, D.: 1998, 'Mars', *Astron. Astrophys. Rev.* **8**, 181–235.

Sprenke, K. F. and Baker, L. L.: 2000, 'Magnetization, Paleomagnetic Poles, and Polar Wander on Mars', *Icarus* **147**, 26–34.

Stevenson, D. J., Spohn, T. and Schubert, G.: 1983, 'Magnetism and Thermal Evolution of the Terrestrial Planets', *Icarus* **54**, 466–489.

Stevenson, D. J.: 1990, in Newsom, H. E., Jones, J. H. (eds), 'Fluid dynamics of core formation. *Origin of the Earth* Oxford University Press, New York, pp. 231–249.

Stevenson, D. J.: 2001, 'Mars Core and Magnetism', *Nature* **412**, 214–219.

Stevenson, D. J.: 2000, 'Core Superheat', *EOS Trans. AGU* **81** (48).

Telford, W. M., Geldart, L. P., Sheriff, R. E. and Keys, D. A.: 1976, *Applied Geophysics*, Cambridge University Press, Cambridge.

Verigin, M. I. *et al.*: 1991, 'Ions of Planetary Origin in Martian Magnetosphere', *Planetary Space Sci.* **39** (1–2), 131–137.

Voorhies, C. V., Sabaka, T. J. and Purucker, M.: 2002, 'On Magnetic Spectra of Earth and Mars', *J. Geophys. Res.* **107** (E6).

Wadhwa, M.: 2000, 'Redox State of Mars' Upper Mantle and Crust From Eu Anomalies in Shergottite Pyroxenes', *Science* **291**, 1527–1530.

Wihelms, D. E. and Squyres, S. W.: 1984, 'The Martian Hemispheric Dichotomy May be Due to a Giant Impact', *Nature* **309**, 138–140.

Williams, Q. and Jeanloz, R.: 1990, 'Melting Relations in the Iron Sulfur System at Ultra-high Pressures: Implications For the Thermal State of the Earth', *J. Geophys. Res.* **95**, 19299–19310.

Wise, D. U., Golombeck, M. P. and McGill, G. E.: 1979, 'Tectonic Evolution on Mars', *J. Geophys. Res.* **84**, 7934–7939.

Zindler, A. and Hart, S.: 1986, 'Chemical Geodynamics', *Ann. Rev. Earth Planetary Sci.* **14**, 493–571.

Yoder, C. F., Konopliv, A. S., Yuan, D. N., Standish, E. M. and Folkner, W. M.: 2003, 'Fluid Core Size of Mars from Detection of the Fluid Tide', *Science* **300**, 299–303.

Zuber, M. T.: 2001, 'The Crust and Mantle of Mars', *Nature* **412**, 220–227.

Zuber, M. T., Solomon, S. C., Phillips, R. J., Smith, D. E., Tyler, G. L., Aharonson, O., Balmino, G., Banerdt, W. B., Head, J. W., Lemoine, F. G., McGovern, P. J., Neumann, G. A., Rowlands, D. D. and Zhong, S.: 2000, 'Internal Structure and Early Thermal Evolution of Mars from Mars Global Surveyor Topography and Gravity', *Science* **287**, 1788–1793.

THE PLASMA ENVIRONMENT OF MARS

A.F. NAGY[1], D. WINTERHALTER[2], K. SAUER[3], T.E. CRAVENS[4], S. BRECHT[5], C. MAZELLE[6], D. CRIDER[7], E. KALLIO[8], A. ZAKHAROV[9], E. DUBININ[3], M. VERIGIN[9], G. KOTOVA[9], W.I. AXFORD[3], C. BERTUCCI[6] and J.G. TROTIGNON[10]

[1]*University of Michigan, Atm Oc & Space Sciences, 1416 Space Research, Ann Arbor, MI 48109, U.S.A.*

[2]*Jet Propulsion Laboratory, California Institute of Technology, 4800 Oak Grove Drive, Pasadena, CA 91109, U.S.A.*

[3]*Max-Planck-Institut für Aeronomie, Max-Planck-Str. 2, 37191 Katlenburg-Lindau, Germany*

[4]*University of Kansas, Lawrence, U.S.A.*

[5]*Bay Area Res Corp, Orinda, CA 94563, U.S.A.*

[6]*Centre d'Etude Spatiale des Rayonnements/CNRS, Toulouse, France*

[7]*Catholic University of America, Department of Physics, Washington, D.C., 20064, USA*

[8]*Finnish Meteorological Institute, Geophysical Research, Helsinki, Finland*

[9]*Space Research Institute, Russian Academy of Sciences, Ul. Profsoyuznaya 84/32, Moscow, Russia GSP-7 117997*

[10]*LPCE/CNRS, Laboratoire de Physique et Chimie, de l'Environnement, 3A, Ave. de la Recherche Scientifique, F-45071 Orleans Cedex 2, France*

(Accepted in final form 14 March 2003)

1. Introduction

When the supersonic solar wind reaches the neighborhood of a planetary obstacle it decelerates. The nature of this interaction can be very different, depending upon whether this obstacle has a large-scale planetary magnetic field and/or a well-developed atmosphere/ionosphere. For a number of years significant uncertainties have existed concerning the nature of the solar wind interaction at Mars, because of the lack of relevant plasma and field observations. However, measurements by the Phobos-2 and Mars Global Surveyor (MGS) spacecraft, with different instrument complements and orbital parameters, led to a significant improvement of our knowledge about the regions and boundaries surrounding Mars.

Figure 1.1 depicts the insights gained, primarily, from the analyses of data from the two missions. Results from theoretical work and numerical simulations helped greatly to place the observations into a self-consistent global framework. The main signatures, defined by the different plasma processes at work, can be summarized as follows:

(1) The bow shock stands off an effective obstacle at a distance expected from gas-dynamic and/or magnetohydrodynamic approximations. The thickness of the

Space Science Reviews **111:** 33–114, 2004.

magnetosheath, the region between the shock and the effective obstacle, is on the order of the solar wind proton gyro-radius, but thinner than the gyro-radius of heavy ions from the planet. Significant mass-loading takes place within the magnetosheath, indicating the existence of an extended hydrogen/oxygen exosphere.

(2) The Magnetic Pileup Region (MPR) is a region dominated by planetary ions. A well-defined boundary, the Magnetic Pileup Boundary (MPB), separates the MPR from the magnetosheath. The MPB is a thin, sharp transition that deflects where the solar wind proton density drops sharply, but not the solar wind electron density nor the solar wind magnetic field. (Previously, the MPB was called the Protonopause, the Planetopause, and various other names.) Within the MPR, on the dayside, the solar wind magnetic field piles up and prominently drapes about Mars. On the dayside, the MPR is limited from below by the ionosphere or the exobase, depending on solar wind conditions. On the nightside, the MPR is bounded by the tail region stretching far behind the planet. The MPR and the MPB are also observed at Venus and comets, and evidence is emerging that they are common features of the interaction of the solar wind with ionospheres of un-magnetized (or weakly magnetized) bodies.

(3) A plasma boundary is observed in the supra-thermal electrons ($>$ 10 eV, MGS data), which hints at the existence of a boundary between the Magnetic Pileup Region and the ionosphere below. However, without thermal ion and electron measurements at the relevant altitudes, it is impossible to determine whether this boundary is an 'ionopause' in the conventional sense. Indeed, it has long been known from Viking measurements, that the ionospheric thermal pressure at Mars is usually insufficient to balance the total pressure in the overlying Magnetic Pileup Region. Thus it is expected that the Martian ionosphere is magnetized, much like Venus' ionosphere during times of high solar wind dynamic pressure and/or low fluxes of ionizing solar radiation. The magnetic pressure associated with this ionospheric field will supplement the ionospheric thermal pressure, possibly enough to balance the overlying MPR pressure. Locally, crustal magnetic fields vastly complicate the topology of the ionosphere and the 'ionopause'.

In subsequent sections of this chapter the magnetosheath and nature of the Magnetic Pileup Region and the Magnetic Pileup Boundary will be discussed, as well as the characteristics of the ionosphere.

2. The Magnetosheath

All solar system objects that are impenetrable obstacles to the solar wind, either by having a sufficiently large intrinsic magnetic field or a dense enough ionosphere, form a bowshock. Because of the geometric aspect of an obstacle such as a planet or comet, the bow shock is detached, and the region between the shock and the

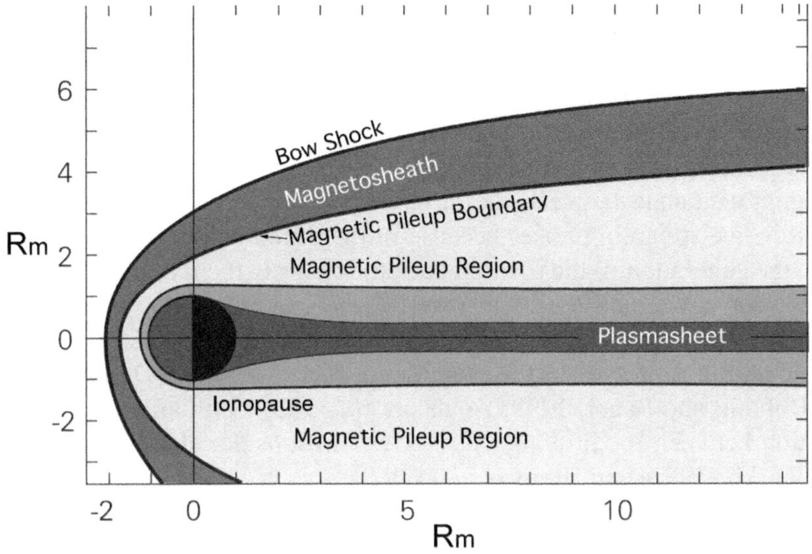

Figure 1.1. A sketch of the structure of the Martian plasma environment, depicting the major boundaries and regions in the equatorial plane. The scales R_M are Mars radii.

obstacle is termed the magnetosheath. The shock-heated and turbulent solar wind plasma and magnetic field signatures of magnetosheaths are similar from object to object, and are readily identified in the data.

In the case of objects with extended exospheres, such as Mars, Venus and comets, heavy ions of planetary or cometary origin significantly modify the overall ion dynamics within the magnetosheath. They appear to be also responsible for the deflection or termination of the solar wind proton flow at the object boundary.

At Mars, remarkable fluxes of planetary ions were observed far from the planet at the terminator bow shock ($\sim 2.8\ R_M$) (Dubinin *et al.*, 1993, 1995). It is not clear whether this observed abrupt increase in the number density of planetary ions is caused only by an increase in the impact ionization rate at the shock front. A sudden appearance of a large number of newly ionized ions (mainly protons) resembles the well-known phenomenon of the ionizing front associated with the Alfvén concept of critical ionization (e.g., Angerth *et al.*, 1962). In ionizing fronts the ionization occurs mainly at the leading edge of a neutral gas as it moves through the plasma. The hydrogen atmosphere of Mars has a large scale height and there are no reasons to expect an appearance of a narrow ionizing front. On the other hand a bow shock and ionizing front have similar substructures. They share charge separation electric fields, and drifts of the electrons in $\mathbf{E} \times \mathbf{B}$ fields, which can destabilize the wave turbulence and lead to the appearance of hot electrons that may initiate avalanche. However, estimates made by Dubinin *et al.* (1993) show that in order to support a self-sustained avalanche the number density of neutrals must be much higher than one may expect at such altitudes.

The Martian magnetosheath is of small spatial extent, with a thickness that is comparable with the gyroradius of the protons. Given the lack of space, the absence of complete thermalization of solar wind and exospheric protons is expected, and is indeed observed (Dubinin *et al.*, 1993). A feature of bi-ion plasma flows is the appearance of bi-ion waves behind the shock. These waves are related to the periodic momentum exchange between the differentially streaming proton and heavy ion fluids. Relative streaming arises because of the difference in inertia, which in turn leads to the generation of the motional electric fields in the reference frames of both streaming ion fluids and therefore results in a strong coupling between the ions. These waves may steepen and give rise to multiple shocklets. Indeed, Phobos-2 and MGS observations provide us the examples of such structures (Dubinin *et al.*, 1996, 1998; Dubinin and Sauer, 1999; Acuña *et al.*, 1998). Similar aspects of turbulent interaction between the protons and newly-born ions are also discussed by Breus and Krymskii (1992) and Breus *et al.* (1992).

Figure 2.1 shows the combined time-energy spectrogram of the proton and oxygen ion fluxes during a long pass of the Phobos-2 spacecraft through the magnetosheath region. The bow shock crossing (00:20 UT) is identified by the sudden heating of the protons. Then a gradual cooling takes place with a subsequent shock-like transition at 01:40 UT, which is accompanied by the appearance of oxygen ion fluxes. The sequence of events, starting with the sudden heating and followed by the drop in the bulk speed and subsequent cooling, is repeated several times.

The variations in the proton and heavy ion velocities are out of phase. Decreases in the proton speed are accompanied by acceleration of O^+ ions. The re-acceleration and the cooling of the protons are accompanied by a decrease of the O^+ velocity, and a broadening of their spectra. The plasma is not fully recovered after the passage of shocklets, indicative of a multiple, step-like, transition to the downstream state.

The heavy ion density increases with decreasing distance to the planet. Since the height of the potential barrier at a shocklet increases with increasing heavy ion density, one may expect that there is a distance from the planet where the solar wind protons will be unable to penetrate further, and a boundary is formed. While this boundary is impenetrable to the protons, the heavies and, importantly, the electrons carrying the magnetic field, go through it. Associated changes of the magnetic field amplitude and direction characterize the apparent border between the magnetosheath and the adjacent region. More details for this scenario of plasma boundary formation are discussed next.

3. The Magnetic Pileup Region (MPR) and Boundary (MPB)

Both Phobos-2 and MGS observations indicate that one or more boundaries exist in the lower magnetosheath across which sharp changes of the plasma and field parameters occur. For example, the magnetometers on Phobos (FGMM, MAGMA)

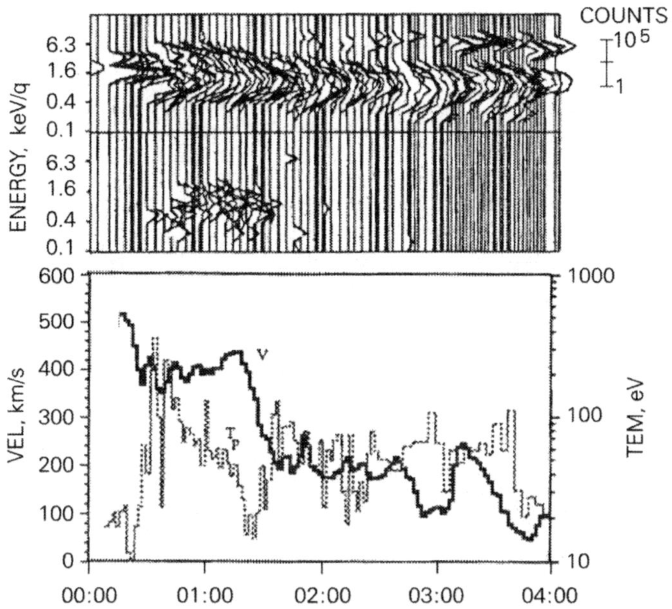

Figure 2.1. Combined energy-time spectrograms of the proton and oxygen ion fluxes measured by sunward and sideward looking sensors along Phobos-2's elliptical orbit on February 15, 1989 (top panel). The bulk velocity and temperature of the solar wind protons are in the bottom panel (from Dubinin *et al.*, 1996).

identified a boundary by a rotation of the magnetic field direction and a decrease in turbulence (Riedler *et al.*, 1989). The boundary, labeled 'Planetopause', was located at a subsolar radius of about 1.25 R_M (R_M: Mars radius). Using the Planetopause as the obstacle boundary that deflects the solar wind flow, a good fit of the observed bow shock location was obtained in gas-dynamic calculations.

Other instrument also detected boundaries. They were introduced as the magnetopause (Rosenbauer et al., 1989; Lundin et al., 1989), the Protonopause (Sauer *et al.*, 1994), and the ion composition boundary (Breus et al., 1991). However, one significant handicap that the Phobos-2 mission suffered under was the large spacecraft periapsis altitude of about 850 km above the surface. The large altitude precluded a decisive statement about the nature of the obstacle. Particularly, no conclusive evidence for the presence or absence of an intrinsic magnetic field could be obtained. The situation fundamentally changed when MGS went down to around 100 km, repeatedly, for many periapses.

Figure 3.1 shows the magnetic field magnitude recorded by MGS during an early elliptical orbit during which the spacecraft explored all the interaction regions from the solar wind to a low altitude periapsis. The bow shock (BS) and an inner 'cavity boundary' (CB) are clearly observed both inbound and outbound. The most prominent signature in this time series of the magnetic field strength is the sudden, strong and sharp jump observed on both sides at the location marked 'MPB' where

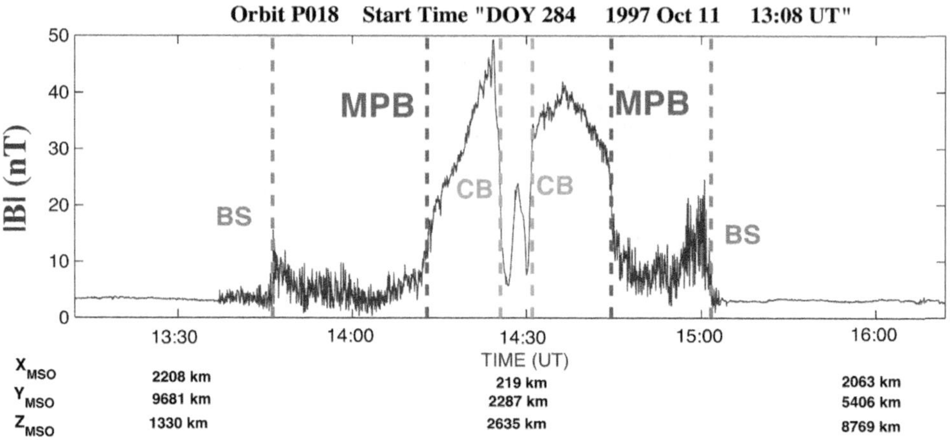

Figure 3.1. Magnitude of the magnetic field recorded by Mars Global Surveyor during an early elliptical orbit (October 11, 1997) around the periapsis (14:28 UT, altitude 120 km, 17.6 local time) displaying the major plasma boundaries symmetrically on both sides (after Mazelle *et al.*, 2002): BS, MPB and CB denote the bow shock, the magnetic pile-up boundary and the magnetic 'cavity boundary', respectively. The horizontal axis shows both time and the spacecraft coordinates in the Mars-centered solar orbital (MSO) system (the X-axis points from Mars to the Sun, the Y-axis points antiparallel to Mars' orbital velocity, and the Z-axis completes the right-handed coordinate system).

MPB stands for the 'Magnetic Pile-up Boundary'. The field magnitude increases dramatically by a factor of about 3 (e.g., outbound) over the very small distance of the order of 100 km. Moreover, the MPB clearly separates two very different regions: a magnetosheath with low amplitude and turbulent magnetic fields, and a region of high pile-up fields, the 'Magnetic Pile-up Region', where the solar wind magnetic field is piled-up and draped about the ionosphere. A similar magnetic pile-up region was found to be present at Venus (Zhang *et al.*, 1991).

The name of the boundary, Magnetic Pileup Boundary, arbitrarily emphasizes the behavior of the magnetic field. Equally correct would have been other names from the literature emphasizing other plasma components, or an instrument-neutral designation such as 'Mantle Boundary' and 'Mantle'. But after long discussions among investigators from both missions, the 'Magnetic Pile-up Boundary' and 'Magnetic Pile-up Region' terminology was adopted. The more important point is that after the analysis of all available plasma parameters from Phobos-2 and Mars Global Surveyor a better understanding of the boundary's physical nature was obtained. Below, a summary of Phobos-2 and Mars Global Surveyor observations is given which will help to clarify that the several boundaries observed in the magnetosheath are one and the same plasma boundary.

3.1. The Magnetic Pileup Boundary

The MPB is always present, even when no 'cavity boundary' is observed. Figure 3.2 illustrates such a case with the bow shock and MPB crossings detected by

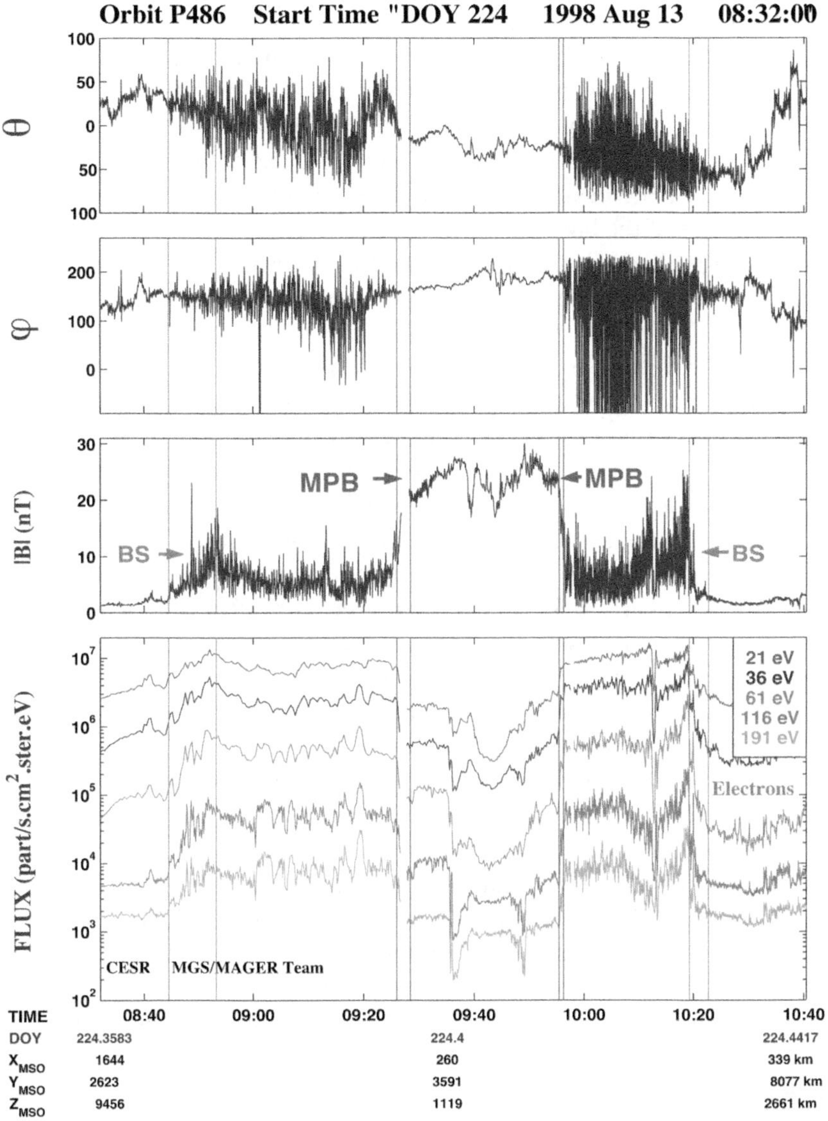

Figure 3.2. Magnetic field and electron data recorded by the Mars Global Surveyor MAG/ER experiment on August 12, 1998 (after Vignes *et al.*, 2000). The three upper panels display the elevation, azimuth angles and the magnitude, respectively, in the MSO system. The lower panel depicts electron fluxes for five energy ranges.

the Mars Global Surveyor magetometer/electron reflectometer (MAG/ER) experiment near the dawn – dusk meridian plane. The coordinate system used is the Mars-centered Solar Orbital (MSO) coordinate system: the X-axis points from Mars to the Sun, the Y-axis points antiparallel to Mars' orbital velocity and the Z-axis completes the right-handed coordinate system. The magnetic field azimuth (with $\varphi = 0$ antisunward here), elevation with respect to the (X, Y) plane and amplitude are plotted in top panels 1, 2 and 3. The last panel shows the electron fluxes for five energy ranges labeled with their geometrical mean energy. Each bow shock (BS) and MPB crossing is indicated between two vertical lines showing the apparent thickness of the boundary. The magnetosheath which can be defined as the region between the bow shock and the MPB is characterized by a huge variability both in the field magnitude and direction, which is particularly obvious on the two upper panels of the figure.

The MPB is revealed as a sharp discontinuity where the magnetic field magnitude suddenly rises by a factor between 2 and 3, where a strong reduction of the field's directional variability occurs, and where the suprathermal electron fluxes dramatically drop by often more than one order of magnitude (see, e.g., the 61 and 116 eV electron fluxes in Figure 3.2). The MPB signature in the field magnitude is sometimes less obvious when observed near noon local time, but both the sharp change in the field direction and the variability, and the change in the electron fluxes are always clearly observed for every crossing from the magnetosheath to the MPR. All the MPB crossings are thus identified in the MGS data by three simultaneous signatures: a more or less sharp increase of the magnetic field magnitude, a correlated reduction of the electron fluxes for energy greater than 10 eV, and a decrease in the fluctuations of the magnetic field.

The Plasma Wave System (PWS) aboard Phobos-2 recorded the spectra of plasma waves in the frequency range 0–150 kHz and also measured the plasma density using a Langmuir probe (Grard *et al.*, 1989a, b). Figure 3.3 shows an example of wave measurements during the fourth elliptical orbit of Phobos-2. The bow shock and the MPB were encountered on February 11, 1989 at 11:03 UT and 11:24 UT, respectively. As can be seen, the signatures of both plasma boundaries are quite clear in the PWS data. The parameters shown are from top to bottom: (1) the dynamic spectrogram of the wave electric field measured by the dipole antenna in the range 3–60 decibels above 1 μV m^{-1}Hz$^{-1/2}$, (2) the root-mean-square of the current collected by the cylindrical Langmuir probe, (3) the high frequency variation of this signal, (4) the electric field averaged over the 100 Hz–6 kHz frequency bandwidth, and (5) its high frequency variation.

The bow shock is followed by a significant increase in the level of the broadband emissions (top panel), which are thought to be electroacoustic noise associated with the dissipation of the solar wind kinetic energy. This wave activity fades away as Phobos-2 approaches the MPB (Grard *et al.*, 1989, 1991, 1993). Panels 3 and 5 (from the top) also reveal that the plasma turbulence usually observed behind the shock decreases and reaches a minimum in front of the MPB. Here the Langmuir

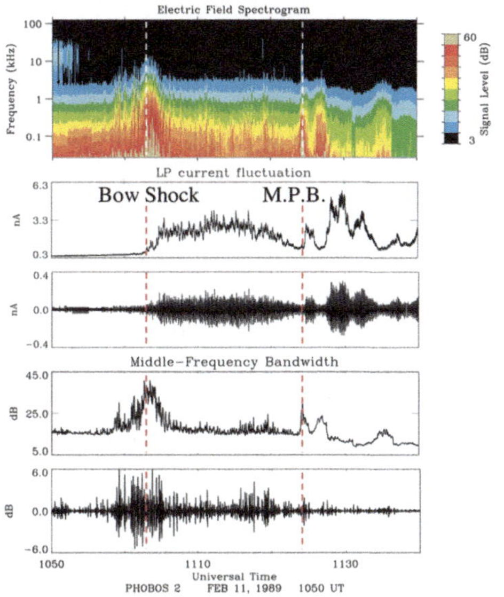

Figure 3.3. Electric field and plasma measurements in the Martian space environment from the plasma wave system onboard Phobos-2. From top to bottom: electric field spectrogram, Langmuir probe current fluctuation, high frequency variation of it, intensity of the electric field averaged over the 100 Hz–6 kHz frequency bandwidth, and high frequency variation of this signal. Crossings of the bow shock and MPB at 11:03 UT and 11:24 UT, respectively, are marked.

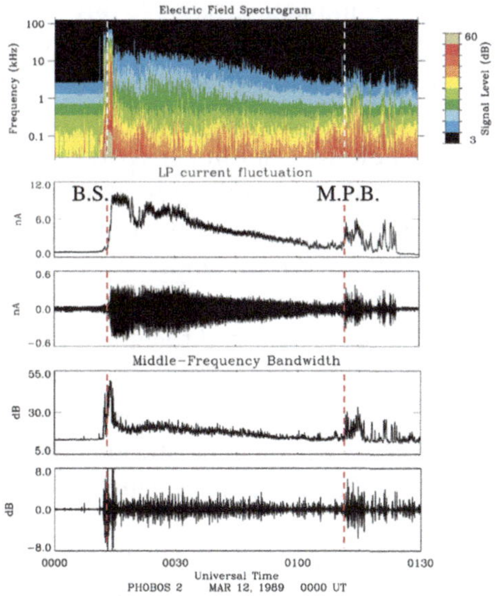

Figure 3.4. Similar to Figure 3.3. The bow shock, and MPB, were encountered along one of the Phobos-2 circular orbits around the red planet at 00:13:30 UT and 01:11:30 UT on March 12, 1989, respectively.

probe current fluctuation is very low (panel 2), indicating a low electron plasma density.

The broadband activity then suddenly increases again as seen in panel 1, and in the turbulence level (panels 3 and 5), as well as in the electron plasma density (panel 2). Thermal electron populations, called plasma clouds, reminiscent of the Venus plasma clouds, are indeed commonly observed behind the MPB. Their density may be as high as 700 cm^{-3} and their temperature is on the order of 10^4 K (Grard *et al.*, 1989, 1991). This is confirmed by the PWS parameters plotted in Figure 3.4, which is similar to Figure 3.3. Here, Phobos-2 crossed the shock at 00:13:30 UT, and the MPB at 01:11:30, March 12, 1989.

The magnetic field measurements are well correlated with the PWS data in identifying both the bow shock and the MPB (Figure 3.5). At the MPB the magnetic field rotation and the drop in the magnetic turbulence level is co-located with the characteristic PWS signatures as described above. The jump in the Langmuir probe current at the MPB indicates a strong increase of the electron density.

Direct measurements of the dynamics of protons and planetary ions by the plasma instruments TAUS and ASPERA are of essential importance for the understanding of the nature of the Magnetic Pile-up Boundary. Figure 3.6(a) depicts the first elliptical orbit of Phobos-2 on February 1, 1989 (Rosenbauer *et al.*, 1989), showing illustratively the inbound crossings of the bow shock and the MPB on the dayside, and outbound crossings in reverse order in the tail. Figure 3.6(b) shows the energy-time spectrogram of ions measured by TAUS along this orbit. The narrow distribution of the solar wind protons (left panels) is suddenly being heated and slowed at approximately 18:15, indicating passage through the bow shock. A short time later, at roughly 18:35, the proton flux is effectively terminated. At the same time planetary ions appear (right panel) (Rosenbauer *et al.*, 1989). The drop of the proton density at about 18:35 UT coincides with the inbound MPB position determined from the magnetic field measurements (Riedler *et al.*, 1989; Sauer *et al.*, 1990, 1992). The planetary ion fluxes continue to a distances of about 10 R_M. The change back to the solar wind (magnetosheath) ion composition occurs at about 23:30 UT, where the MPB is crossed outbound.

The outbound MPB crossing is also shown in Figure 3.7, where ASPERA density measurements and TAUS data are plotted together with the MAGMA magnetic field variation (Lundin *et al.*, 1989). The MPB is marked by the steep decrease in magnetic field amplitude and the strong change of the B_x component. Crossing this boundary from the MPR towards the magnetosheath side, the density of planetary ions (thin line) decreases, whereas the proton density (thick line) simultaneously increases.

A last example of multi-instrument detection of the MPB is given in Figure 3.8. Measurements are shown during one of the Phobos-2 closest approaches to the dayside of Mars, on February 8, 1989. The top panel represents the magnetic field magnitude. The bow shock was crossed at 05:35 UT. The decrease in the floating potential (second panel), which is positive in the solar wind due to photoelectrons,

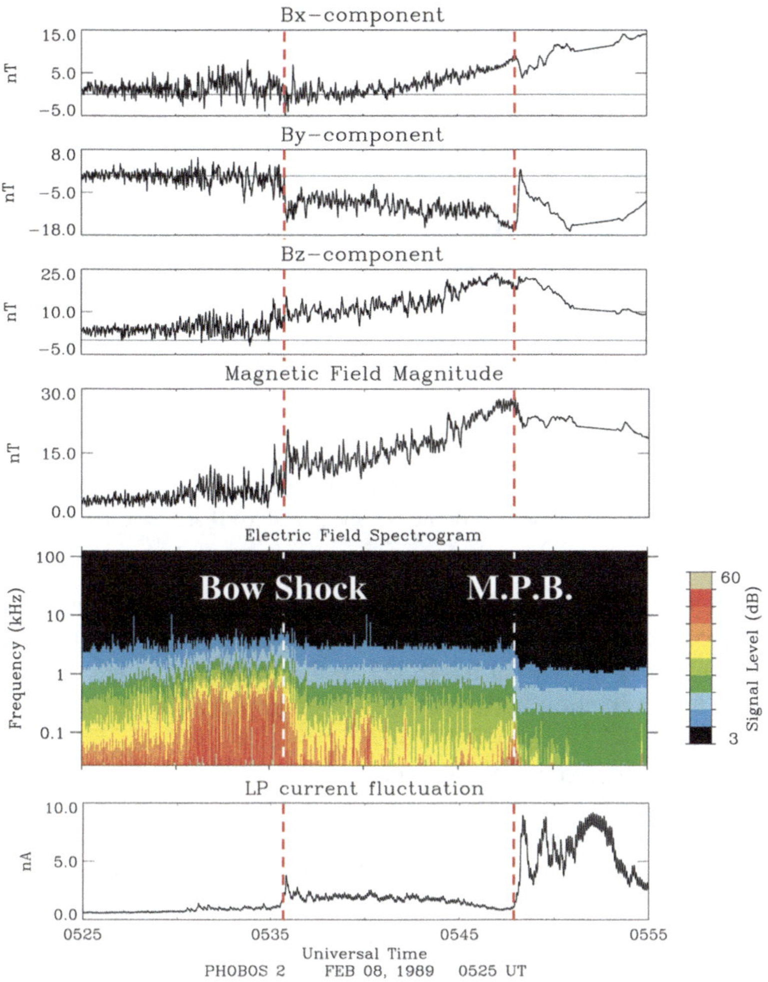

Figure 3.5. Magnitude and orientation of the magnetic field vectors measured during the third Phobos-2 elliptical orbit (Riedler *et al.*, 1989) (top four panels). Electric field dynamic spectrogram and root-mean-square of the current collected with the PWS Langmuir probe (bottom two panels). The crossings of the bow shock (05:35:50 UT) and MPB (05:48 UT) are marked by vertical dashed lines.

is caused by an increase in the density of ambient electrons. The electron number density (third panel) is derived from the measurements of the potential difference between the conducting spacecraft at floating potential and the negatively biased electric field probe. The results of the Langmuir probe, which could measure only a dense plasma ($n_e > 10\ \mathrm{cm}^{-3}$), are shown in the fourth panel. The next two panels depict the proton number density and the bulk speed measured by the ASPERA. At 05:48 UT, at the altitude of 880 km, the magnetic field reached a maximum and the proton fluxes dropped. However, in contrast to the entry into a true Earth-like magnetosphere, a sudden increase in the electron number density was observed.

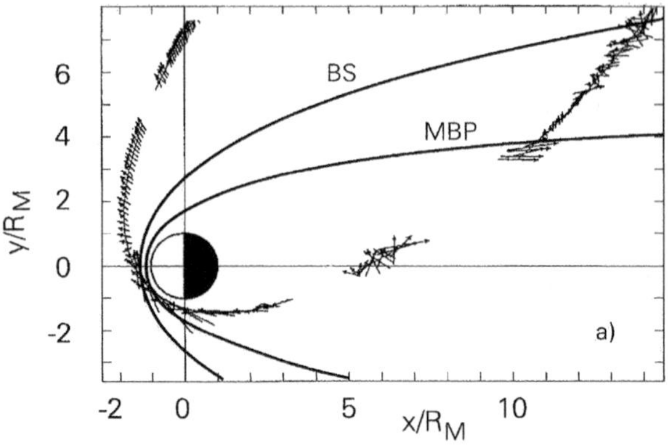

Figure 3.6. (a) View of the plasma boundaries in the equatorial plane, with arrows depicting the magnetic field vectors measured by Phobos 2 along the elliptical orbit on February 1, 1989. (b) Energy-time spectrograms of ions measured by TAUS during that orbit. The fluxes of the solar wind protons, left panel, and heavy ions ($m/q > 2$) of planetary origin, right panel, are shown (from Rosenbauer *et al.*, 1989). The crossings of the bowshock and MPB are indicated at the left.

This increase in n_e corresponds to the appearance of the planetary ions (the seventh panel). The bottom panel shows the wave activity in three frequency ranges. The increase of the wave activity at 05:48 UT in the frequency range of 5–50 Hz was attributed to the interaction between the two ion populations (Grard *et al.*, 1991).

3.2. The MPB Location and Shape

Attempts to characterize the location and shape of the boundary in the lower magnetosheath were first carried out using the Phobos-2 measurements. The major obstacles to a clear characterization were (1) the basic nature of the boundary (e.g., magnetopause versus Venus-like, etc.), and (2) the relative small number of boundary crossings by the spacecraft. Although there was no consensus of what kind of boundary it was, there were little doubts about its existence.

These problems were resolved by the Mars Global Surveyor mission, which showed that there is no global intrinsic magnetic field (and therefore no magnetopause), and which crossed the boundary many thousand times. With these measurements it is clear that this is a plasma boundary, that is formed by the interaction of the solar wind with the Martian exosphere/ionosphere. Furthermore, this kind of boundary may in fact be a general characteristic associated with non-magnetic solar system bodies which are surrounded by a gaseous envelope.

The MPB crossing locations inferred from the Phobos-2 PWS observations are shown in Figure 3.9 (diamonds). The off-center models of the bow shock (solid line) and MPB (dotted line) derived by Trotignon *et al.* (1993, 1996) are also displayed. On the vertical axis, $(Y'^2+Z'^2)^{1/2}$ is the distance of the Phobos-2 spacecraft from the aberrated tail X' axis, where (X',Y',Z') denote the aberrated Mars solar orbital system. It is worth noting that the orbital motion of the planet makes the

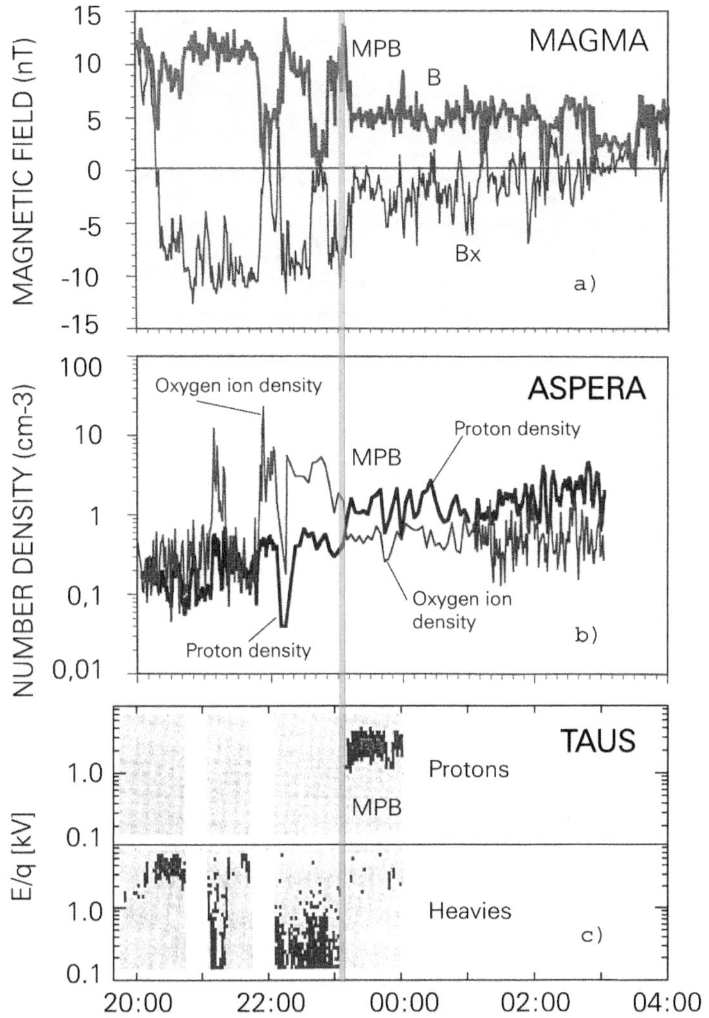

Figure 3.7. For the outbound crossing of the MPB at about 23:30 UT on February 1–2, 1989 seen in Figure 3.6, the magnetic field magnitude and x-component are shown in the top panel. The number density of protons (thick line) and planetary ions (thin line) are displayed in the center panel. The bottom panel displays the energy/charge measurements for this interval.

apparent direction of the solar wind flow deviate from the anti-sunward direction, therefore a 4° aberration angle has to be considered. Moreover, a better fit to the observations is obtained by allowing the foci of the conic, used to derive the bow shock and MPB models, to lie along the X' symmetry axis, as opposed to being fixed on the origin, i.e., the planet center (Slavin and Holzer, 1982).

Figure 3.9 also shows that there are only four observations on the dayside, and none near the sub-solar point. For this reason two conic sections were used to best fit the observed MPB locations: one on the night side of Mars, for $X' < 0$, and the

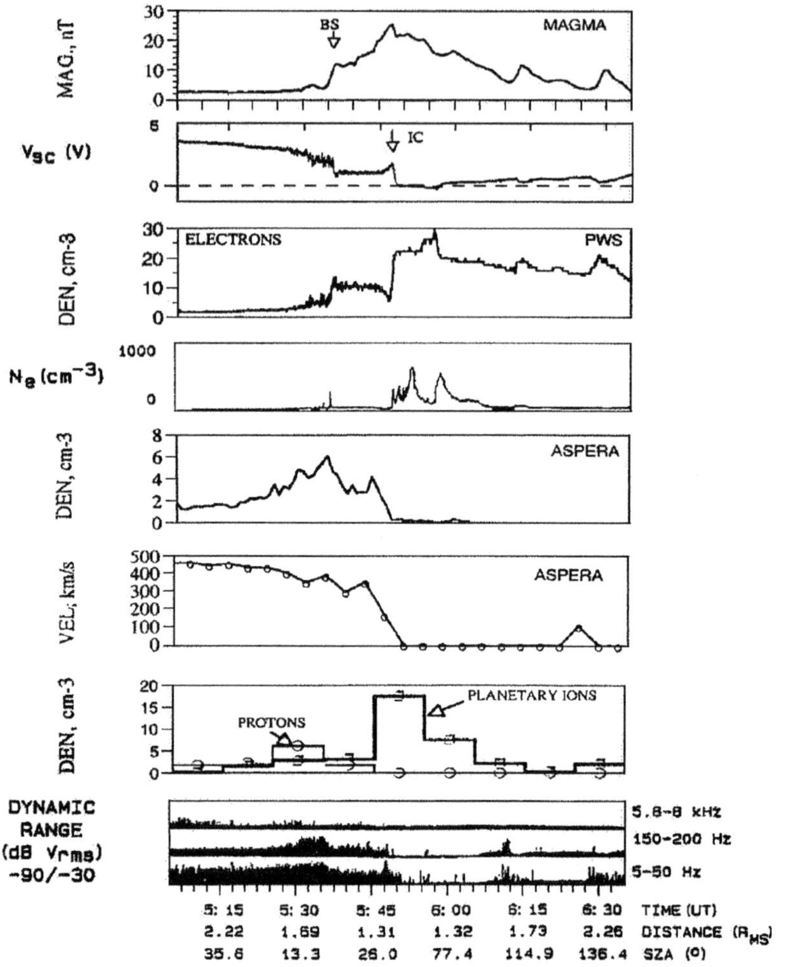

Figure 3.8. From top to bottom are the magnetic field magnitude, the spacecraft potential, the electron number density derived from the potential measurements, the x-component of the proton velocity, the density of planetary ions, and the electric field fluctuations in three frequency ranges along Phobos-2's third elliptical orbit on February 8, 1989 (from Dubinin *et al.*, 1996).

other on the dayside, for $X' > 0$. The night side fit was calculated first, because most of the crossing events occurred there. The dayside conic was then obtained by ensuring the continuity in amplitude and derivative between the two conics at $X' = 0$ (Trotignon *et al.*, 1996). The solid line in Figure 3.9 is the result of this process.

From this analysis the average planetocentric standoff distance of the MPB was found to be about 1.2 R_M (4050 km) at the subsolar point, and 1.4 R_M (4640 km) at the terminator, where Mars' radius R_M was taken to be 3390.5 km. These values are in good agreement with the ones given by Riedler *et al.* (1989), and Verigin *et al.* (1993). The Martian tail diameter was found to be 4.2 R_M at $X' = -2.3\ R_M$,

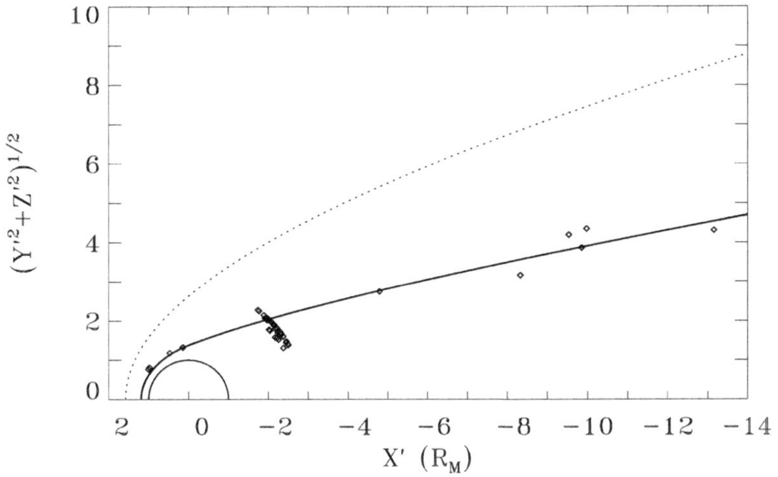

Figure 3.9. Crossings of the Magnetic Pile-up Boundary identified from observations by the plasma wave system aboard Phobos-2. There are 45 crossings (diamonds), of which only 4 are on the dayside. Models of the bow shock (dotted line) and the MPB (solid line), as derived by Trotignon *et al.* (1993, 1996). (X',Y',Z') are the aberrated Mars solar orbital coordinates in units of Mars radii R_M ($1 R_M = 3390.5$ km); A 4° aberration angle was considered.

and 7.6 R_M at $X' = -10\ R_M$. The Martian tail diameter (normalized by the planet's radius) appears to be about twice as large as the width of the Venus' induced magnetotail. Some interpreted the large diameter of the tail, reasonably so, as evidence for the presence of an intrinsic global magnetic field (e.g., Verigin *et al.*, 1993).

The Mars Global Surveyor measurements have now anchored many of the Phobos-2 observational uncertainties. The MPB was crossed by MGS during most of the elliptical orbit phases of the mission. During this phase of the mission, the highly inclined orbital plane explored a large range of local times and solar zenith angles, although the coverage was not uniform (for instance there is no data for MPB crossings below 20° solar zenith angle).

Figure 3.10(a) displays 488 MPB crossings by MGS from September 12, 1997 to August 29, 1998, in aberrated cylindrical coordinates (from Vignes *et al.*, 2000). The overall shape of the MPB was obtained by fitting an axi-symmetric conic section to the observed boundary crossings. The post-terminator crossing positions are much more variable than the ones on the dayside, and are not always easily identifiable. The magnetic signature of the MPB often appears to be weak and gradual at $X' < -2\ R_M$. This is at least in part due to the crossing geometry, since MGS crosses the boundary more tangentially in the post-terminator regime. The difficulties in identification may be reflected in the increased scatter of the data points far from the terminator (cases with poorly defined crossings were not included in this study).

The average planetocentric stand-off distance of the MPB was found to be about 1.3 R_M (4400 km) at the subsolar point, and about 1.5 R_M (5000 km) at the ter-

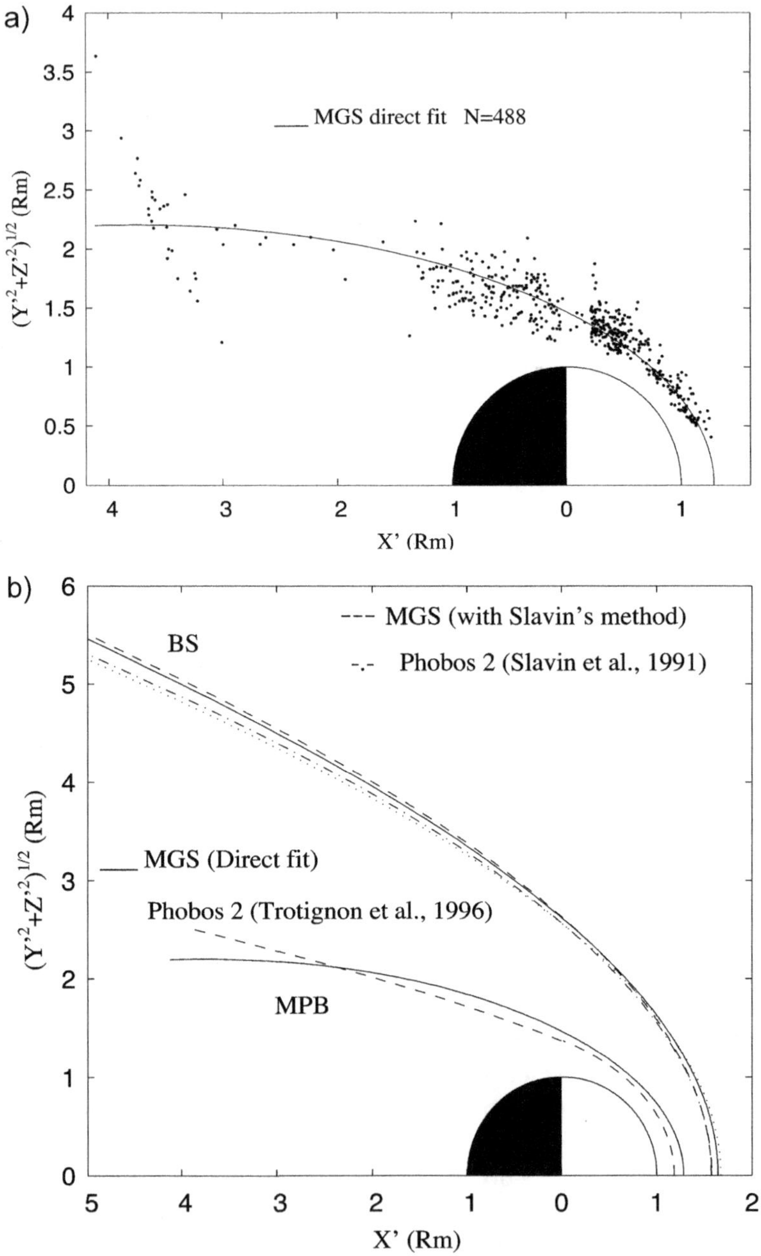

Figure 3.10. (a) MPB fit from MGS crossings during the first year of the mission in aberrated Mars-centered Solar Orbital (MSO) coordinates; (b) Comparison between Martian BS and MPB fits from MGS and Phobos-2 in aberrated MSO system (from Vignes *et al.*, 2000).

minator, using $R_M = 3390$ km. These values are in reasonably good agreement with the ones given by Trotignon *et al.* (1996) from a similar study using the 45 Phobos-2 'Planetopause' crossings in the equatorial plane (see Figure 3.9).

Figure 3.10(b) from Vignes *et al.* (2000) compares the shape and location results of the MPB inferred from these two studies and also helps to visualize the relative location of the MPB compared to the bow shock. The resulting fits for the MPB are very close, especially in view of the fact that Trotignon *et al.* (1996) used different criteria to define the boundary crossings, and had only 4 boundary crossings available on the dayside.

These results show that Phobos-2 and MGS explored the same plasma feature in the environment of Mars, the MPB, and that the MPB is a permanent plasma boundary. The location of the MPB is highly variable on a day-to-day basis. Its altitude seems to depend on planetary latitude especially in the southern hemisphere, which is the region that contains the strongest crustal magnetic fields (Crider *et al.*, 2002). But the high variability can also be produced by plasma processes associated with mass loading, such as the variation of the local convective electric field. This is the case for the bow shock, which is sensitive to variations of the upstream IMF orientation (Vignes *et al.*, 2002). The Phobos-2 and MGS missions occurred during very different phases of the solar cycle, but Figure 3.10(b) reveals that the mean bow shock locations appear very consistent.

Taking a subset of the MGS MPB crossings to filter out the effects of the crustal magnetic fields in the southern hemisphere and the possible pickup ion asymmetry in IMF coordinates due to the electric field direction, the distance of the MPB at the terminator is 241 km (or 20%) closer to Mars on average when the solar wind pressure exceeds 1.1 nPa, as shown in Figure 3.11. The response of the MPB to solar wind pressure may be indicative of a residual/leakage magnetic field resulting from the anomalies. Krymskii *et al.* (2002) discuss such a leakage of magnetic flux from the southern hemisphere.

3.3. Magnetic Field Draping

The magnetic field morphology of the solar wind interaction with Mars, just like the solar wind interaction with Venus and comets, is dominated by the 'draping' of interplanetary magnetic field (IMF) lines around the planet. In fact, most of the current understanding of magnetic field draping comes from investigations of Venus and comets starting with Alfvén's (1957) theory and subsequent applications (e.g. Luhmann, 1986; Schwingenschuh *et al.*, 1987; Cloutier *et al.*, 1999; Mazelle *et al.*, 2002). Figure 3.12 is a sketch depicting the magnetic field lines, and the parameters defining the draping geometry.

Locally, the field is expected to be predominately horizontal relative to a spherical surface on the dayside, or $|B_\phi| \gg |B_r|$, where B_ϕ is the azimuthal component of the magnetic field, and B_r is the radial component in spherical coordinates. The angle that the magnetic field makes to the local horizontal is the inclination

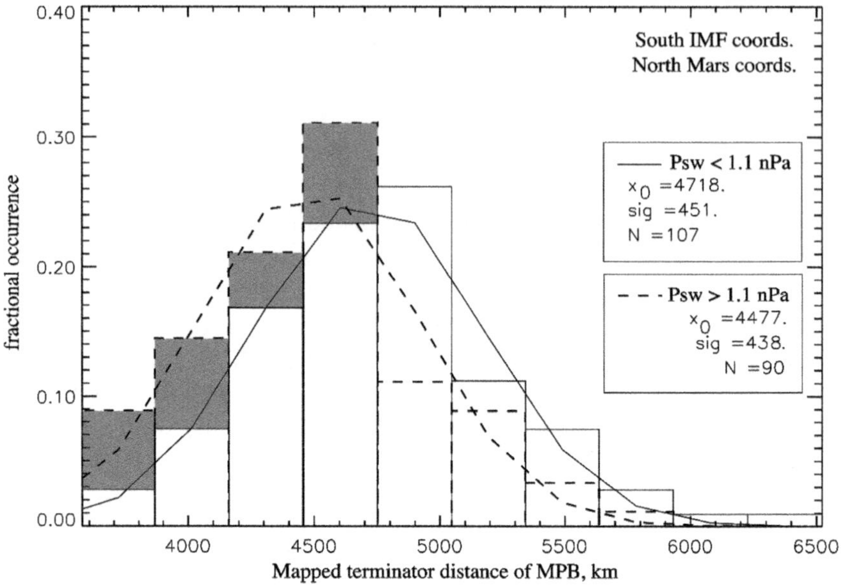

Figure 3.11. Histograms showing the distance of MGS MPB crossings mapped to the terminator plane. The shaded histogram includes all crossings in which the upstream solar wind ram pressure was greater than 1.1 nPa. The white histogram shows the lower solar wind pressure crossings. The curves are the Gaussian fits to the histograms. The associated parameters are shown in the boxes. x_0 is the average MPB distance (mapped to terminator plane) in km, sigma is the width of the gaussian, N is the number of points included in the distribution (from Crider *et al.*, 2003).

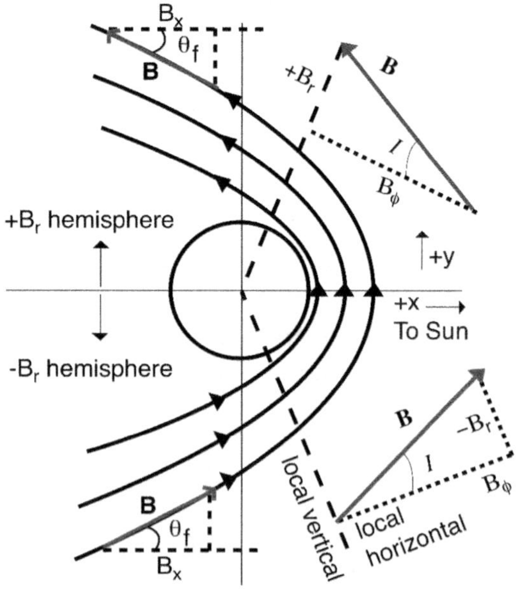

Figure 3.12. Sketch depicting the geometry of draped magnetic field lines at Mars. Positive x is toward the Sun, $+z$ is up from the ecliptic plane.

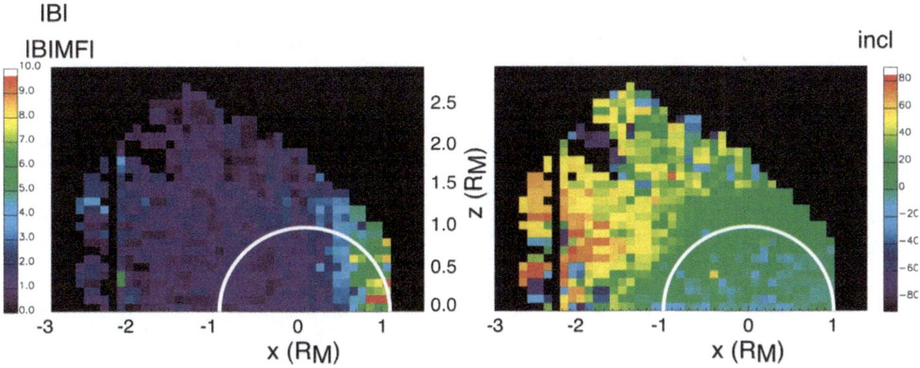

Figure 3.13. The left-hand panel is the MGS composite of the magnetic field magnitude around Mars. The right side shows the inclination angle as a function of position (see text).

angle, or $i = \arcsin(B_r/|\mathbf{B}|)$. In the dayside magnetosheath, the inclination angle is assumed to be small, and to have its sign vary with hemisphere. That is, in the upper half of Figure 3.12 the radial component is small and points away from the planetary center, giving rise to a positive angle i. In the lower half of the figure, B_ρ is towards the planet, yielding a negative i. The magnetic field in the tail is oriented more nearly radial than it is on the dayside. One determines the average direction of the induced magnetotail using the flare angle, θ_f. The flare angle is defined as the angle the magnetic field makes with the x-axis, $\theta_f = \arccos(B_x/|\mathbf{B}|)$.

The most comprehensive data set regarding magnetic field draping comes from the Mars Global Surveyor spacecraft owing to its extensive coverage in altitude and local time during its elliptical orbit mission phase. Crider *et al.* (2001, 2002) used the MGS data to determine the average magnetic field configuration resulting from the solar wind interaction with Mars. The data are grouped in spatial bins that are cubes with length of 0.1 R_M per side. Each segment of an MGS orbit that falls within a bin contributes one magnetic field value to the bin average. The images in Figure 3.13 are the projection of these bins onto a single half-plane. The left panel is the magnitude of the field and the right panel is the inclination angle.

The average draping parameters found for certain regions of interest are summarized in Table 3.1. The magnetic field values in this survey are all taken from positions far away from known crustal magnetic sources. Therefore, the topology is due primarily to the solar wind interaction.

The left panel in Figure 3.13 shows that the magnetic field magnitude ($|\mathbf{B}|$) is highest in front of the planet. $|\mathbf{B}|$ decreases with increasing solar zenith angle and increasing altitude. The right panel in Figure 3.13 shows a large green area close to the planet where the magnetic field is primarily horizontal. The average inclination angle, $\langle i \rangle$, increases from 0° at local noon to 10° at the terminator. Note that these values are adjusted in sign by hemisphere such that a positive value is expected (Crider *et al.*, 2001).

TABLE 3.1

Draping parameters from the Mars Global Surveyor magnetometer data.

Parameter	Average	Standard dev.
$\|\mathbf{B}\|$ (day)	20.7	±14.7
$\|\mathbf{B}\|$ (night)	11.6	± 7.6
θ_f (day)	57.6	±26.6
θ_f (night)	40.0	±27.8
i (day)	5.6	±13.6
i (night)	12.5	±33.6

Crider *et al.* (this issue) also find that the inclination angle increases with altitude on the dayside. On the night side, the magnetic field tends to be highly inclined to the local horizontal. This is the magnetic field geometry that is expected in the magnetotail (e.g., McComas *et al.*, 1986). Nearly radial magnetic fields, such as these, have strong implications for the existence of night side ionospheric holes at Mars, as were observed at Venus (Marubashi *et al.*, 1985). However, the possibility remains that the magnetic field in the Martian wake contains a significant contribution from magnetic flux of planetary origin that leaks out into the tail (Krymskii *et al.*, 2002).

Bertucci *et al.* (2003a) studied the variation of draping across the MPB, using the simple approach shown in Figure 3.14. As the MGS trajectory crosses field lines closer and closer to the planet , it is clear that $|B_x|$ should tend to increase due to draping. However, one must consider the 3-dimensional field geometry. Figure 3.14 shows a $y - z$ projection at some $x > 0$ of the IMF transverse component $\mathbf{B_T}$ with its two cylindrical component B_r and B_t, as well as the projected MGS trajectory. A consideration of successive slices at different x positions helps to show that in a draping regime, the changes in B_x will be accompanied by a variation of $\mathbf{B_T}$ such that its radial cylindrical component B_r and B_x will be correlated. In the Martian magnetosheath, the magnetic field direction is highly variable and, despite the incipient draping, no correlation would be expected. However, in the magnetic pileup region, the magnetic field becomes more regular and stable, and in general, $|B_x|$ is smaller than the transverse component $|B_T| = (B_y^2 + B_z^2)^{1/2}$. The latter is a necessary condition for draping to exist.

The results of the correlation analysis between B_x and B_r are summarized in Figure 3.15 for one representative MGS orbit. In the Martian magnetosheath, upstream of the MPB (panel on the left), the lack of correlation shows that the draping is undetectable. In contrast, in the magnetic pileup region (downstream of the boundary), the very high linear correlation coefficient reveals a dramatic

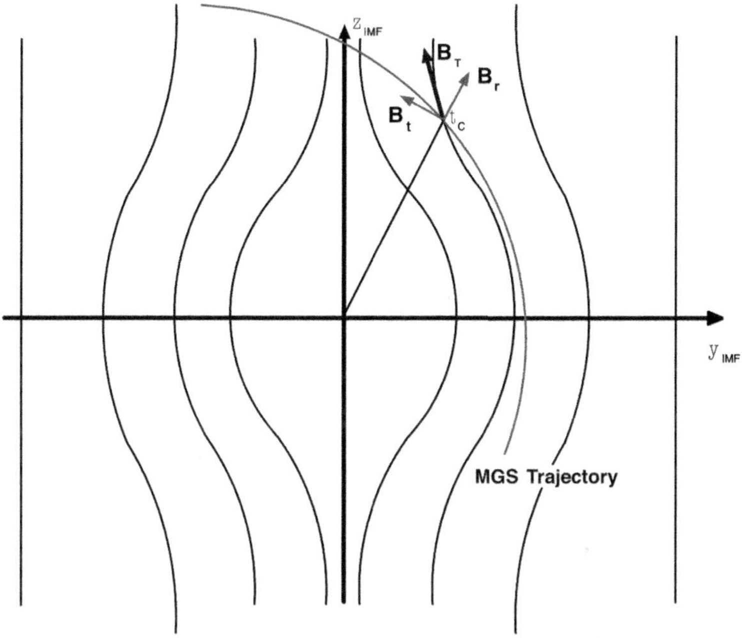

Figure 3.14. Slice perpendicular to V_{SW} of the 3D field line pattern and projection of the MGS trajectory, containing the spacecraft position at some $x_{IMF} > 0$. B_r is the radial cylindrical component of the transverse magnetic field B_T at the spacecraft position. The draping at Mars revealed by the correlated variation between $B_{x'}$ and B_r (Bertucci *et al.*, 2003a).

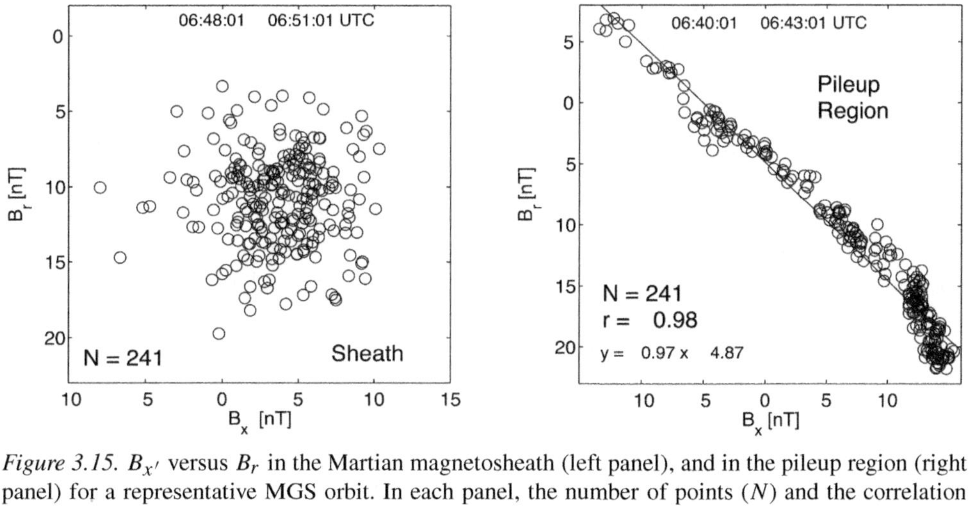

Figure 3.15. $B_{x'}$ versus B_r in the Martian magnetosheath (left panel), and in the pileup region (right panel) for a representative MGS orbit. In each panel, the number of points (N) and the correlation coefficient (r) corresponding to the best linear least-square fit is shown (Bertucci *et al.*, 2003a).

increase in the draping as the magnetic field configuration becomes regular (right panel). This striking result shows that the Martian MPB represents the outer edge of the region where the draping effect becomes prominent. Furthermore, this feature can be used to identify the MPB, especially when the signature on the magnetic field intensity is ambiguous.

The draping enhancement across the MPB is a common feature also at comets. In particular, Israelevich *et al.* (1994) have used this same method with the Giotto spacecraft magnetometer data and obtained very similar results at comet P/Halley. The MPB (crossed on the dayside) separates the cometary magnetosheath, where there is no evidence of significant draping, from the magnetic pile-up region where there are strongly draped fields. Similar properties have also been observed for the MPB crossing (also on the dayside) by the Giotto spacecraft at comet P/Grigg-Skjellerup (Neubauer *et al.*, 1993), and across the magnetic tail of comet P/Giacobini-Zinner by the ICE satellite. Slavin *et al.* (1986) report turbulent, weakly draped magnetic fields in the ionosheath, while at the interior of the tail boundary the magnetic field increases in magnitude and adopts a draped configuration. The latter observation enforces the idea that not only at comets, but also at Mars, the magnetotail boundary and the MPB are one and the same plasma boundary (Neubauer, 1987; Mazelle *et al.*, 2002). Recently, Bertucci *et al.*, (2003b), using the same method as Bertucci *et al.*, (2003a), reported a strong enhancement of the magnetic field draping at the outer part of the magnetic barrier region at Venus as a clear evidence for a magnetic pileup boundary.

3.4. Observations of Low Frequency Waves at the MPB

Compared to the broad spectrum of waves usually observed in the magnetosheath, the vicinity of the Martian MPB often contains well-defined compressive, low frequency waves. This should not be surprising, since around the MPB plasma properties, such as the plasma $\beta(= 8\pi nkT/B^2)$, and the particle anisotropy, change significantly. Figure 3.16 shows MGS MAG/ER data for a $\sim$ 10 a.m. local time orbit (Bertucci *et al.*, 2002). The MPB is crossed a few seconds after 04:36 UTC, at $\sim$ 700 km altitude. On both sides of the boundary the magnetic field, steady in direction, displays high-amplitude compressive fluctuations. The timescale of these fluctuations is of the order of a few tens of seconds (typically, 20s), well below the local proton gyrofrequency on both sides of the boundary. On the upstream side, the oscillations in $|B|$ are 'tooth shaped', i.e., a series of dips superimposed on a nearly constant background value (type '1'). Inside the MPB, the waves are quasi-monochromatic and much more coherent (type '2'). An analysis of the MAG data together with the ER data at both sides of the MPB is very useful in order to identify the modes to which these waves are associated. An examination of the suprathermal electron fluxes shows that they are anticorrelated with $|B|$ upstream from the MPB, while downstream, the two quantities are correlated (vertical point-

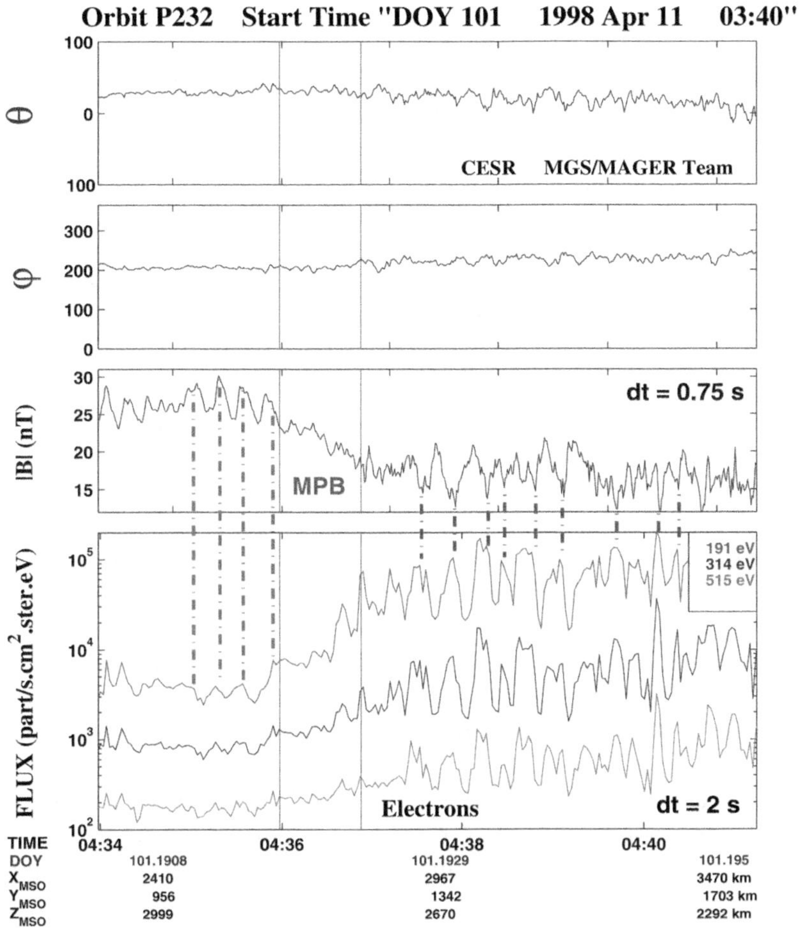

Figure 3.16. Observations by MAG/ER of compressive, linearly polarized low frequency waves around the MPB for an orbit for $a \sim 10$ a.m. local time orbit. Point-dashed lines emphasize the anti-correlation (correlation) between the electron fluxes and $|B|$ upstream (downstream) from the boundary, revealing mirror mode-like (fast mode-like) waves, respectively (from Bertucci *et al.*, 2002).

dash lines). In this figure, only the 190–520 eV range is depicted for clarity, but the same behavior is observed in all suprathermal energies ($E > 10$ eV).

For each wave event, an ambient field vector B_0 is computed by time-averaging the individual vector measurements over the event duration. By defining a mean field coordinates system with one axis along B_0, one can separate the wave magnetic field component along this direction (the compressive component $B_{\|}$), from the perpendicular components. On both sides of the MPB, the oscillatory signal is primarily in the $B_{\|}$ component. This means that the waves are linearly polarized along $\mathbf{B}_0$. The minimum variance analysis (MVA) gives wave vectors $\mathbf{k}$ nearly perpendicular to $\mathbf{B}_0(\theta_{KB} = 89°$ and $88°$, upstream and downstream from the MPB respectively). Computation of the aflvénic magnetic field wave components in the direction $\mathbf{k}\times\mathbf{B}_0$, and magnetosonic wave components along $\mathbf{k}\times(\mathbf{k}\times\mathbf{B}_0)$, also shows that the oscillations are fully reproduced in the magnetosonic component alone. For numerous MGS orbits the low frequency waves observed in close vicinity of the MPB are always compressive and linearly polarized, with propagation nearly normal to $\mathbf{B}_0$. Such waves are observed neither in the magnetosheath proper (closer to the bow shock) nor in the solar wind. The occurrence of these waves is another characteristic feature of the Martian MPB. As a result of a survey of 282 MPB crossings by MGS, type 1 waves occur in at least 48%, type 2 waves in at least 27%, and both modes simultaneously in at least 18% of the observations. Finally, at least 11% of the observations show neither type 1 nor type 2 waves.

The characteristics of the waves upstream from the boundary (linearly polarized, quasi-perpendicular propagation, anti-correlation between $|B|$ and the suprathermal electron density) are those of mirror mode waves. This purely kinetic mode is stationary in the plasma frame ($\omega_r = 0$) for a homogeneous medium. It is driven by anisotropies in the plasma pressure ($\beta_\perp/\beta_{||} > 1 + 1/\beta_\perp$, where $\beta_\perp$ and $\beta_{||}$ are the perpendicular and parallel plasma β), especially when β is high. At Mars these waves have length scales of several upstream proton gyroradii, which is consistent with theoretical properties. One of the possible sources for generating anisotropies is the heating of the ion population (mainly perpendicular to $\mathbf{B}$) downstream from quasi-perpendicular shocks. However, mirror mode waves at Mars appear to be independent of the shock's geometry. On the other hand, they appear always attached to the boundary, as if they are generated upstream from the MPB, grow and are convected down to the MPB in the decelerated flow. Distributions (likely highly non-gyrotropic) of newborn pickup ions (especially heavies, e.g., O^+) also contribute to large $\beta_\perp$ values upstream from the MPB, where the β is high since the ambient magnetic field has a low magnitude on average, and the thermal pressure of the shocked solar wind is high. In the case of comets, mirror-mode waves have been reported on the outer side of the magnetic tail boundary of P/Giacobini–Zinner (Tsurutani *et al.*, 1999). This is more evidence of the connection between the dayside MPB and the magnetic tail boundary.

The properties of the waves downstream from the MPB (type 2) coincide with those of magnetosonic fast-mode waves, and their nature is very different from that of mirror mode waves. Moreover, the β strongly decreases at the MPB since the magnetic pressure is much higher in the MPR, and the hot solar wind particle populations are replaced by colder populations of planetary origin. Thus, downstream of the MPB, the plasma becomes stable to the mirror mode, and the mirror mode waves are replaced by magnetosonic fast-mode waves. However, in light of their large amplitude and the small scale of the MPR, it is difficult to explain the fast mode waves as arising from locally growing plasma microinstabilities (especially ones with perpendicular propagation). Alternatively, the waves can be interpreted as stationary, bi-ion waves standing downstream of the MPB (e.g., Sauer *et al.*, 1990). Interestingly, the same relationship of mirror mode and fast magnetosonic mode (Mazelle *et al.*, 1989) was reported by Mazelle *et al.* (1991) at both sides of comet P/Halley MPB. In a related study, Glassmeier *et al.* (1993) suggest an equivalent interpretation for the waves at the MPB. However, further studies still need to be done in this respect.

3.5. THE NATURE OF THE MPB/MPR – DISCUSSION

The extensive magnetic field measurements by MGS down to altitudes below 100 km established that at present there is no planetary dynamo operating in Mars. While the remnant fields are locally strong, they are too weak to stop the solar wind, on a global scale, at the observed distances. Thus the interpretation of the 'obstacle boundary' as a magnetopause is, in the classical sense, inappropriate.

Further, since the first spacecraft measurements around Mars it has been known (or at least strongly suspected), that the ionospheric pressure is also typically insufficient to stop the solar wind at the observed distances. Therefore, an alternate explanation of the interaction is required.

If one combines the results of Phobos-2 and MGS data analyses, the signatures of the Magnetic Pile-up Region and the Magnetic Pile-up Boundary can be summarized as follows:

(1) Moving from the magnetosheath to the MPR, the magnetic filed piles up, accompanied by magnetic field rotation.
(2) There is a drop in the magnetic field fluctuations.
(3) There is a drastic change of ion composition: the proton density decreases, commensurate with an increase of planetary ion density.
(4) There is a sharp increase of the electron density.
(5) The electron temperature drops.
(6) Localized high-frequency emissions appear.
(7) The MPB is a permanent feature of the interaction.
(8) Two interaction mechanisms that potentially could account for all or part of these characteristics are charge exchange, and mass loading.

Breus *et al.* (1989), Ip (1992) and Ip *et al.* (1994) have suggested that charge-exchange of the solar wind protons with an oxygen-rich atmosphere may cause the termination of the proton fluxes. However, Lichtenegger *et al.* (1997) have shown that this process can only lead to a significant depletion of the solar wind in a small region above the subsolar region at low altitudes. The absorption by oxygen leads to a significant effect only below about 1500 km, while neutral hydrogen has only a small influence on the solar wind loss at Mars.

A complication in the charge-exchange calculation is the relatively large gyroradius of the protons compared with scale of the system. Consideration of their finite gyroradius indicates that the path through the absorption region is elongated, effectively increasing the losses. Nevertheless, the total absorption near the terminator in the altitude region between 500 and 1500 km is only about 1% to 6% (similar to Venus (Gombosi *et al.*, 1981)). It is worth noting that the calculations by

Lichtenegger *et al.* (1997) were made for solar-maximum conditions, which were appropriate for the Phobos observations.

Lichtenegger and Dubinin (1999) reexamined the mechanism of charge-exchange, motivated by the MGS results and in anticipation of Nozomi mission results. They used the model by Krasnopolsky and Gladstone (1996) to calculate the solar wind absorption for solar minimum and solar maximum conditions. At solar minimum, when the density of hydrogen constituents (H and H_2) is expected to be much higher, the absorption was found to reach more than 20%. The neutral densities are too low at the required distances, therefore it was thought unlikely that charge-exchange processes are dominantly responsible for the observed proton flux termination.

Sauer *et al.* (1990, 1992, 1994, 1995) have suggested that the sharp boundary terminating the access of the solar wind is caused by the direct interaction of the solar wind protons with heavy ions of planetary origin, a mass-loading process. In addition to the features related to the extended atmosphere, the importance of the ion gyroradius effect is unique at Mars. The pick-up gyroradius of the O^+ ions significantly exceeds the characteristic size of the system upstream of the bow shock, and it becomes comparable with the magnetosheath width at closer distances. In such a configuration a relative streaming of the different ion species is possible. The existence of a second ion population leads to an additional coupling between the ions and electrons through the Lorentz forces and the charge neutrality requirement. Even a small admixture of heavy ions into a proton-electron plasma significantly modifies the flow properties (Dubinin and Sauer, 1999). These issues are discussed further in the Section 6, 'Theory and Modeling'.

4. The Tail Region

The first measurements in the Martian tail were obtained during the Mars-5 mission (Gringauz *et al.*, 1975; Dolginov *et al.*, 1976a; Vaisberg *et al.*, 1975). The measurements led to a qualitative demonstration of the compressibility of the Martian magnetotail with increasing solar wind ram pressure. A boundary layer was revealed, and measurements of ion spectra with unusually high energies also led to speculations concerning the existence of the magnetotail plasma sheet, even though Mars-5 did not reach the inner region of the Martian tail (Dolginov *et al.*, 1976b; Gringauz *et al.*, 1976).

The plasma investigations of the Phobos mission for the first time measured directly the features of the deep tail region, as well as the plasma characteristics in the equatorial region of the Martian tail (the MGS orbits did not include the tail region to an appreciable extend).

4.1. Hints of a Planetary Magnetic Field

It is well known that the typical distribution of the electrons in the solar wind is asymmetrical in the wings, because of halo electrons streaming out from Sun along the interplanetary magnetic field. Dubinin *et al.* (1994) have applied the measurements of the electrons in the wings as tracers of the field lines, because streaming electrons have to replicate field line kinks associated with the draping of the IMF around Mars. This approach provides a very sensitive tool for the study of magnetic field topology. Figure 4.1 illustrates this method.

In the solar wind, due to the halo component, the antisunward/sunward asymmetry of the electron fluxes is positive. The draping of the IMF around Mars changes the field direction in some regions and therefore changes the anisotropy of the electron fluxes. In the region where the B_x-component changes sign, because of draping, the sunward flux exceeds the antisunward flux.

While the B_x-component changes sign many times during the tail crossing, the variations in the anisotropy of the halo electrons generally replicates the variations in B_x well, thus indicating that the flux tubes encountered by the spacecraft were populated by halo electrons and thus connected to the IMF. This is an indication that the magnetic field around Mars is primarily induced. On the other hand, there are a number of tail crossings with anomalous characteristics. Figure 4.1 (bottom) shows one such example.

The magnetic field shows a rather stable orientation with respect to the current sheet separating the two lobes. Suprathermal electrons trace the field variations except in the region bounded by the dashed lines. The drop in the electron flux by a factor 3–5 is an interesting feature. A violation of the electron tracing suggests that this region of the tail may be magnetically disconnected from interplanetary space. These observations were considered as indirect evidence of an intrinsic magnetic field. The measurements made by Phobos-2 only covered the near equatorial region. The more recent MGS observations (Acuna *et al.*, 1998) have demonstrated that the regions with strong crustal magnetization, those that can give rise to the observed anomalies in the tail, are mainly localized in the mid and high latitudes of the southern hemisphere.

Further indirect evidence of the presence of a Martian magnetic field came from radio observations of the night-time ionosphere. The single frequency radio-occultations during the primary and extended Mariner 9 missions did not detect its existence (Kliore *et al.*, 1972, 1973). Subsequent and more sensitive dual-frequency Mars 4,5 radio experiments revealed the existence of the Martian night-time ionosphere, with an electron density peak of about 5×10^3 cm^{-3} (Vasiliev *et al.*, 1974) (also, see the Ionosphere section below). However, Viking 1 and 2 showed that this peak occurs only about 40% of the time (Zhang *et al.*, 1990b). Phobos-2 measured typical electron fluxes of $\sim 2 \times 10^8$ cm^{-2} s^{-1} in the magnetotail, and Verigin *et al.* (1991a;b) and Haider *et al.* (1992) suggested that these fluxes are responsible for the measured ionospheric densities (Figure 4.2). Further, they

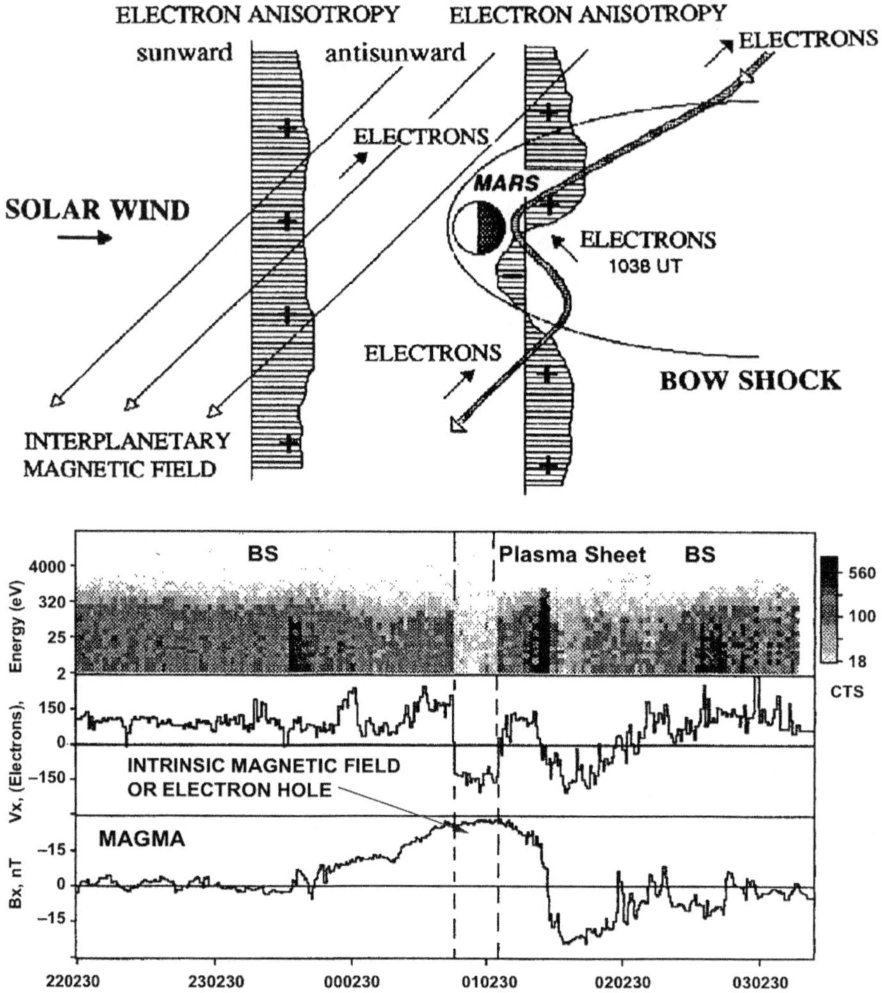

Figure 4.1. (a) A sketch depicting how the halo component of the solar wind electron flux streaming along the interplanetary magnetic field lines traces the magnetic field topology near Mars. Electron fluxes measured by sunward and antisunward looking sensors of the ASPERA instrument in the solar wind will see the antiparallel streaming (left), even in the region where the B_x component of the IMF changes sign because of draping around Mars (right) (from Dubinin *et al.*, 1994). (b) Phobos-2 orbit on March 2, 1989. The interruption of the electron flux in the time interval marked by dashed lines, and the fact that the change in flux occurs without a correlated change in B_x, gives a hint that Phobos-2 encountered flux tubes that are not connected with the IMF (from Dubinin *et al.*, 1994).

proposed that a weak intrinsic magnetic field, partially screening the Martian nightside atmosphere, might frequently prevent magnetotail electrons from reaching the ionosphere.

Finally, straightforward analysis of Martian magnetotail boundary crossings by the Phobos 2 orbiter indicated that equatorial crustal magnetization appears to in-

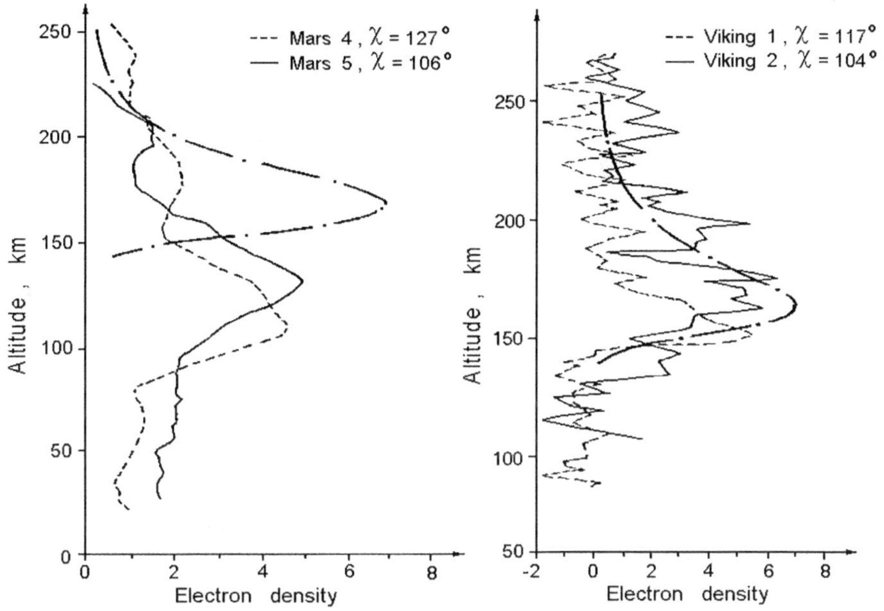

Figure 4.2. The electron density (in units of 10^3 cm^{-3}) profiles in the Martian night-time ionosphere as measured with Mars 4 and 5, and Viking 1 and 2 radio-occultations. Also shown is the calculated electron density (dash–dot). χ is the zenith angle of the point where radio beam was tangent to planet (from Verigin et al., 1991b).

fluence its location (Verigin *et al.*, 2001). Qualitative analysis demonstrated that the magnetotail thickness can be affected significantly by a region of enhanced crustal magnetization, if this region is located in the vicinity of the planetary terminator. The smoothed dashed line in Figure 4.3, which is passing through the MPB (called 'magnetopause' in the reference) positions observed under intermediate ρV^2 conditions (5×10^{-9} dyn cm^{-2} $< \rho V^2 < 1.4 \times 10^{-8}$ dyn cm^{-2}), separates the observations corresponding to low and high ram pressures. This curve has definite bulges, corresponding to enhanced magnetotail thickness, in the ranges of areographic longitudes of about 30°–80°, 120°–160°, 210°–270°, and 300°–340°. These four longitude ranges correspond well to the regions of enhanced magnetization of the Martian crust as reported by Acuña *et al.* (1999). Specific local displacement of the magnetotail boundary downstream of these regions were estimated to be ~ 500–1000 km.

4.2. Magnetic Field Topology

A good insight into the formation of the Martian magnetotail was obtained by observations during the multiple crossings of the tail by the Phobos 2 orbiter. Yeroshenko *et al.* (1990) found that the magnetic field in the magnetotail can be organized into two tail lobes with 'to' and 'away' from Sun directions of the

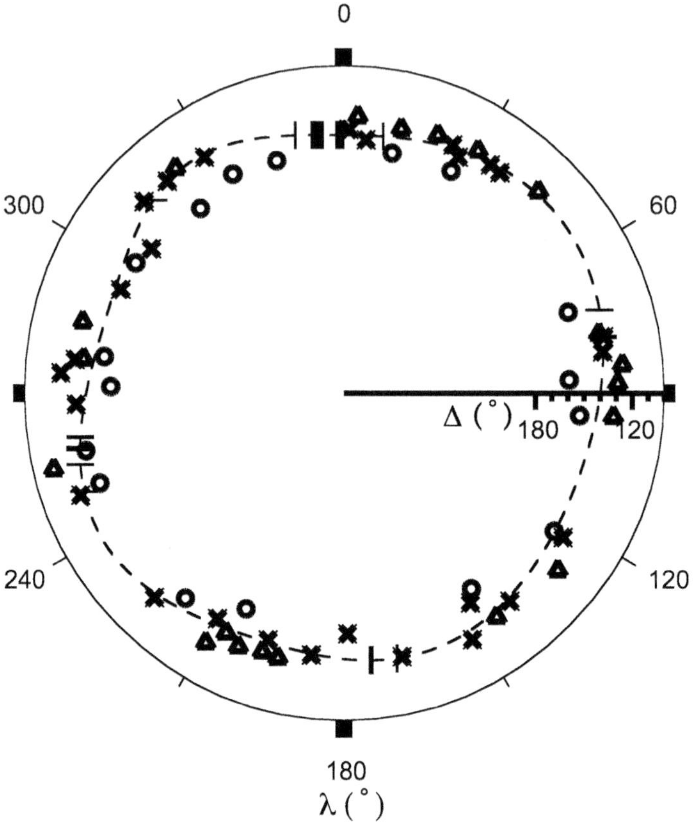

Figure 4.3. Dependence of the tail boundary zenith angle (Δ), observed by Phobos 2, on the east areographic longitude, λ. Boudary locations (i.e., Δ) observed during intermediate ρV^2 conditions (5×10^{-9} dyn cm^{-2} $< \rho V^2 <$ 1.4×10^{-8} dyn cm^{-2}) are denoted by x. The boundary locations observed during low and high solar wind pressures are denoted by triangles and circles, respectively. The dashed line is smoothly drawn through the intermediate locations. The bulges occur at longitudes of strong crustal fields (from Verigin et al., 2001).

magnetic field (Figure 4.4), separated by a thin magnetic neutral sheet. This ordering was achieved by rotating the original reference frame around the solar wind velocity vector until the perpendicular component of the interplanetary magnetic field was aligned with the Y-axis of the rotating reference frame. Presence of the two tail lobes in the rotating reference frame, confirmed later by Schwingenschuh *et al.* (1992a) with the use of a wider data set, provided evidence for the inductive character of the Martian magnetotail.

Similar to the geomagnetic tail, the magnetic field intensity, B_t, in the Martian magnetotail provides pressures balance, with the external solar wind ram pressure ρV^2 and, hence, B_t should be a function of this pressure. Figure 4.5 presents experimental confirmation of this expected dependence, obtained with the use of TAUS

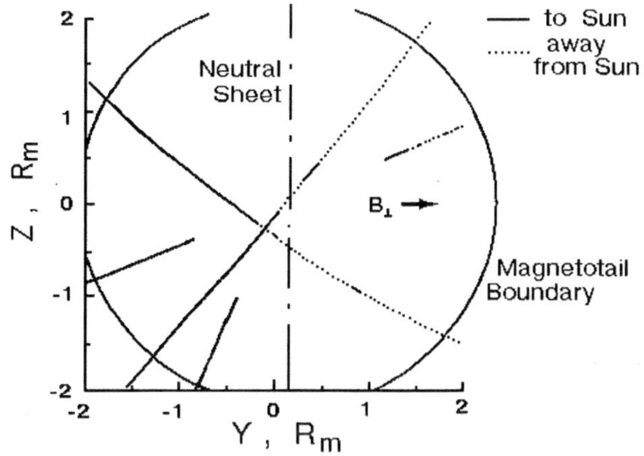

Figure 4.4. The magnetic field in the tail is separated into two lobes (see text) (from Yeroshenko et al., 1990).

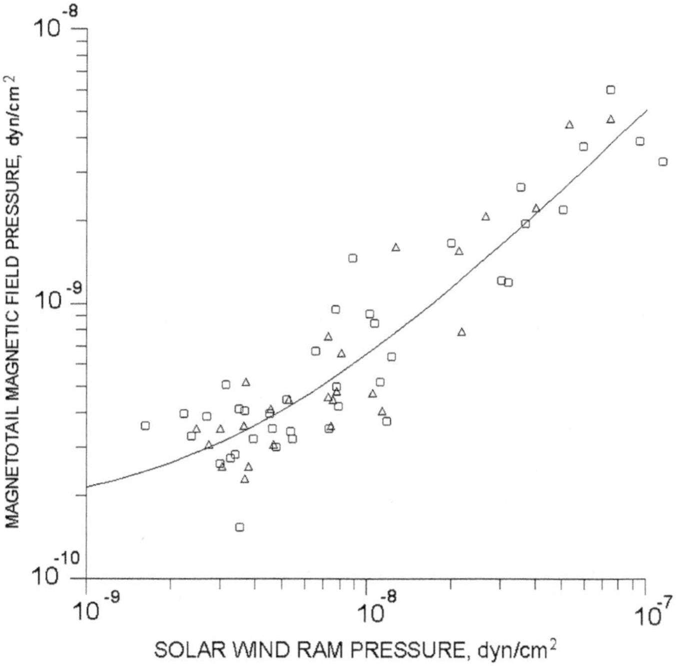

Figure 4.5. Dependence of the magnetic field pressure in the Martian tail on the solar wind ram pressure. Entry and exit of Phobos 2 into and out of the magnetotail are denoted by the dash and Δ symbols, respectively (from Rosenbauer et al., 1994).

and MAGMA data (Rosenbauer *et al.*, 1994). The smooth curve in this figure is the theoretical dependence, given as:

$$B_t^2/8\pi = \rho_p V^2 \sin^2\alpha + p. \tag{4.1}$$

The best-fit coefficients are, $sin^2\ \alpha \approx 0.049 \pm 0.004$ and $p \approx (1.7 \pm 0.3) \times 10^{-10}$ dyne cm^{-2}. Coefficient $\alpha \approx$ arc $\sin(\sqrt{0.049}) \approx 13°$ is just the average flaring angle of the Martian magnetotail boundary at the Phobos 2 orbit. This flaring angle turned out to be considerably less than the geomagnetotail flaring angle at comparable distances ($\approx 22°$), but exceeded the one for the purely induced Venusian magnetotail ($< 9.5°$). Thus, at that time, it was concluded that both intrinsic and induced magnetic fields play an important role in the solar wind interaction with Mars (Rosenbauer *et al.*, 1994).

More detailed information on the shape of the Martian magnetotail, under different solar wind ram pressures, can be found in the tables presented by Verigin *et al.* (1997). The tail region and its correlation with upstream solar wind parameters (e.g., IMF and dynamic pressure) was also studied by Slavin *et al.* (1984), and Luhmann *et al.* (1991), and modeled by Lichtenegger *et al.* (1995).

There are several properties which make Mars a unique object for studying the interaction between the solar wind and a planetary body. This is partly due to the crustal magnetic field (Connerney *et al.*, 1999), which is relatively small and localized so that it cannot play a major role in the interaction as do the magnetic fields of planets like Earth, Jupiter or Saturn. In this respect, Mars is more similar to Venus than it is to Earth.

The induced magnetotail forms as a result of the atmospheric mass loading and subsequent draping of passing magnetosheath flux tubes that sink into the wake. In general, planetary bodies without intrinsic magnetic fields, but with substantial atmospheres, are known to possess such cometlike induced magnetotails.

Phobos-2 observations, as well as MGS observations on the dayside (see relevant section regarding draping, above), have shown that the magnetotail of Mars consists of two lobes of sunward and antisunward fields, their polarity being controlled by the orientation of the cross flow component of the interplanetary magnetic field (Yeroshenko *et al.*, 1990; Schwingenschuh *et al.*, 1992a). Although these measurements suggest an induced nature of the Martian magnetotail (like that of Venus), the lack of observations in the polar regions of the magnetotail cannot exclude the possible existence of a hybrid-like tail for Mars (Dubinin and Podgorny 1980; Axford, 1991), emanating from the magnetic anomalies discovered by Mars Global Surveyor (Acuna *et al.*, 1998; Connerney, 1999).

The hybrid or combined character of the solar wind interaction with Mars is also supported by the results of a spectral analysis of the Phobos-2 magnetic field data (Möhlmann *et al.*, 1991) and the modulation of the tail diameter found in the plasma data (Verigin *et al.*, 2001). The flaring of the Martian tail was investigated (Zhang *et al.*, 1994) and it was found that the magnetotail is similar to the Earth's tail in its dependence on the pressure of the solar wind. The draping at low altitudes

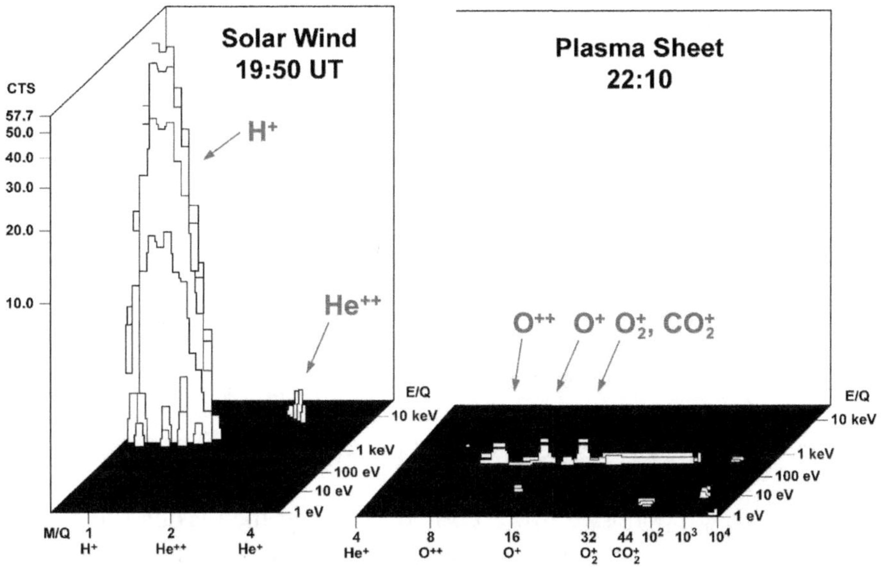

Figure 4.6. $E/q - M/q$ matrix measured by the ASPERA instrument on March 25, 1989 in the solar wind and plasma sheet. The plasma sheet contains mainly ions of the planetary origin. The planetary ions gain the energy (E/q) of about 1 keV q^{-1} (from Dubinin *et al.*, 1993a).

has been investigated using MGS data and was found to be better defined above the ionosphere than in the ionosphere (Crider *et al.*, 2001).

4.3. Plasma Acceleration Processes in the Martian Tail

The central tail is the region where energy of the magnetic field, accumulated from the flux of solar wind kinetic energy, is transferred back to the plasma. So far the Phobos 2 orbiter has provided the only plasma measurements in the deep planetary tail. Plasma instruments TAUS (Rosenbauer *et al.*, 1989) and ASPERA (Lundin *et al.*, 1989) revealed that the tail of the Martian magnetosphere has a significant population of planetary ions, primarily oxygen ions. The most intense fluxes of tailward streaming heavy ions (ions with mass/charge ratio $M_i/q_i > 3$) were observed in the magnetotail plasma sheet (Rosenbauer *et al.*, 1989; Verigin *et al.*, 1991a). The velocity and intensity of heavy ion fluxes reach their peak values in the vicinity of the magnetic neutral line. The energy of these ions varies from $\sim$ 100 eV up to $>$ 6 keV. Ion dynamics in the Martian tail were discussed by Dubinin *et al.* (1993a). The remarkable feature of the ASPERA observations is that the energy of the different heavy ions in the plasma sheet at $\sim 2.8\ R_M$ is close to the bulk energy of the protons in the solar wind. Figure 4.6 shows $E/q - m/q$ ion matrixes from the measurements in the solar wind and in the plasma sheet.

It is seen that the O^+ ions and the heavier molecular ions (O_2^+ or CO_2^+) have approximately the same energy/charge as the solar wind protons. Another interesting feature is that ions with $m/q = 8$ (O^{++}) have also approximately the same energy/charge. These features indicate, that to a large degree electric fields must be responsible for the acceleration of these ions. This is not really suprising, because both O^+ and molecular ions have large gyroradii ($r_{O+} \sim 3000$ km in a magnetic field of $B_\perp \sim 5$ nT and $W \sim 1$ keV), which are comparable or larger than the width of the plasma sheet. Dubinin *et al.* (1993a) suggested that the ion acceleration is due to the action of magnetic field stresses. The $\mathbf{j} \times \mathbf{B}$ force is the strongest in the center of the tail:

$$(\mathbf{B} \cdot \nabla)\mathbf{B}/4\pi \cong 1/4\pi(B_\perp \, dB_x/dl_\perp) \tag{4.2}$$

where $B_\perp$ and B_x are the transverse and longitudinal components of the magnetic field and $l_\perp$ is the distance between the spacecraft and the current sheet. The B_x component in the tail can be well modeled by $B_x = B_{ox} \tanh(l_\perp/\delta)$, where δ is a characteristic size of the field reversal region ($\delta \sim 2000$ km). Then the magnetic field stresses are proportional to $K/\cosh^2(l_\perp/\delta)$, and the plasma sheet geometry with respect to the spacecraft can be roughly estimated from the IMF orientation.

The peak ion energy variations are similar to the changes in magnetic field stresses. The important feature of the near-Mars tail is that the heavy ions are not magnetized. This means that the $\mathbf{j} \times \mathbf{B}$ force accelerates the electron fluid and the ions are accelerated by the related ambipolar electric field $E_x \sim (\mathbf{j_e} \times \mathbf{B})_x/(n_e ec)$. The energy gained by the ions in the electric field is on the order of:

$$W \sim qLB_\perp B_{xo}/2\pi n_e \delta \, \cosh^2(l_\perp/\delta). \tag{4.3}$$

Thus $W \sim 1 - 2$ keV if $n_e \sim 5$ cm^{-3}, $L/\delta \sim 5 - 10$, $B_{xo} \sim 20$ nT, $B_\perp \sim 5$ nT. This simple model leads to a reasonable agreement with the observations and explains why oxygen and molecular ions have approximately the same energy and the energy of double-ionized oxygen ions is two times higher. Hence a chain of the processes $n_{\text{sw}} m_p V_{\text{sw}}^2 \rightarrow B^2/8\pi \rightarrow n_{O+} m_{O+} V_{O+}^2$ explains that $W_{\text{sw}}(H^+) \sim W_{\text{tail}}(O^+)$ if $n_{\text{sw}} \sim n_{\text{tail}}$. It is also clear that ions with smaller Larmor radii will gain less energy, and that is also in agreement with the observations (Dubinin *et al.*, 1993a).

The measurements made by the TAUS instrument (Rosenbauer *et al.* 1989, Kotova *et al.*, 1997a) have revealed another interesting feature of the ions in the Martian plasma sheet. 2D ion spectra collected during the elliptical orbits showed a supersonic, highly anisotropic distribution function of heavy ions (Rosenbauer *et al.*, 1989). The 3D distribution function of heavy ions, taken while the spacecraft was spinning, has a 'mushroom cap' shape similar to the shape of proton distributions in the terrestrial plasma sheet boundary layer (Kotova *et al.*, 1997a).

Several acceleration mechanisms were invoked for the explanation of such distributions in the Earth's magnetotail (Eastman *et al.*, 1986). A simple field aligned acceleration in an electric field and the adiabatic deformation of the distribution function due to changes in the magnetic field magnitude were examined by Kotova

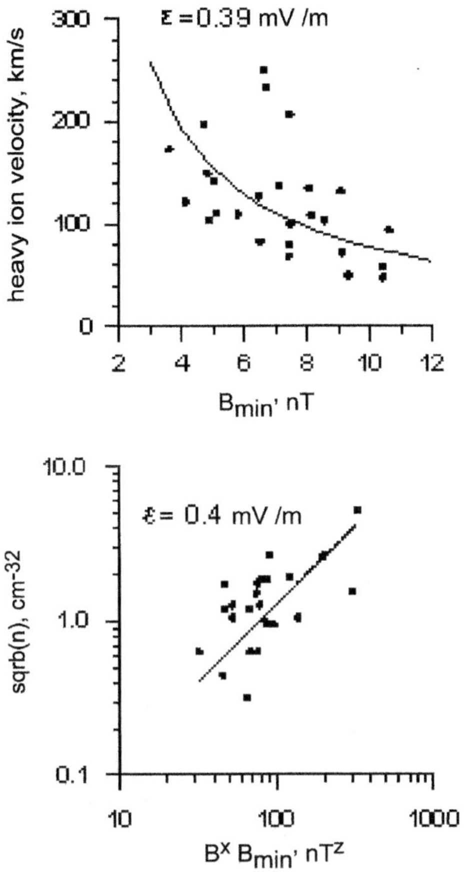

Figure 4.7. Correlations of plasma sheet ion and magnetic field parameters showing evidence for acceleration to occur in the current sheet.

et al. (1997a). It turns out that this model requires the existence of some process that accelerates the ions to velocities of several tens of km s^{-1}. This pre-acceleration may be the result of the direct interaction of ionospheric ions with magnetosheath plasma in the 'polar' regions of the Martian magnetosphere (Verigin *et al.*, 1991a).

Another acceleration mechanism that could account for the 'mushroom cap' shape of the heavy ion spectra, is a classical acceleration in the cross-tail current sheet (Speiser, 1965; Shabanskiy, 1972). The radius of the Martian magnetotail is comparable to the gyroradius of the heavy ions in the plasma sheet. Thus ion kinetic effects in the Martian magnetotail are more important than in the Earth's magnetotail. However, the inverse correlation between the maximum heavy ion velocity (v_{max}) and the minimum magnetic field (B_{min}) in the tail (Figure 4.7) points to the possible existence of a cross-tail electric field in the Martian magnetotail (Kotova *et al.*, 1997a, 2000a). Taking into account that:

$$v_{\max} \approx 2c\varepsilon / B_{\min},$$

where ε is a cross tail electric field in the Martian magnetosphere, c is the velocity of light, the average electric field during the time of the Phobos 2 flight can be estimated as $\varepsilon \approx 0.4$ mV m^{-1}.

Based on the equation (Eastwood, 1972; Hill, 1975):

$$\varepsilon = \frac{B B_{\min}}{\sqrt{4\pi n M_i}}$$

the average electric field can also be estimated from the correlation of $\sqrt{n}$ (n is the heavy ion density) and $B \cdot B_{\min}$ (B is the magnitude of the magnetic field in the tail lobes). The correlation is shown in Figure 4.7 (lower panel), and the average electric field is again $\varepsilon \approx 0.4$ mV m^{-1}.

4.4. Plasma Loss

The initial efforts to estimate the loss rate of the Martian ions and neutral gas began about thirty years ago. These first theoretical calculations by Michel (1971), Cloutier *et al.* (1973) and Wallis (1978) indicated a particle loss rate of several grams per second, which we now know is far too low. After the Mars 5 ion measurements at the edge of the Martian magnetotail, Vaisberg (1976) evaluated the planetary heavy ion loss rate through its boundary layer as $\sim 10^{25}$ s^{-1} or ~ 250 g s^{-1}. The absence of a mass-spectrometer aboard this orbiter did not allow the identification of the ion mass and Bezrukikh *et al.* (1978) interpreted their measurements as observations of a proton boundary layer.

The Phobos 2 plasma measurements confirmed the existence of the proton boundary layer. The Phobos 2 instruments had the capability to identify the different masses and that led to the discovery of the magnetotail plasma sheet (Rosenbauer *et al.*, 1989) and to an estimate of the planetary ion loss rate, $\Phi_{esc} \sim 1$–3×10^{25} s^{-1}, through this structure, (Rosenbauer *et al.*, 1989; Lundin *et al.*, 1989; Lundin and Dubinin, 1992). Similar to the terrestrial magnetotail, the Martian plasma sheet surrounds a narrow current sheet (Figure 4.8). In contrast to the terrestrial case, however, the Martian plasma sheet consist mainly of planetary heavy ions ($m/q > 3$) (Verigin *et al.*, 1991a). Ions with a variety of masses were detected, although oxygen ions appear to be prevalent (Lundin *et al.*, 1989; Dubinin *et al.*, 1993a) with supersonic, highly anisotropic, mushroom cap like distribution functions (see previous section).

The measured values of the ion fluxes in the tail are reliable, but using these values to arrive at an estimate of the total escape flux is associated with significant uncertainties. The uncertainties are mainly associated with the highly variable position of the Martian plasma sheet, which is schematically shown in Figure 4.9, which is based on Phobos 2 TAUS observations (Verigin *et al.*, 1991a). The innermost magnetopause positions are marked in this figure by thin vertical bars.

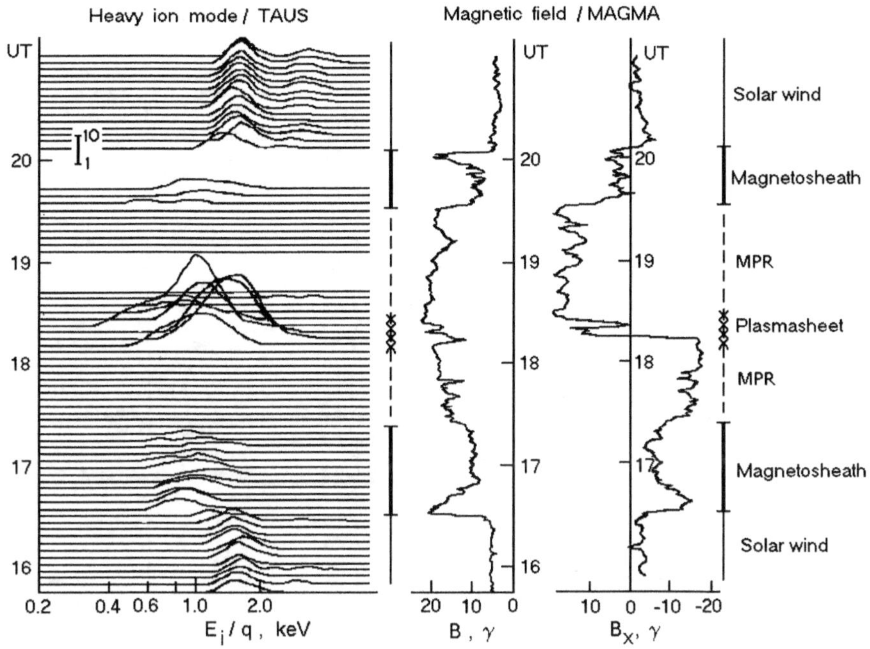

Figure 4.8. Left: Phobos 2/TAUS energy-time spectrogram, which contains heavy ions and protons (admixture) in the Martian magnetotail; Right: total magnetic field and the B_X component (adopted from Verigin *et al.*, 1991a; The authors used the term 'Magnetosphere' for the region labeled MPR). It demonstrates the co-location of a dense plasma population and magnetic field reversal in the Martian magnetotail.

Observations of multiple plasma sheet crossings in some orbits may be caused by plasma sheet motion.

Using information from a variety of observations, Verigin *et al.* (1991a) estimated that the diameter of the Martian magnetotail, D, is approximately 1.5×10^4 km, that the average heavy ion flux in the plasma sheet, F, is about 2.5×10^7 cm^{-2} s^{-1}, and that the plasma sheet thickness, d, is $\sim 0.1D$. The estimate of the plasma sheet thickness was based on the percentage of the time that Phobos 2 spent in the plasma sheet, while crossing the magnetotail (Figure 4.9). The average escape rate of planetary oxygen through the tail is then $\Phi_{esc} \sim F \cdot d \cdot D = 5 \times 10^{24}$s^{-1}. This estimate is in general agreement with other esitmates; the mean suggested escape rate is around $\Phi_{esc} \sim 10^{25}$ s^{-1}, with an uncertainty of about a factor of two.

The ion loss process, through the Martian plasma sheet, is the only one that was observed by in-situ measurements so far, though it is not likely to be the only important planetary oxygen loss process. Another potential escape mechanism is the tailward flow of ionospheric thermal ions (see Section 6.1.2). (Ma *et al.*, 2002; 2004). This tailward flowing plasma may be the source of some of the ions measured in the tail. The hot neutral oxygen densities and escape fluxes at Mars have been calculated, using two different modeling approaches (Kim *et al.*, 1998;

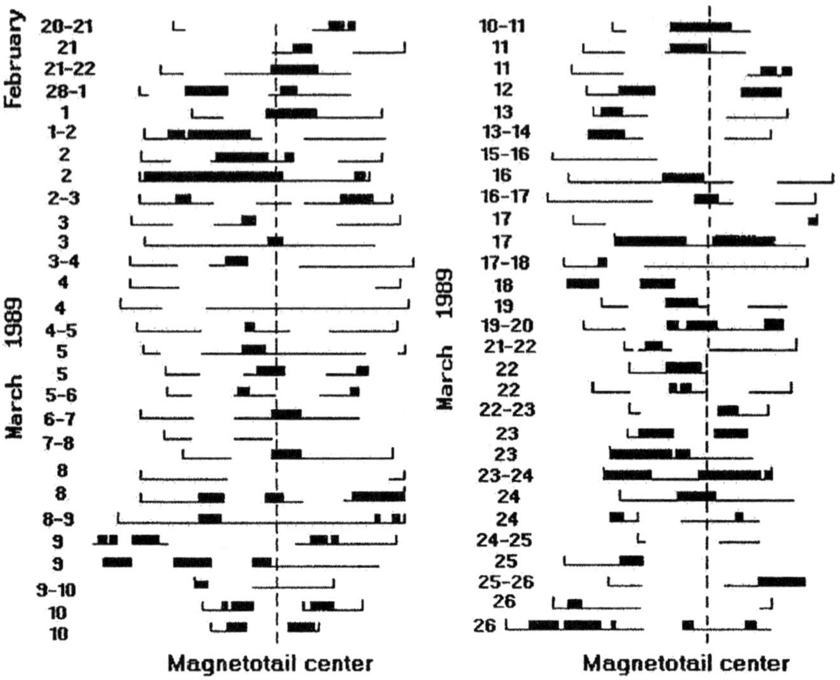

Figure 4.9. Plasma sheet location in the Martian magnetotail as observed during Phobos 2 circular orbits. The dashed line corresponds to the magnetotail center (from Verigin et al., 1991a).

Hodges, 2000). The main source of the hot oxygen atoms was found to be the dissociative recombination of the O_2^+ (see Section 6.3). The estimated escape fluxes vary over the solar cycle; during solar cycle maximum the calculations indicated escape rates, via this mechanism, is in excess of 2×10^{26} s^{-1}. Finally, pick up ion escape and sputtering from the Mars atmosphere also needs to be considered (e.g., Leblanc and Johnson, 2002).

5. The ionosphere

An ionosphere was first detected on Mars by the Mariner 4 spacecraft in 1965 with the radio occultation experiment (Fjeldbo and Eshleman, 1968) and was followed by measurements with Mariners 6 and 7 (Rasool and Stewart, 1971) and Mariner 9 (Kliore *et al.*, 1972). The peak densities were observed to be about 10^5 cm^{-3} and the structure and extent was very different from terrestrial expectations (see Bauer, 1973, for an interesting discussion of these early developments in our knowledge of the ionospheres of Mars and other planets).

The first *in situ* thermospheric and ionospheric measurements ever made for a planet other than Earth were made by instruments onboard the Viking 1 and 2 landers. Neutral composition was measured with mass spectrometers (Nier and

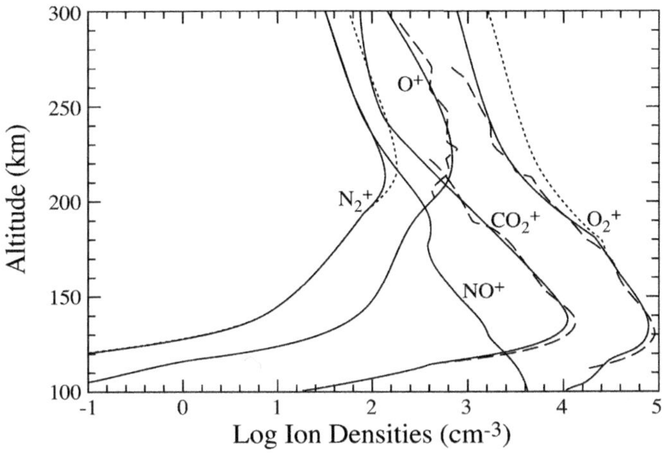

Figure 5.1. Ion densities from a chemical/vertical transport model of the Martian ionosphere for zero outflow upper boundary condition (dashed line) and maximum outflow (solid line). The dotted line are measured profiles from the Viking RPA experiment (Hanson *et al.*, 1977). From Fox (1997).

McElroy, 1977), although not atomic oxygen, and the major ion composition by the retarding potential analyzers (RPA) (Hanson *et al.*, 1977). The measured ion density profiles for CO_2^+, O_2^+ and O^+ are shown in Figure 5.1, illustrating that O_2^+ is indeed the major ion species throughout the Martian dayside ionosphere (c.f., see discussion in Schunk and Nagy, 2000). The observed O^+ density increases with altitude, up to the measurement limit of about 300 km, but never quite becomes the major ion unlike the ionospheres of Earth and Venus. Electron and ion temperature profiles, and photoelectron fluxes, were also measured by the Viking RPA experiments (Hanson *et al.*, 1977; Hanson and Mantas, 1988). These Viking measurements remain our only in situ source of information on 'thermal' ions and electrons at Mars, although the Mars Global Surveyor (MGS) mission has provided important *in situ* measurements of the magnetic field and superthermal electron fluxes (Acuña *et al.*, 1998) and also provided additional radio occultation electron density profiles (Bougher *et al.*, 2001).

Even though *in situ* measurements of the ionosphere are very limited, an extensive set of electron density profiles are available from a number of missions, obtained with the radio occultation technique, over many portions of the solar cycle (Mariners 4, 6, 7, and 9; Viking 1 and 2; the Soviet Mars 4 and 5 missions; MGS). A review of this radio occultation data and a comparison with ionospheric profiles for Venus (mostly from the Pioneer Venus mission) can be found in Zhang *et al.* (1990c) and Kliore (1992). Figure 5.2 shows some of the Viking radio occultation profiles.

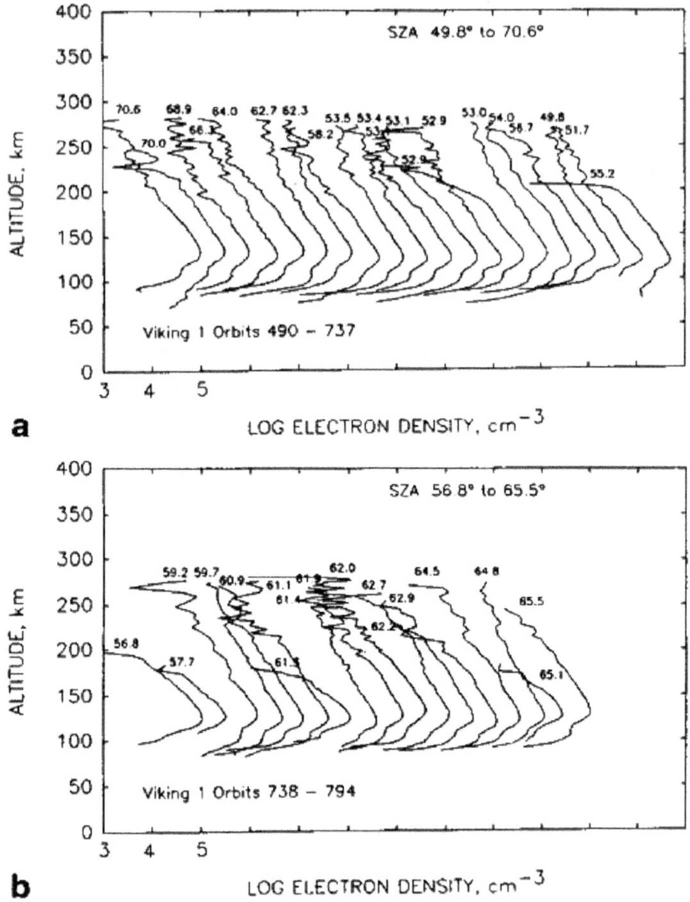

Figure 5.2. Electron density profiles measured on the Martian dayside by the Viking radio occultation experiment on the Viking mission. (a) Orbits 490–737, (b) Orbits 738–794. From Kliore (1992).

5.1. Dayside

The behavior of the ionospheric peak at both Venus and Mars is well-described by Chapman theory, with a peak density at Mars that varies with solar zenith angle, χ, as $n_{\text{emax}} = 2.3 \times 10^5 \text{ cm}^{-3}(\cos\ \chi)^{1/2}$, where all the data has been normalized to an $F10.7$ flux of 150, and where χ is given in degrees (Kliore, 1992). The altitude of the peak is at about 125 km near $\chi = 60°$, and in agreement with Chapman theory this altitude increases with increasing χ. However, the ionospheric peak at 55° was several km higher during the first 71 Mariner 9 orbits than it was at other times. This was attributed to the atmospheric expansion associated with a large dust storm (Stewart and Hanson, 1982).

Other information can also be derived from the radio occultation profiles. As noted by Kliore (1992), with only a few exceptions, all the electron density profiles exhibit great uniformity in their topside scale heights; the topside scale height is about 35 km. Assuming a photochemical control of the topside ionosphere, the electron density scale height should be twice the neutral CO_2 scale-height ($h_e = 2\ h_n$), which allows the neutral temperature to be remotely monitored (Bauer and Hantsch, 1989; Bauer, 1999). The data is consistent with a Martian thermospheric temperature of $\approx$ 200 K, which also agrees with MGS atmospheric drag measurements (Keating *et al.*, 1998).

The ionopause is a feature almost always observed in the ionosphere of Venus either by the radio occultation technique, or in situ by experiments onboard the Pioneer Venus Orbiter (Kliore, 1992; Brace and Kliore, 1991). The ionopause is marked by sharp decrease in the electron density at the top of the ionosphere. However, no clear ionopause signatures have been observed in the radio occultation profiles at Mars (see Figure 5.2). Interpretation of these observations is considered later.

5.2. Nightside

A nightside ionosphere was observed at Mars by the radio occultation experiments on the Mars 4 and 5 spacecraft (Savich *et al.*, 1979) and by the Viking mission (Zhang *et al.*, 1990b). Zhang *et al.* (1990b) concluded that the nightside ionosphere at Mars is generally weaker than that at Venus and the peak altitude is higher. A well-developed ionosphere is present near the terminator. However, its existence deeper on the nightside is more problematic as the following quote from Kliore (1992) notes: '…it appears that the nightside ionosphere of Mars more than 10° from the terminator is a rather ephemeral phenomenon. In contrast, on Venus the nightside ionospheric peak is a robust and ubiquitous presence, regardless of solar activity…'.

Two mechanisms are known to maintain the nightside ionosphere of Venus: (1) flow of plasma from the dayside topside ionosphere into the nightside, followed by downward flow and recombination, and (2) ionization on the nightside by precipitating electrons (e.g., Brannon *et al.*, 1993; Dobe *et al.*, 1995). The first mechanism is dominant when the dayside ionosphere is unmagnetized, which it ordinarily is, at least during periods of high solar activity. Electron precipitation is able to produce an ionosphere near the peak region, but not at higher altitudes. As a later section will discuss, the dayside ionosphere of Mars is generally magnetized, due to either processes related to the solar wind interaction (Cloutier *et al.*, 1999), or to crustal magnetic fields (Acuña *et al.*, 1998). Therefore it is not surprising, perhaps, that the nightside is not well supplied by plasma from the dayside. However, this is an area that requires both more data and theoretical study.

5.3. Ionospheric energetics and airglow

Dayside electron and ion temperatures generally exceed the neutral temperature in all planetary ionospheres (cf., Schunk and Nagy, 2000) and Mars is not an

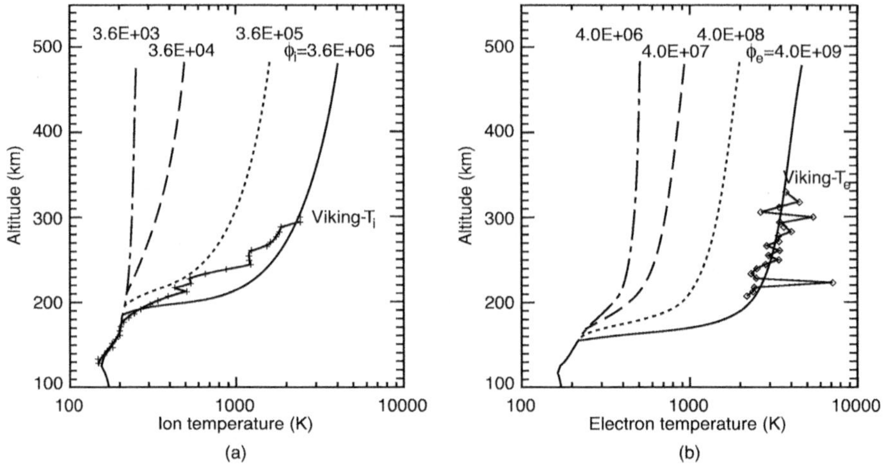

Figure 5.3. Electron and ion temperatures measured by the Viking 1 Retarding Potential Analyzers (Hanson and Mantas, 1988) and some calculated ion and temperature profiles for different topside heat fluxes from (Choi *et al.*, 1998).

exception. The only measured temperature profiles are from the Retarding Potential Analyzer (RPA) experiment on the Viking landers (see Figure 5.3). The electron temperature is about 3000 K for altitude above 200 km and the ion temperature increases from a few hundred K at 200 km up to 3000 K at 300 km. As discussed in Section 6.3, a variety of models (e.g., Rohrbaugh *et al.*, 1979; Choi *et al.*, 1998) have been developed to explain these temperatures and they all require some combination of heat input from the solar wind and/or modification of the thermal conduction coefficients by ionospheric magnetic fields (cf. Nagy and Cravens, 1997).

A key energy input into the thermal ionospheric plasma comes from collisions of this thermal plasma with superthermal electrons. Superthermal electrons have energies significantly in excess of thermal energies, $E > 1$ eV, and have non-Maxwellian energy distributions (cf., Schunk and Nagy, 2000). Two sources contribute to the superthermal electron population, both at Mars and at Venus (e.g, Spenner *et al.*, 1980; Gan *et al.*, 1990): photoelectrons produced by the photoionization of atmospheric neutrals by solar radiation, and solar wind electrons (actually magnetosheath electrons), which make it into the ionosphere along magnetic field lines linking the ionosphere and magnetosheath. Apparently, both superthermal electron sources are present at Mars. The Viking RPA measured superthermal electron fluxes (Mantas and Hanson, 1979), as did the HARP spectrometer on Phobos (Shutte *et al.*, 1989) and the MGS electron reflectometer (Mitchell *et al.*, 2000).

Some superthermal electron flux spectra from MGS are shown in Figure 5.4, as well as a theoretical photoelectron spectrum from Fox and Dalgarno (1979). According to Mitchell *et al.* (2000), spectra in the ionosphere-proper appear to be dominated by the photoelectron contribution, whereas in a boundary layer near the

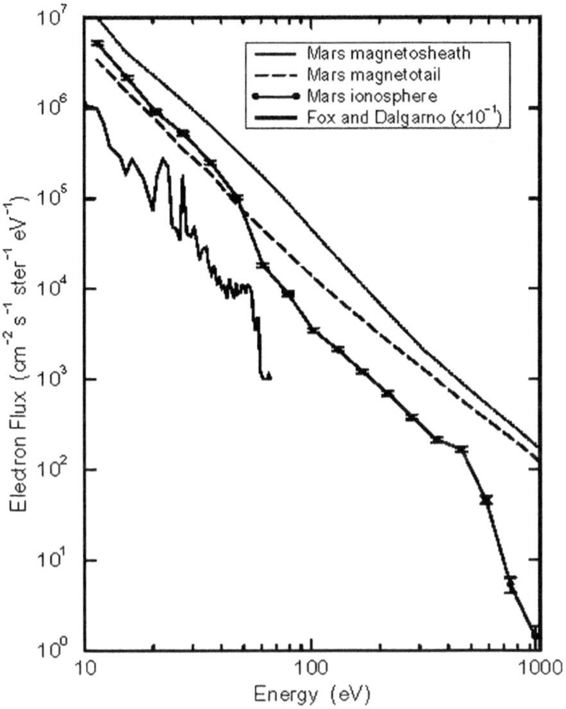

Figure 5.4. Superthermal electron flux versus energy measured by the MGS electron reflectometer in the magnetosheath, in the magnetotail, and in the ionosphere. Also show is a theoretical photoelectron spectrum calculated by Fox and Dalgarno (1979). From Mitchell *et al.* (2000).

top of the ionosphere the spectra take on a more magnetosheath-like appearance. Figure 5.4 from Mitchell *et al.* (2001) also shows this boundary layer which is more apparent for the higher energy electrons. A similar boundary layer was present at Venus and was evident in the measured superthermal electron distributions at that planet (Spenner *et al.*, 1980; Gan *et al.*, 1990).

The ionospheric spectrum for electron energies below about 100 eV is what one would expect from theoretical calculations (see Figure 5.4 again) if the energy resolution is lowered to instrumental values. However, a feature is apparent in the measured ionospheric spectra near 500 eV. This feature is interpreted as being due to Auger electrons produced by the ionization from the K-shell of O atoms, primarily residing in CO_2 (Mitchell *et al.*, 2000). This population of superthermal electrons is associated with the soft X-ray part of the solar spectrum, and it is a useful diagnostic indicating atmospheric origin rather than solar wind origin. Thus it is helpful in mapping out the transition from the ionosphere to the 'solar wind'.

Airglow measurements provide another useful diagnostic of ionization and superthermal electron processes in planetary ionospheres Measurements were made at Mars by the Mariner 6, 7, and 9 missions and also by the Mars 5 mission (cf., Paxton and Anderson, 1991). Prominent CO_2^+ and CO^+ airglow features appear in the spectrum and are diagnostics of the ionization processes occurring in the ionosphere (see Fox, 1993). The Cameron bands of CO also have a significant contribution from photoelectron excitation. At Venus, measurements of the oxygen 1304 line have been used to deduce energy inputs into the nightside ionosphere by 'auroral' electrons from the magnetotail (Fox and Stewart, 1991), which have also been invoked as one possible source of the nightside ionosphere.

5.4. Solar Wind Interaction with the Ionosphere

For many years it was debated whether or not a significant intrinsic magnetic field existed at Mars, and only recently, with the arrival of MGS, has this issue been settled. MGS is the only spacecraft carrying a magnetometer that has encountered the ionosphere (Acuña *et al.*, 1998). However, we have long realized that magnetic fields induced by the solar wind might be present within the ionosphere (e.g., Cloutier and Daniell, 1979). Prior to MGS, the nature and strength of the Martian magnetic field was estimated indirectly in several ways (cf., Luhmann *et al.*, 1987; Krymskii, 1992), including methods based on observed properties of the ionosphere (Luhmann *et al.*, 1987; Mahajan and Mayr, 1989; Shinagawa and Cravens, 1989, 1992; Zhang *et al.*, 1990b,c; Krymskii, 1992). Overall, these studies suggested that a small, largely horizontal magnetic field (whether induced or intrinsic) of no more than $\approx$ 50 nT was possible. The ionosphere, particularly during solar minimum time periods, was apparently in an 'overpressure' state in which the solar wind dynamic pressure exceeds the maximum ionospheric thermal pressure.

Our understanding of the Martian ionosphere has greatly benefited from the Venus analogy. Extensive studies of the Venus ionosphere, using the large PVO database and numerous theoretical models (cf., Luhmann and Cravens, 1991) demonstrated that an ionosphere of a planet lacking a large intrinsic magnetic field, such as Venus or Mars, should be permeated by large-scale horizontal magnetic fields, extending downward from the 'magnetic barrier' (or MPB), during the overpressure case. On the other hand, the Venus ionosphere was observed to be unmagnetized when the maximum ionospheric pressure significantly exceeded the solar wind dynamic pressure. In this case, a clear ionopause was present and was located at relatively high altitudes. Statistical analyses have shown that the solar wind dynamic pressure exceeds the maximum ionospheric pressure approximately 85% of the time at Mars (Phillips *et al.*, 1984; Zhang *et al.*, 1990c), so that the Martian ionosphere should generally be found to be in the overpressure case, and hence magnetized.

The MGS magnetometer measurements indeed show that the magnetic field in the ionosphere are largely induced (except in regions with significant crustal anomaly fields) (Acuña *et al.*, 1998; Connerney *et al.*, 1999; Cloutier *et al.*, 1999). This is illustrated in Figure 5.5, which shows the measured magnetic field strength and orientation in the ionosphere. Note that the magnetic field is mainly horizontal

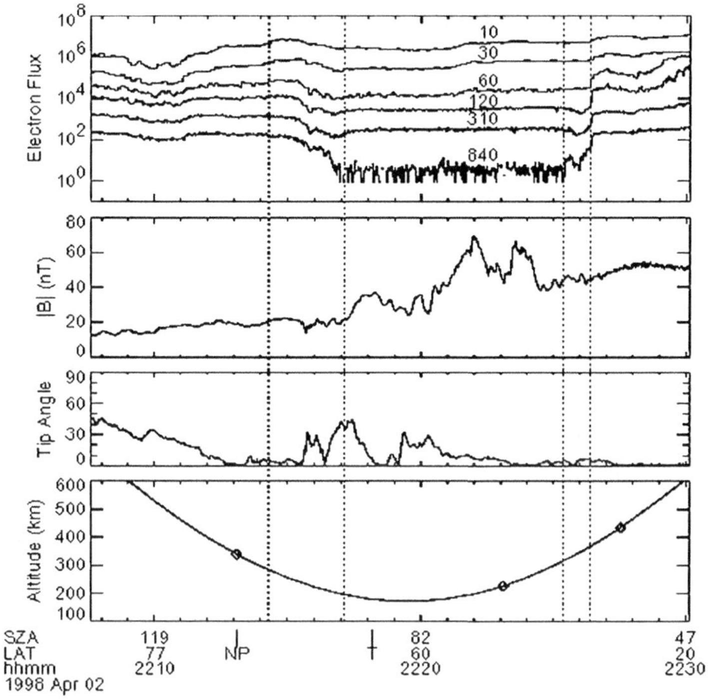

Figure 5.5. Time series of the superthermal electron flux for several energy channels from the MGS electron reflectometer for a April 2, 1998 pass. Units are cm^{-2} s^{-1} $ster^{-1}$ eV^{-1}. Also shown is the magnetic field strength (in nT) and the tip angle of the field (angle from the horizontal). From Mitchell *et al.* (2000).

with a strength of roughly 50 nT or less (Mitchell *et al.*, 2000), as expected for fields induced by the solar wind interaction (Shingawa and Cravens, 1989; Krymskii, 1992; Cloutier *et al.*, 1999). A boundary layer is evident in the superthermal electron fluxes, as discussed above, yet there is no obvious signature of this layer in the magnetic field. Unfortunately, the thermal electron density was not measured *in situ*, so that the overall plasma density structure in this boundary layer is not known. Cloutier *et al.* (1999) notes that a magnetic field rotation occurs near this boundary layer and is analogous to similar field rotation observed near the Venus ionopause.

The observed topside scale height in the Martian ionosphere is only about 35 km, as discussed earlier, and appears to be ‘photochemical’ rather than the larger scale height that a free (‘diffusive’) vertical transport would yield. Plasma scale heights from many radio occultation profiles obtained during the Mariner 9 and Viking missions were analyzed and confirm this statement (Izakov and Roste, 1996; Breus *et al.*, 1998; Ness *et al.*, 2000). Horizontal magnetic fields, such as the ones that have now been observed in the ionosphere, will indeed inhibit vertical transport (Shinagawa and Cravens, 1989; Krymskii, 1992).

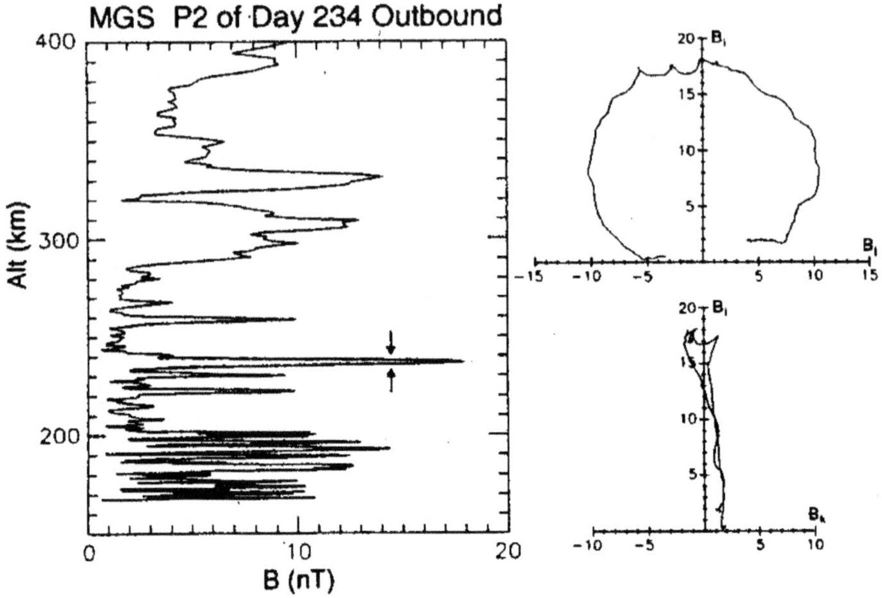

Figure 5.6. On the left: magnetic field strength versus altitude as measured by the MGS magnetometer experiment for Day 234 of 1998. On the right: hodogram of a magnetic flux rope, in principal axis coordinates. From Cloutier *et al.* (1999).

On a few occasions, when the Martian ionosphere appeared to have relatively low magnetic field strength, perhaps associated with low solar wind dynamic pressure conditions, MGS observed spatially narrow features with strong localized field surrounded by low background field (Cloutier *et al.*, 1999). One example of such conditions is given in Figure 5.6 for day 234 of 1998. Cloutier *et al.* (1999) produced a hodogram for one of the magnetic structures and demonstrated that it was a 'force-free' magnetic flux rope. Magnetic flux ropes with similar helical structures were commonly observed by the PVO magnetometer in 'unmagnetized' ionospheres of Venus (Elphic *et al.*, 1980; Luhmann and Cravens, 1991). The formation of magnetic flux ropes at either Venus or Mars is not really understood yet, although much work has been carried out on this topic (cf., Cravens *et al.*, 1997). Some promising suggestions have been made by Krymskii *et al.* (2000) for Venus, in which ropes form, not at the ionopause as some previous theories had suggested, but due to instabilities in the magnetic 'belt', which has been observed by the PVO magnetometer at low altitudes in the ionosphere. Perhaps similar ideas can be applied to flux rope formation in the Martian ionosphere(see Vignes *et al.*, 2002).

5.5. Role of Crustal Anomalies

A key difference in the characteristics and dynamics of the ionospheres of the two nonmagnetic planets, Venus and Mars, does exist due to the crustal magnetic anomalies that were discovered by the MGS magnetometer at Mars (Acuña *et al.*, 1998, 1999; Connerney *et al.*, 1999; Ness *et al.*, 1999). The local, but fairly widespread crustal magnetic field anomalies, observed by MGS, may create an effective large-scale magnetic field structure in the southern hemisphere (Ma *et al.*, 2002). The major fraction of the planetary magnetic field flux is concentrated near the surface of Mars. The rest of the flux 'escapes' from the planet and, in the context of the solar wind Mars interaction process, can act as an effective dipole. The anomalies can also create regions with effective horizontal (as well as radial) magnetic fields at ionospheric altitudes, where the radio occultation measurements were made (Krymskii *et al.*, 2002).

'Mini-magnetospheres' are formed in regions of strong crustal field (Mitchell *et al.*, 2001) that should strongly affect ionospheric processes, partly by acting as a more effective obstacle to the solar wind, holding it off at altitudes exceeding 400 km (Brain *et al.*, 2003; Ma *et al.*, 2002). For example, the outermost lines of crustal magnetic field should be able to re-connect with IMF lines draped around the mini-magnetospheres, so the region outside the 'closed' magnetosphere is covered with a boundary layer of 'opened' field lines, which connect the Martian surface and the solar wind. Brain *et al.* (2003) have recently studied the re-connection of Martian magnetic fields to the solar wind and found that the re-connection regions cover about 7% of the Martian surface. Outside of the mini-magnetospheres the Martian ionosphere should be qualitatively similar to the ionosphere at Venus as discussed earlier. On the other hand, numerous cusp-like regions should also form above the crustal anomalies through which charged particles can penetrate into the ionosphere and neutral atmosphere resulting in heating and ionization (Brain *et al.*, 2003; Ness *et al.*, 2000).

The crustal anomalies should also affect the nightside ionosphere. Neither day-to-night flow nor electron impact ionization caused by precipitating energetic electrons can repopulate the nightside ionosphere within an intense crustal magnetic field region (mini-magnetosphere) except in the 'cusp' regions. This means that the ionospheric plasma can be denser outside a mini-magnetosphere region than inside it, because day-to-night plasma transport and/or impact ionization caused by energetic electrons precipitating from the magnetotail can keep it populated. Mitchell *et al.*, (2001) have detected plasma voids at around 400 km on the nightside, which correspond to regions of crustal magnetization.

6. Theory and Modeling

In the preceding sections the observed properties of the dayside bowshock, magnetosheath and ionosphere were summarized. Here in this section the theoretical model calculations, associated with these regions are described. The primary challenge of the model calculations is, of course, to explain the details of the interactions, subject to the constraints imposed by the observations (see Introduction).

First we describe three current global models of the solar wind interaction with Mars. This is followed by an outline of the model of hot atom production around Mars and a review of various ionospheric models.

6.1. Global Models of the Solar Wind/Mars Interaction

Before beginning a discussion of the modeling tools used to simulate Mars and some of the results that have been published concerning the solar wind interaction with Mars, it is instructive to consider why one would resort to such simulations. There are a variety of answers to this question. One of them is that the visualization of the full system, provides leading questions, occasionally provides answers, and often provides insight. Also the various models are based on a variety of different assumptions/simplifications and thus can lead to a better understanding of the relative importance of the different mechanisms, controlling the interaction processes. The theoretical models of the solar wind interaction with Mars can be placed in two general categories: fluid and kinetic approaches. The hydrodynamic model calculations of Spreiter and colleagues starting back in the 1960's (Spreiter and Briggs, 1962; Spreiter and Stahara, 1980) provided groundbreaking new insights into the basic, large scale interaction processes between the supersonic solar wind and solar system objects. These hydrodynamic results are still used in some applications (e.g. Zhang *et al.*, 1993a), but to a large degree have been superseded by more sophisticated models during the last decade*. The three current models in use are:

(i) The semi-kinetic model developed by Brecht;
(ii) The multi-species, single fluid MHD model developed by Nagy and colleagues
(iii) The two-fluid MHD model developed by Sauer and colleagues.

The challenge for the numerical models is to reproduce, mathematically, the observations described earlier – the boundaries and regions and their characteristics now heuristically understood from the data. In particular, the MPB and MPR, the ionospheric characteristics, the shock, and the unusually large upstream wave activity should be identifiable in the model results. Of course, not all model approaches encompass the entire interaction problem. Therefore the reader will find that one model will be better at explaining the characteristics of the shock, while an other may excel in describing the cause and effect of the MPR and MPB, or the workings of the ionosphere. Indeed, a rather considerable problem in itself is to understand the limitations of the individual model approaches, and their domain of applicability, which typically are not heralded by the model architects ('If the best tool in one's toolbox is a hammer, solutions tend to appear as nails.').

* The results from a new 3D, single fluid, non-ideal MHD model have just been published (Harnett and Wingler, 2003).

6.1.1. *The Semi-Kinetic (Hybrid) Model*

The kinetic simulations most often used in space research are the so-called hybrid particle simulations, where the ions are treated as individual particles, whereas the electron are considered as a massless neutralizing fluid (cf., Brecht *et al.*, 1993). The hybrid model allows easy inclusion of multiple ion species, because they are simply particles following the basic Lorentz force laws and are automatically kinetic in their behavior. This approach to treating the ions permits direct simulation of ion gryo-motion effects and drift motion due to electric and magnetic fields as well as density gradients. The electron momentum equation, given below as Equation (6.5) is equivalent to Ohm's Law, containing the Hall, $\mathbf{J}_e \times \mathbf{B}$ and ∇p_e terms. The Hall term breaks the symmetry of the equations and permits, among other things, filamentation of the plasma, and the propagation of whistler mode waves. The pressure and resisitive terms in the Ohm's law also contribute to the electric field.

The rich suite of electromagnetic modes resulting from the model equations, as well as, the inclusion ion gyromotions, means that the physics of collisionless shock formation are directly simulated in this model. This includes parallel waves and the subsequent acceleration of the ions in the shock structure. One should note that the ion pressure is not needed in any of the equations. This is a fluid dynamic concept and is not valid within the kinetic approach. However, the particle distributions and moments can be calculated in order to investigate the shape of the distribution functions and the ion pressure tensor. Indeed it is the anisotropic pressure, created by the reflected ions at the shock, that leads to the electromagnetic modes within the shock. These wave modes result in two effects. First, they lead to a very structured shock surface. Second, they carry away energy in the form of whistler, Alfvén-Ion-Cyclotron (AIC) waves. This is what leads to shock overshoots seen at the various planets (Russell *et al.*, 1982; Mellot and Livesey, 1987).

The charge neutrality assumption and the neglect of the transverse part of the displacement current (called the Darwin approximation) truncate Maxwell's equations. The charge neutrality removes high speed plasma oscillations, while the charge separation and Darwin approximations remove the light waves. This allows the use of larger cell sizes and larger time steps making the numerical solution of this model easier. Given the above described formulation and approximations the controlling equations are as follows:

$$\nabla \times \mathbf{B} = \mathbf{J}, \tag{6.1}$$

$$\nabla \times \mathbf{E} = -\frac{\partial \mathbf{B}}{\partial t}, \tag{6.2}$$

$$m_i \frac{\partial \mathbf{v}_i}{\partial t} = q_i\{\mathbf{E} + \mathbf{v}_i \times \mathbf{B}\} - q_i \eta \mathbf{J}, \tag{6.3}$$

$$\frac{\partial \mathbf{x}_i}{\partial t} = \mathbf{v}_i, \tag{6.4}$$

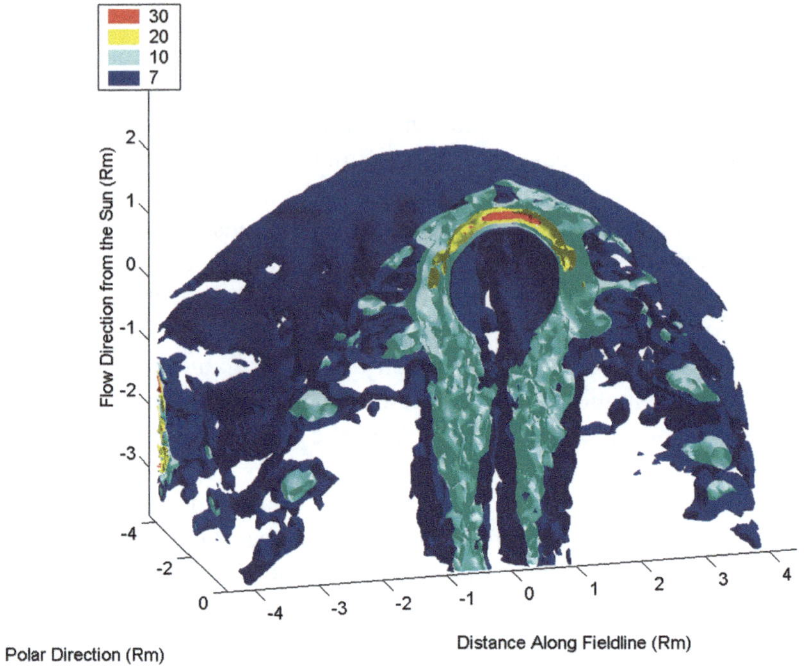

Figure 6.1. A 3-D surface plot of the Martian bow shock cut through the ecliptic plane. The magnetic barrier is symmetric, the shock is symmetric and the two lobes of the magnetotail of Mars are clearly seen. One, can see evidence of the wave structure of the region below the shock. In this perspective the solar wind come from the top of the picture and the IMF is in the plane of the cut through the volume.

$$-en_e\mathbf{E} + \mathbf{J}_e \times \mathbf{B} - \nabla p_e + en_e\eta\mathbf{J} = 0, \tag{6.5}$$

$$\sum_i n_i = n_e, \tag{6.6}$$

where the charge neutrality given in (6.6) holds for scale lengths greater than the Debye length, η is the plasma resistivity, and the subscripts e and i denote electron and ion parameters, respectively. These equations are solved with a predictor-corrector method, first published by Harned (1982). Full details of this hybrid formulation, including the equations solved can be found in Brecht and Thomas (1988).

This approach to global modeling is often considered to be the most complete approach. However, it has certain limitations, the main practical one being that it is very computer intensive. The time steps must be small enough to resolve the fastest of the allowable electromagnetic modes, the whistler wave. The spatial cell sizes must be small enough to resolve the ion gyroradii and the Hall effects. In the case of Mars, this means cells of the order of 100's of kilometers.

This code has been applied to simulations of the solar wind interaction with Mars. In particular, the magnetic topology of Mars in full 3-D was investigated

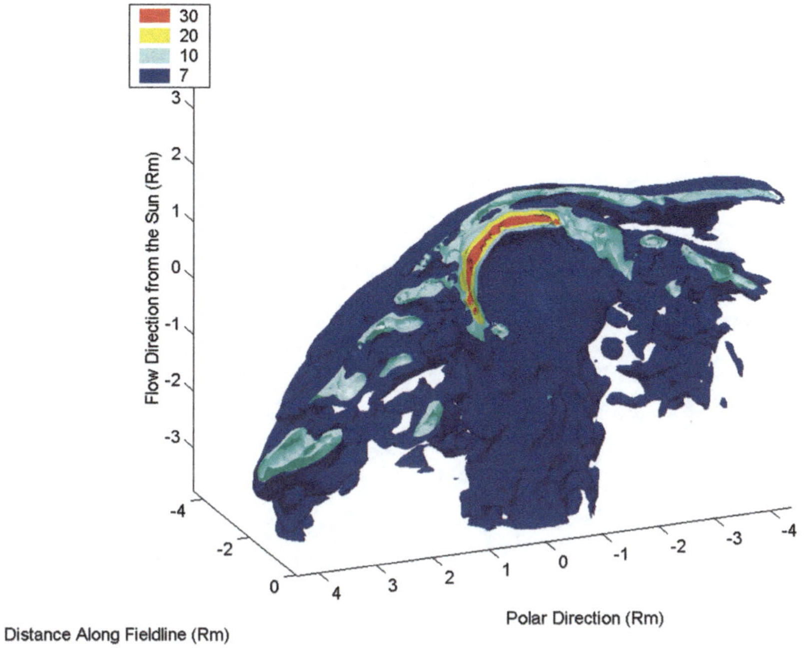

Figure 6.2. A 3-D surface plot of the Martian bow shock cut through the polar regions. The magnetic barrier is highly asymmetric, the shock is also asymmetric and one can see considerable difference between the two polar regions. In this perspective the solar wind comes from the top of the plot. The convection electric field is in the plane of the cut through the volume and points in the positive direction.

in order to see the effects of the kinetic interaction on the shock (Figure 6.1 and Figure 6.2). It was found that the bow shock did not thermalize the solar wind plasma as occurs on other planets (Brecht *et al.*, 1993) and that the shock was asymmetric in the polar directions with the normal solar wind IMF configuration (Brecht and Ferrante, 1991; Brecht *et al.*, 1993; Brecht, 1997a).

It was also suggested by the hybrid simulations that the solar wind deposits a significant amount of energy into the planets ionosphere and atmosphere (Brecht, 1997b). Finally, Phobos 2 orbits have been flown through the hybrid simulations as a way of comparing the results of the simulations with the data obtained by this spacecraft. The results were encouraging and agreed well with the data (Brecht *et al.*, 1993). More recently, as the research has evolved, it has been found that adding an ionosphere and exosphere to the simulations leads to the appearance of flux filaments (Figure 6.3). They only occur in the North Pole of the planet where the IMF is in the ecliptic plane, as is typical for the IMF. Indeed the preponderance of the flux seems to be transported preferentially over the North Pole. Currently the model has started to be used to investigate the role of the crustal magnetic fields (Acuña *et al.*, 1998, 1999) on the shape of the shock, and scavenging of the Martian ionosphere, caused by the solar wind interaction with Mars.

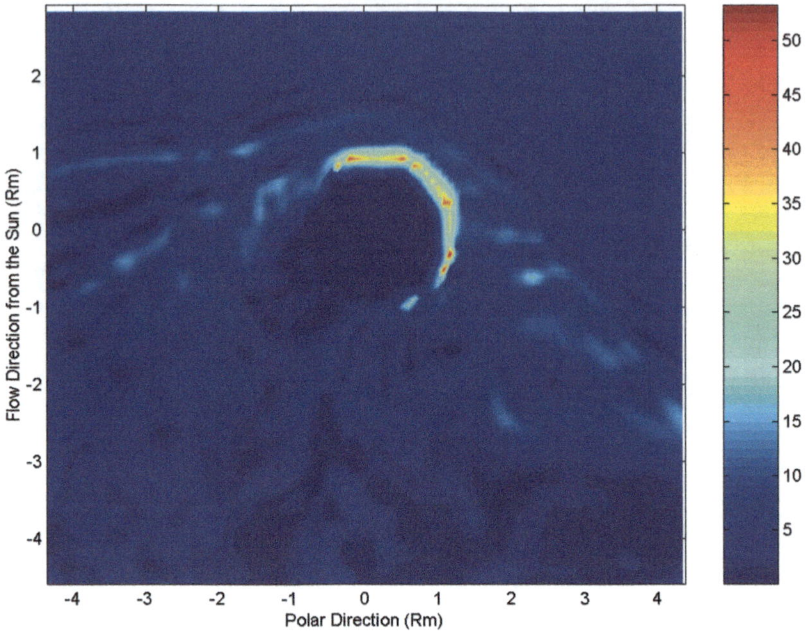

Figure 6.3. A 2-D slice through the simulations shown in Figure 6.1 and Figure 6.2. This plane contains the solar wind flow and the convection electric fields. The magnetic barrier is highly asymmetric and strongest over the North pole of the planet. One, can also see the tendency of the simulation to develop strong regions of magnetic field within the barrier region, suggesting that flux ropes may well be more favorably formed in the North pole region as well.

6.1.2. *The Multi-Species, Single Fluid MHD Model*

It has been argued that the use of semi-kinetic models is especially appropriate for Mars, where the ion gyroradius is of the same order as the planetary radius, which is true mainly outside the bowshock. However, it is important to note that the ideal MHD equations have been found to be successful even in such situations. The possible reason for this maybe the presence of significant wave activity and turbulence, which lead to a wide variety of wave particle interactions, which in turn act as pseudo-collisions. The wave activity and turbulence were observed by the Phobos-2 wave instrument (Grard *et al.*, 1989), and the MGS magnetometer (e.g., Acuña *et al.*, 1998; Cloutier *et al.*, 1999), and was predicted by the semi-kinetic code described above.

Further, there are many questions of interest that are truly fluid in their behavior. This is especially true when one is addressing the influence of the solar wind interaction on the ionosphere and atmosphere. In these regions even the classical collision frequencies are high and the velocities small, resulting in effective gyroradii which are small compared to the scale sizes of the system.

The MHD calculations described next (Ma et al., 2002; 2004) solve the coupled set of equations consisting of a number of continuity equations, one momentum equation, one energy equation, and one magnetic field equation. The justification for using a single momentum equation is that often the fluids of interest are closely coupled where they exist together (however, see Section 6.1.3). The actual equations (assuming three species) are:

$$\frac{\partial \mathbf{W}}{\partial t} + \{\nabla \cdot \mathbf{F}\}^T = \mathbf{Q}, \tag{6.7}$$

where the total energy density is defined as

$$\varepsilon = \frac{1}{2}\{\rho_1 + \rho_2 + \rho_3\}u^2 + \frac{1}{\gamma - 1}p + \frac{B^2}{2}, \tag{6.8}$$

the state vector, flux tensor, and source vector in Equation (6.7) are

$$\mathbf{W} = \begin{pmatrix} \rho_1 \\ \rho_2 \\ \{\rho_1 + \rho_2 + \rho_3\}\mathbf{u} \\ \mathbf{B} \\ \varepsilon \end{pmatrix}, \tag{6.9}$$

$$\mathbf{F} = \begin{pmatrix} \rho_1 \mathbf{u} \\ \rho_2 \mathbf{u} \\ \rho_3 \mathbf{u} \\ \{\rho_1 + \rho_2 + \rho_3\}\mathbf{uu} + \left\{\frac{B^2}{2}\right\}\mathbf{I} - \mathbf{BB} \\ \mathbf{uB} - \mathbf{Bu} \\ \mathbf{u}\{\varepsilon + p + \frac{1}{2}\} - \{\mathbf{B} \cdot \mathbf{u}\}\mathbf{B} \end{pmatrix}, \tag{6.10}$$

$$\mathbf{Q} = \begin{pmatrix} 0 \\ S_2 - L_2 \\ S_3 - L_3 \\ \{\rho_1 + \rho_2 + \rho_3\}\mathbf{g} - \{\rho_1 + \rho_2 + \rho_3\}\nu\mathbf{u} - \mathbf{u}L_2 - \mathbf{u}L_3 \\ 0 \\ Q_6 \end{pmatrix}, \tag{6.11}$$

where the source term of the energy equation, Q_6, is

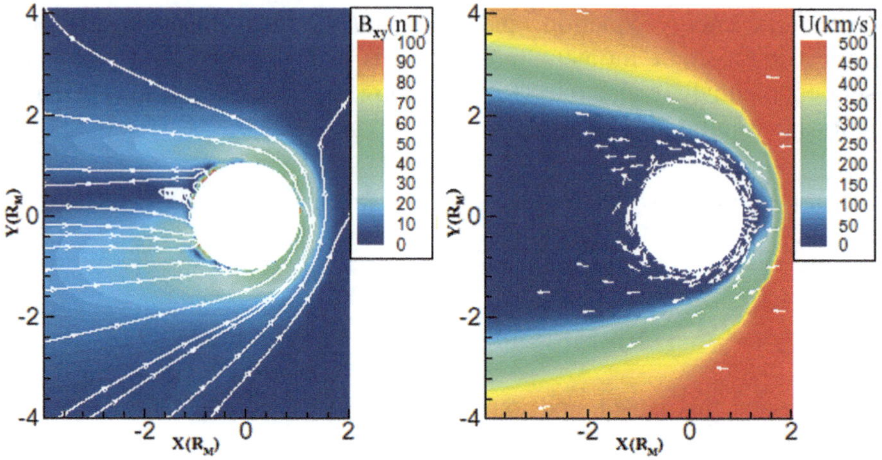

Figure 6.4. The component of the calculated magnetic field (left) and velocity magnitudes (right) in the *xy* plane. The color plots show the magnitudes; the white lines marked with arrows indicate the vector direction of the magnetic field and the arrows the direction (not the magnitude) of the velocity (from Ma et al., 2002).

$$
\begin{aligned}
Q_6 = \; & \{\rho_1 + \rho_2 + \rho_3\}\mathbf{u} \cdot \mathbf{g} - \frac{1}{2}u^2\{L_2 + L_3\} - \{\rho_1 + \rho_2 + \rho_3\}\nu u^2 - \\
& - \frac{1}{\gamma - 1}\frac{L_2 p}{\{32\rho_1 + \rho_2 + 2\rho_3\}} - \frac{1}{\gamma - 1}\frac{L_2 p}{\{16\rho_1 + 0.5\rho_2 + \rho_3\}} + \\
& + \frac{1}{\gamma - 1}S_2\frac{k}{m_2}T_o + \frac{1}{\gamma - 1}S_3\frac{k}{m_3}T_o
\end{aligned}
\tag{6.12}
$$

and where ρ_1, ρ_2 and ρ_3 are the H^+, O_2^+ and O^+ mass densities, respectively, S_2, S_3 and L_2, L_3 are the O_2^+, and O^+ mass source and loss rates, respectively, p is the total thermal pressure of the plasma, $\mathbf{u}$ is the velocity of the plasma, ν is the ion neutral collision frequency (taken to be $4 \times 10^{-10}\{(O) + (CO2)\}$ s^{-1}, T_o is the temperature of the newly produced ions and the other symbols have their usual definition.

The computational domain defined by $-24\ R_M \leq x \leq 8\ R_M$, $-16\ R_M \leq y$, $z \leq 16\ R_M$, where $R_M = 3396$ km is the radius of Mars, was used in these calculations and the inner boundary was taken to be 140/100 km above the Martian surface. The numerical approach used in solving the equations is described in Powell *et al.* (1999) and the boundary conditions adopted can be found in Ma *et al.* (2002; 2004). A 60-order harmonic expansion for the crustal magnetic field (Arkadi-Hamed, 2001), which reproduces the observed fields very well, was used.

In the calculations published by Ma et al. (2002) the location of the largest crustal field (180 longitude) was pointing sunward, and the solar wind flow direction was parallel with the equator of the planet. Figure 6.4 shows the calculated magnetic field and velocity components in the equatorial plane, while Figure 6.5

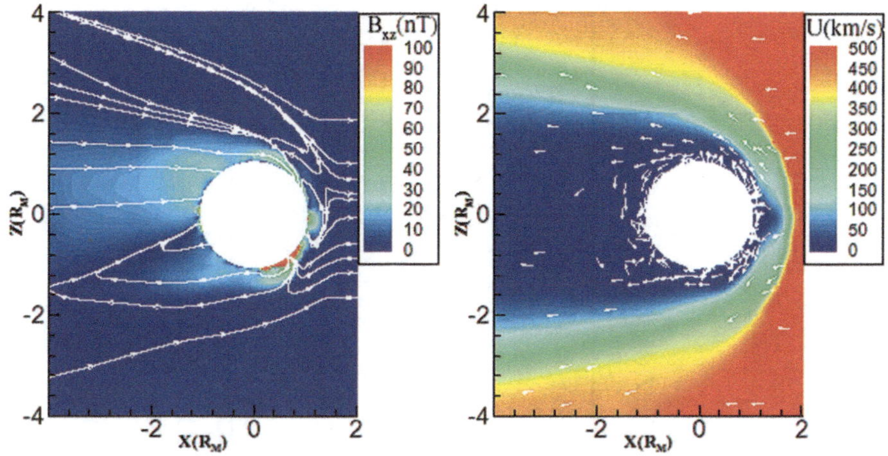

Figure 6.5. The same as Figure 6.4, but in xz plane (from Ma et al., 2002).

shows the field and velocity in the noon-midnight meridian plane. The color plots show the magnitude of the field and velocity components; the white lines show the field directions and the white arrows the direction (not the magnitude) of the velocities. The draping of the IMF is clearly visible in the equatorial plane as indicated in Figure 6.4 and the small 'bends and kinks' are the result of the crustal fields. There also appears to be a flux rope like feature; flux ropes have been observed by the MGS magnetometer (Vignes *et al.*, 2002). The general trend of the plasma flow is upward and toward the terminator on the dayside and either downward into the ionosphere (helping to maintain the nightside ionosphere) or down the tail (and escaping from the planet) on the nightside. Figure 6.5 shows the presence of closed field lines (mini 'magnetocylinders'), the result of the 'merging' of the crustal and IMF fields. The presence of such mini-magnetospheres have been inferred by the electron reflectometer carried aboard the Mars Global Surveyor (MGS) spacecraft (Mitchell *et al.*, 2001). Figure 6.6 shows the plot of the dayside location of the calculated 9 nT values (three times the assumed IMF field) as the proxy for the bowshock position. In order to compare these results with the observed bowshock locations (which have been presented assuming symmetry about the X-axis) the calculated curves are shown both in the $X - Y$ and $X - Z$ planes. The agreement between the calculations and observations (Vignes *et al.*, 2000) are good, especially considering the fact that there is a scatter, of the order of one Mars radius, in the observed locations.

Given the strong spatial variations in the crustal field the ionopause locations are not expected to be spherically symmetric around the planet. MGS results have demonstrated the positive correlation between the ionopause height and the strength of the magnetic field (Mitchell *et al.*, 2002). 'Ionopause' in this case is defined by the change in character of the superthermal electron spectrum (the 'photoelectron boundary'), and does not necessarily coincide with a sharp change

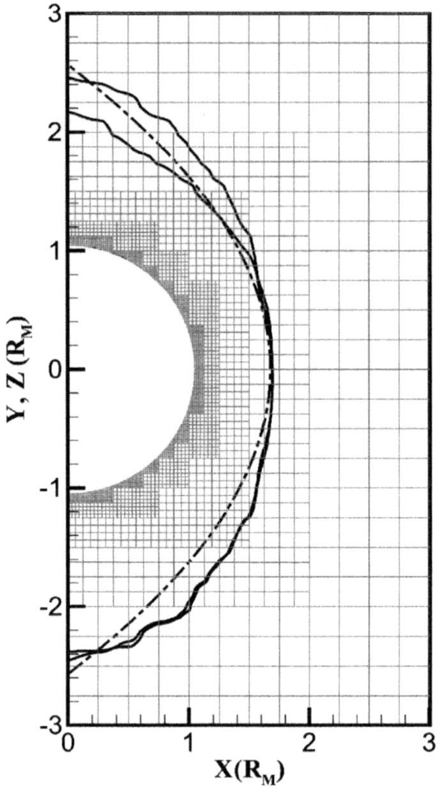

Figure 6.6. Plot showing where the magnetic value is equal to 9 nT in equatorial plane (blue line), and in meridian plane (red line); This value of magnetic field is used as a proxy for the location of the bow shock. Also shown is the mean value of the observed bowshock location (green line) from Vignes *et al.*, 2000 (from Ma et al., 2002).

in the thermal electron density, as was originally meant by this term for the Venus ionosphere.

The model calculations of Ma *et al.*, (2002) also provide some indication of such a situation. They determined the locations where the pressure balance between the kinetic pressure of the ionospheric ions equals the sum of the magnetic pressure and the total (kinetic plus dynamic) solar wind proton pressure. The position of this pressure balance is a reasonable indicator of the ionopause location. They compared the locations of these pressure balances, which they calculated for the the non-magnetic case with those calculated in the presence of crustal fields. They found that the presence of the crustal magnetic fields does raise the ‘ionopause’ altitude; the largest outward movement is near the strongest field location and corresponds to an altitude increase of about 370 km.

Ma *et al.*, (2002; 2004) also calculated the trans-terminator and tailward escape fluxes. They found that these fluxes decrease in the presence of the crustal field and that they change with solar cycle, as expected. Their calculations indicate that these ionospheric fluxes make up a significant fraction of the total escape flux estimated from the Phobos-2 measurements (Lundin *et al.*, 1989; Rosenbauer *et al.*, 1989). The calculated trans-terminator flux is also consistent with the 'upper-limit' estimates of Fox (1997). Ma *et al.* (2004) have modified their earlier model and are now using a spherical grid stucture, which leads to much better radial resolution. The ionospheric resolution that they have obtained is about 10 km, considerably less than the plasma scale height, and thus leading to meaningful results in this region.

6.1.3. *Two-Fluid, Multi-Species MHD*

The major initial motivation for constructing an MHD model with separate ion momentum equations was observations made in the different plasma environments of comets, Venus and Mars. In addition, the active experiments in space in which barium and lithium were released into the solar wind creating artificial comets (Bryant *et al.*, 1985) provided further stimulus. These observations indicated that the solar wind protons and the heavy ions of the obstacle have their own separate dynamics. The first of such a model was a one dimensional MHD model, with two momentum equations (Sauer *et al.*, 1990; Baumgärtel and Sauer, 1992). While the classical one-fluid MHD models failed to reproduce a number of important aspects of the observed signatures of the proton flow, the main features of the interaction can be described by an MHD model in which protons and heavy ions develop their own interconnected dynamics (Sauer *et al.*, 1990).

The continuity and momentum equations used in the model are given below for the protons (the same equations are also used for the heavy ions; thus interchanging the subscripts p and h, lead to the heavy ion equations):

$$\frac{\partial n_p}{\partial t} + \nabla \cdot \{n_p \mathbf{u}_p\} = 0, \tag{6.13}$$

$$\frac{\partial \{n_p \mathbf{v}_p\}}{\partial t} + \nabla \cdot \{n_p \mathbf{v}_p \mathbf{v}_p\} = \frac{1}{m_p}\frac{n_p}{n_e}\left\{e n_h[\mathbf{v}_p - \mathbf{v}_h] \times \mathbf{B} - \nabla\left[\left(p_e + \frac{B^2}{2\mu_o}\right)\mathbf{I} - \frac{\mathbf{BB}}{\mu_o}\right]\right\}, \tag{6.14}$$

$$\frac{\partial p_e}{\partial t} + \nabla \cdot \{\mathbf{v}_e p_e\} + \{\gamma - 1\} p_e \{\nabla \cdot \mathbf{v}_e\} = 0, \tag{6.15}$$

$$\frac{\partial \mathbf{B}}{\partial t} - \nabla \times \left\{\frac{1}{n_e}\left[n_p \mathbf{v}_p + n_h \mathbf{v}_h - \frac{1}{\mu_o}\nabla \times \mathbf{B}\right] \times \mathbf{B}\right\} = 0, \tag{6.16}$$

where n_p, n_h and n_e are the proton, heavy ion and electron number densities, n_c is the charge density, $\mathbf{B}$ is the magnetic field, p_e is the electron pressure and $\mathbf{v}$ denotes the relevant velocities.

At first a relatively simple model for analyzing the interaction of a flowing plasma with a heavy ion source (which is representative of a cometary or planetary atmosphere), was developed by Baumgärtel and Sauer (1992). This stationary 1D model shows that a sub-sonic/Alfvénic flow penetrating the 'ionospheric'

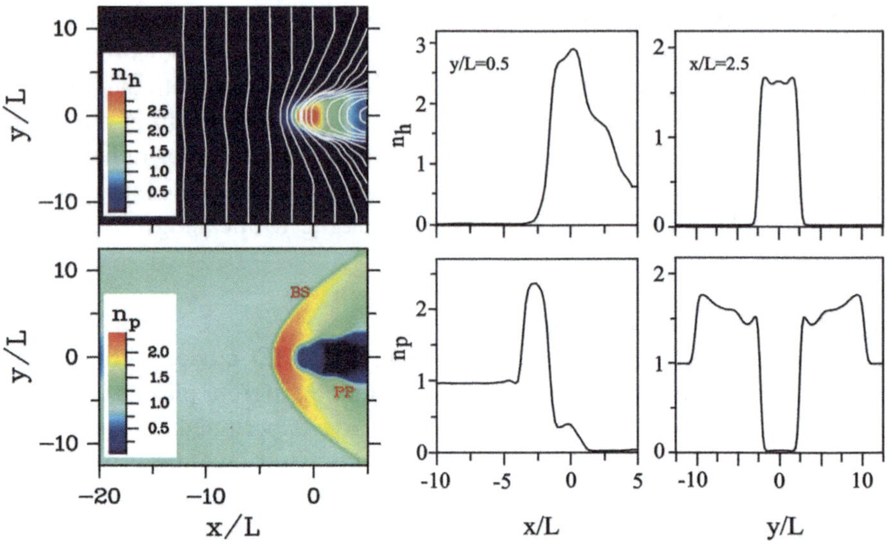

Figure 6.7. Results of 2D bi-ion fluid simulations of solar wind interaction with a heavy ion source. The momentum coupling is provided solely by the ambipolar electric field (the magnetic field is not included in the dynamics calculation). The upper three panels show (from left to right) the spatial distribution of the heavy ion density (n_h), and cuts along $x(y/L = 0.5)$ and $y(x/L = 2.5)$. In the lower three panels the same is shown for the proton density (n_p). A remarkable feature is the formation of a boundary where protons are replaced by heavy ions (MPB). The magnetic field lines shown as white lines in the upper color plot show the enhanced draping at this boundary (adapted from Sauer and Dubinin, 2003).

plasma behaves like a gas in a Laval nozzle. On the front side of the heavy-ion cloud the protons are accelerated and then decelerated after reaching the center of the 'obstacle'. At a 'critical heavy-ion density', which is less than the density of the incoming protons the flow becomes choked. This phenomenon arises from the proton flow becoming sonic before reaching the center of the heavy ion cloud. In reality of course the flow is not one-dimensional and in two dimensions one is lead to expect the formation of a proton flow boundary.

A similar magnetic field-free model, in which a source of heavy ions is included, shows that the self-consistent interaction between the proton flow and the heavy ions leads to the formation of a sharp boundary between both ion fluids. Figure 6.7 shows the spatial dependence of both the proton and heavy-ion densities as color plots (left), together with cuts through the simulation box, representing crossings along the flank ($y/L = 0.5$) and through the tail ($x/L = 2.5$) of the interaction region. The coordinates x and y are normalized by the characteristic scale length, L, of the source function. Since the magnetic field is not included in the dynamics calculation, the coupling between the protons and heavy ions is exclusively provided by the electric field $E \sim \text{grad} p_e$, where p_e is the electron pressure. At the MPB (called protonopause in Sauer *et al.*, 1994), the proton density drops sharply to near zero, forming a proton cavity. Simultaneously, the heavy-ion density increases sharply. Throughout this boundary, momentum is transferred from the protons to the heavy ions, which, because of their higher mass, move slowly inside the proton cavity (MPR). It is important to note that two wave modes exist in such a bi-ion plasma (McKenzie *et al.*, 1993): one is related to the protons ('proton mode') and the other to the heavy ions ('heavy-ion mode').

The magnetic field is calculated from the electron dynamics, like in gasdynamic models (Spreiter and Stahara, 1980), so that the sharp drop of the electron velocity at the MPB leads to enhanced draping of the magnetic field, and an abrupt increase of its magnitude. This effect is seen in the upper color plot of Figure 6.7, where the magnetic field lines are plotted together with the heavy ion density distribution. The collocation of the draping feature and the change of the ion composition is in agreement with the observation (see Section 3.3).

Results from a 2D simulation, using the above equations, in which the solar wind magnetic field is self-consistently included, have been published recently (Sauer and Dubinin, 2000). The source of heavy ions was taken to be an exospheric/ionospheric density distribution of oxygen, based on reasonable parameters from the literature (the planetary body itself was ignored). The main aim of these studies was to elucidate the momentum and energy coupling between the proton and heavy-ion fluids. By excluding the high-density part of the ionospheric source distribution, no stagnation of the electron flow (determined by both the proton and heavy-ion flows) appeared in the results,. This implies the absence of a stagnation point of the magnetic field flux, or, in other words, the absence of an ionopause. The grad p_e term plays the dominant role in the momentum coupling between the two ion species. A sharp boundary is formed, the MPB, which separates the upwind protons from heavy ions. This occurs at a location where the heavy-ion densities are relatively low compared to the undisturbed proton density. Simulations with different orientations of the solar wind magnetic field shows that the shape of the MPB becomes asymmetric, being rather sharp on one side, with a more gradual transition on the other side (also see Plate 1 by Sauer *et al.*, 1994). The magnetic field piles up at that boundary where the ion composition changes. Thus there is a general feature consisting of a 'magnetic pile-up boundary' (MPB) coinciding with an 'obstacle boundary' for the proton flow. Moreover, this transition is associated with a drop in the electron temperature, which is an additional signature of the bi-ion transition.

Figure 6.8 shows a comparison between Phobos-2 measurements along (a) the elliptical and (b) the circular orbits with corresponding cuts through the appropriate simulation boxes (Sauer and Dubinin, 2000). Overall, there is good agreement. The variations of both the proton and electron densities demonstrate the efficiency of the bi-ion fluid approach. Whereas the proton density drops down at the MPB, the electron density jumps simultaneously (by virtue of charge-neutrality), indicating a change in the ion composition. Also noteworthy is the sharp change in the temperature at the MPB (Figure 6.8b) shown in both the Phobos-2 measurements and the simulations.

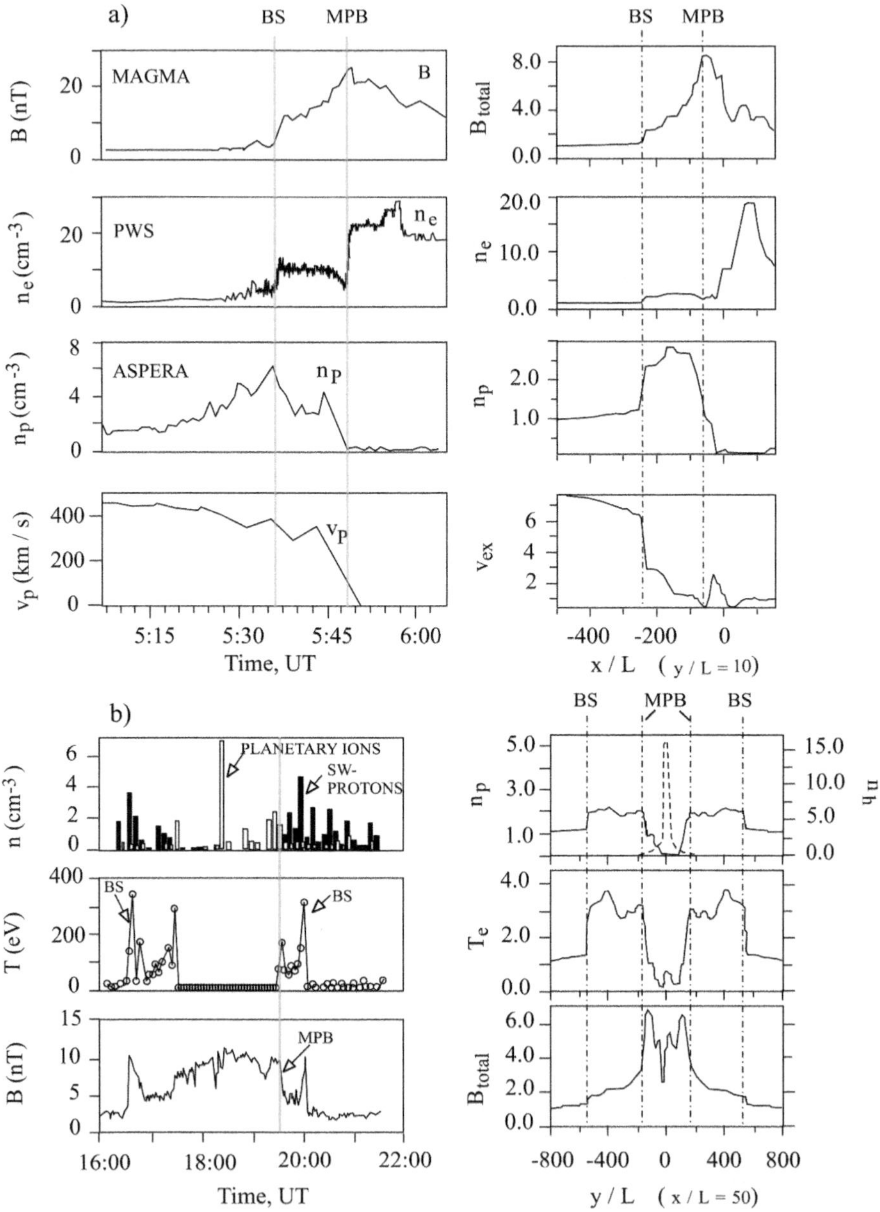

Figure 6.8. Comparison between Phobos-2 measurements (left panels) and bi-ion fluid simulations (right panels) for the third elliptical orbit (a) and one circular orbit (b), after Sauer and Dubinin (2000).

A summary of the insights gained fro the bi-ion fluid simulations is as follows. When the dynamic pressure of the solar wind is not too high, it will be balanced by the ionospheric pressure. Then, three main plasma boundaries exist: the bow shock (BS), the magnetic pile-up boundary (MPB), and the ionopause. The latter may be deformed locally by magnetic anomalies. The MPB is the 'obstacle boundary' for the solar wind protons, determining the position and the shape of the bow shock. In the opposite case of high dynamic solar wind pressure, the ionospheric pressure is insufficient to stand-off the magnetospheric plasma flow, and no ionopause is formed. In this case, the magnetospheric plasma interacts with the localized crustal magnetic fields to create mini-magnetospheres along the flanks of the tail. The MPB is formed in the magnetosheath region, where the exospheric ion density becomes comparable with the solar wind proton density. The other parameters have only a weak influence. This may explain why the bow shock position at Mars does not show the usual solar cycle variation found at Venus.

6.1.4. *Pros and Cons of the Model Approaches*

The solar wind interaction with Mars is a complex physical problem, thus it is appropriate and desirable to have a number of different models available. Given the practical limitations of numerical models described above, each has certain strength and weaknesses. Inter-comparing the results of such calculations, as well as examining agreements and disagreements with observations, advances our understanding of the controlling processes.

In principle, the hybrid kinetic model (e.g., Brecht *et al.*, 1993) should describe the macro and micro-physics of the interaction processes best, and it has been successful in reproducing some of the observed features of the modeled interaction. However, for practical reasons a number of simplifications and assumption must be introduced in order to run the model, thus it also has a number of shortcomings. For example, if one is attempting to perform detailed chemistry calculations then many particles per cell are needed in order to obtain a complete 6 degree of freedom distribution function. This is very costly, and frankly can probably be done as well or better with a fluid approach, unless the distributions are expected to be rather non-maxwellian. Next, the code is complex and requires the developer to be extremely careful in checking the results. This occurs, because particle codes can have numerical instabilities that resemble very closely real plasma instabilities. Further, with the large suite of electromagnetic modes possible in this code care must be given to resolving them both temporally and spatially. This leads to major concerns. Hybrid simulations are computationally intensive and therefore, are not easily run. However, with newer, bigger, and faster machines, even PC's, may be able to run reasonable simulations soon.

On the positive side the MHD model of Ma *et al.* (2002; 2004) is 3D, considers the solar wind protons and the major ionospheric ions separately, via separate continuity equations, and does include the crustal magnetic fields. On the negative side, it assumes only one ion velocity and one plasma temperature. Given all this, it still provides very good global insight into the interaction, and models bowshock and ionopause locations quite well, consistent with observations. It also predicts the presence of mini-magnetocylinders and gives plasma fluxes, which are also in reasonably good agreement with other estimates and observed values.

The MHD model of Sauer and Dubinin (2000) has, on the positive side, separate momentum equations for the solar wind protons and the heavy planetary ions, allowing for separate velocities. It also allows for momentum transfer by electromagnetic fields from the protons to the heavy ions, which appears to be the fundamental process for the formation of a MPB. In this respect, the use of a 2D model, as long as the electron flow is not stopped, is not a limitation. There is very good agreement between the calculations and the spacecraft observations of the MPB. However, on the negative side, it is a 2D model at this time, and it requires either the assumption of no solid obstacle or axisymmetry. It also neglects ion pressures (temperatures), which result in sharper boundaries, because the cold ions are easy to 'stop' in the simulations. Despite the various shortcomings of the different models, as indicated above, they all make important contributions towards advancing our comprehensive understanding of the interaction processes.

6.2. Hot Atoms and X-ray Models

The co-existence of solar wind plasma and the planetary neutral gas, results in a strong interaction between them. One of the fundamental collisional interactions is the charge-exchange (CX) process between an energetic ion and cold atmospheric atom, which produces an energetic neutral atom and an ionized gas particle. Examples of CX processes of special importance at Mars are:

$$\mathrm{H_{SW}}^{+} + \mathrm{H} \rightarrow \mathrm{H_{SW}} + \mathrm{H}^{+}, \tag{6.17}$$

$$\mathrm{H_{SW}}^{+} + \mathrm{O} \rightarrow \mathrm{H_{SW}} + \mathrm{O}^{+}, \tag{6.18}$$

$$\mathrm{H_{pickup}}^{+} + \mathrm{H} \rightarrow \mathrm{H_{pickup}} + \mathrm{H}^{+}, \tag{6.19}$$

$$\mathrm{O_{pickup}}^{+} + \mathrm{H} \rightarrow \mathrm{O_{pickup}} + \mathrm{H}^{+}. \tag{6.20}$$

The first three processes shown above (6.17), (6.18) and (6.19) produce fast hydrogen ENA's, (H-ENA's), while reaction (6.20) produces fast oxygen atoms (O-ENA's). The subscript SW refers to solar wind protons and pickup refers to particles originating from the planetary atmosphere or exosphere. The important feature of these CX processes is that the velocity of the newly formed ENA is about the same as the velocity of the initial energetic ($E \gg 1$ eV) fast ion. Thus the ENA's carry information about the properties of the ions from the site at which they were formed.

The solar wind also contains heavy, multiply charged ions, for example, O^{6+} ions. The charge-exchange with a multiply charged ion, say an n times charged ion X^{n+}, and a neutral, M, is a source of X-ray photons, $h\nu$:

$$X^{n+} + M \rightarrow X^{*(n-1)+} + M^{+} \rightarrow X^{(n-1)+} + M^{+} + h\nu, \tag{6.21}$$

where the superscript * refers to an exited state. The CX processes that produce ENAs and X-rays are of importance from the point of view of potential observations, because they provide opportunities to obtain global images of the source ions.

The Martian atmosphere and exosphere alters the incoming energetic solar wind by (a) 'mass loading' the solar wind with newly created ions produced mainly by photoionization, and solar wind electron impact ionization of the atmospheric gases, and (b) undergoing charge-transfer (charge-exchange) interactions with the solar wind ions. According to some models, a fraction of the solar wind ions (mainly H^+ and He^{++}) directly impact the Martian upper atmosphere near its exobase (at $\sim$ 180 km altitude) (Brecht, 1997b; Kallio and Janhunen, 2001). Others undergo charge-exchange reactions with ambient exospheric and thermospheric neutrals, particularly hydrogen and helium at high altitudes ($h >$ 500 km) and oxygen at lower altitudes, and then impact the exobase as ENA's (Kallio *et al.*, 1997). In both cases, solar wind energy is directly deposited into the upper atmosphere, resulting in increased ionization rates and UV emissions. (It should be noted that there is no observational evidence that the solar wind protons required for charge exchange penetrate through the MPB to reach the ionosphere.)

When the effects of such H-ENA precipitation were studied by Monte Carlo simulations, the precipitating hydrogen atoms were estimated to increase the ionization rate by about 1% compared to the ionization rates due to extreme ultraviolet radiation under typical solar wind conditions (Kallio and Barabash, 2000, 2001). This effect is comparable to, or even stronger than similar effects caused by the O^+ and H^+ precipitation (Luhmann and Kozyra, 1991; Brecht, 1997b; Kallio and Janhunen, 2001). Calculations also indicate that a substantial fraction of the precipitating H-ENA is scattered back from the Martian atmosphere, resulting in an H-ENA albedo (Kallio and Barabash, 2001; Holmström *et al.*, 2001). Imaging these H-ENA's would visualize the spots or regions of the most intense ENA precipitation.

All the model calculations described in this section were obtained by using the empirical flow model that is based on Phobos-2 measurements (Kallio, 1996; Kallio and Koskinen, 1999). For all three simulations the atomic and the molecular hydrogen densities were taken from Krasnopolsky and Gladstone (1996) and the cold and hot oxygen densities came from Zhang *et al.* (1993b).

While the energy transfer associated with the proton or ENA precipitation exceeds the one from the O^+ precipitation, the oxygen ions are the ones that cause massive sputtering of the atmosphere (Luhmann and Kozyra, 1991; Luhmann *et al.*, 1992; Kass and Yung, 1995, 1996). The O and CO_2 sputtered by O^+ ions has been estimated to range from 3×10^{23} to 4.7×10^{24}atoms s^{-1}, small compared to the non-thermal escape of the hot oxygen atmospheric component (Kim *et al.*, 1998; Hodges, 2000).

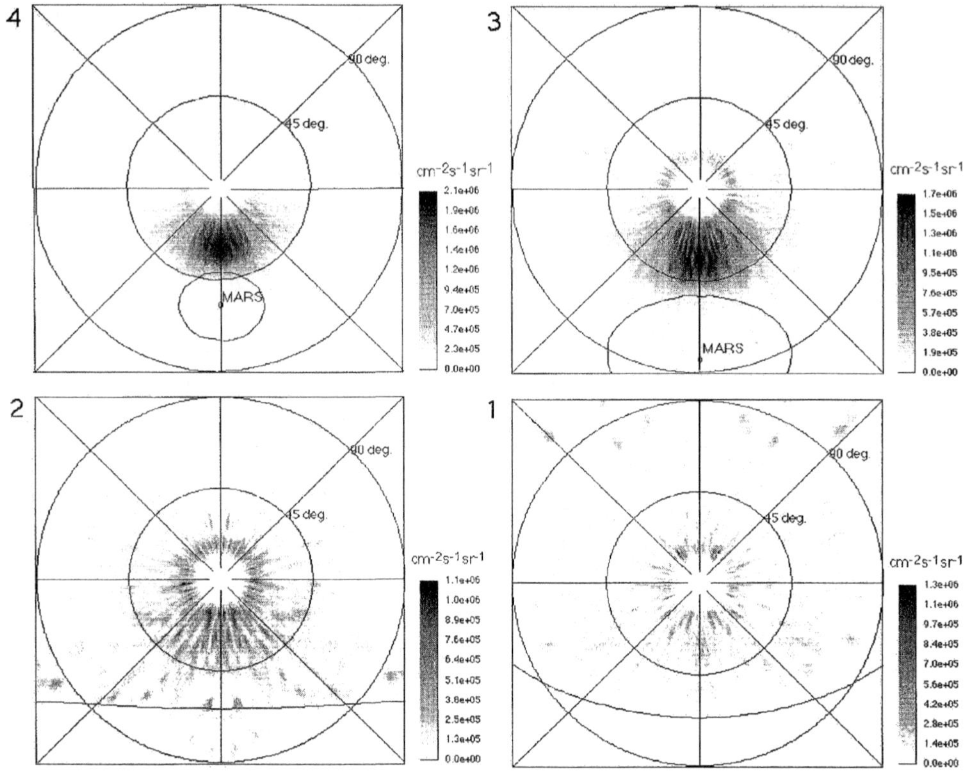

Figure 6.9. Simulated hydrogen ENA particle fluxes at four vantage points No. 4-1 from the apocenter (point No. 4) to the pericenter (point No. 1). The look direction (the polar coordinate axis) is toward the Sun, that is at the center of the images. The inner (outer) circle around the Sun represents the polar angle, θ of 45° (90°). The lines originating from the center of the figures represent constant azimuth angles, φ. The ENAs in the region $\theta < 15°$ are blocked. The solar wind parameters are $n_{sw} = 2.5$ cm^{-3}, $U_{sw} = 400$ km s^{-1}.

The supersonic solar wind upstream of the bow shock can experience charge-exchange with the Martian hydrogen exosphere (Equation (6.17)) over very long distances resulting in a narrow ($\sim$ 10°) anti-sunward beam of ENA's with the energy of the bulk flow of the solar wind. Figure 6.9 shows simulated directional hydrogen ENA fluxes produced by solar wind protons. The H-ENA flux was integrated over energy as a function of two spherical angles at four positions (1–4) along the orbit of a virtual spacecraft, whose orbit is similar to that of Mars Express. In this fish-eye projection, the look direction (the polar coordinate axis) is toward the Sun. In Figure 6.9 the distance from the Sun (at the center of each of the four images) gives the polar angle, θ. The azimuth angle, φ, measured around the polar axis, was chosen to be zero for the center of Mars. The ENA fish-eye pictures illustrate the integrated fluxes in several directions for the energy range $E > 100$ eV, based on an axially symmetric model for the solar wind proton flow around Mars. Note that in Figure 6.9 the ENA flux at polar angles less than 15° was set to zero to mimic ENA observations (an ENA instrument cannot look directly toward the Sun because of the EUV radiation from the Sun).

Figure 6.9 illustrates how an H-ENA image evolves when the vantage point moves from the apocenter (position 4) toward the pericenter (position 1). The maximum fluxes are obtained at $\varphi = 0°$, that is, between the Sun and the center of Mars. The spread of the distribution to non-zero φ depends on the proton temperature: a zero proton temperature would give non-zero fluxes only along the $\varphi = 0°$ line, that is, on the line between the Sun and Mars. Non-zero proton temperatures give, instead, non-zero fluxes at other φ angles as well, as is the case in Figure 6.9. Note that the plasma model used is axially symmetric and therefore the solution on the hemisphere $\varphi = 0° - 180°$ and $\varphi = 180° - 360°$ is symmetric. The value of the maximum ENA flux depends on the flux of the solar wind protons, the shape and the distance of the bow shock and the magnetopause, as well as the proton parameters downstream of the bow shock. The ENA fluxes produced by the shocked solar wind are most sensitive to the neutral hydrogen distribution controlled by the exobase temperature and the position of the boundary separating the solar wind and planetary plasmas (Holmström *et al.*, 2002). The ENA images display the entire interaction region and can be converted into global distributions of the proton flow and neutral gas using extracting inversion methods similar to the one developed for Earth (Roelof and Skinner, 2000).

The Martian hydrogen corona provides another source of fast H^+ ions. The accelerated planetary protons, originating from ionization of the hydrogen corona, also charge-exchange with the exospheric gas, resulting in planetary hydrogen ENA emissions (Equation (6.19)). The H-ENAs produced from these pick-up H^+ ions differ from the H-ENAs produced from the shocked solar wind protons in energy, because the pick-up protons can gain an energy up to four times the solar wind energy (Lichtenegger *et al.*, 2002).

The ASPERA/Phobos-2 observations of the plasma inside the Martian magnetosphere have shown the existence of two basic ion populations. The tail beams of H^+ and O^+ with energy 1–3 keV, and the outflowing ionospheric ions, with energy 10–100 eV near the tail flanks (Lundin *et al.*, 1993). The related ENA flux has been estimated to be 10^3 cm^{-2} s^{-1} keV^{-1} for the energy range 1–10 keV, and up to 10^5 cm^{-2} s^{-1} keV^{-1} for lower energies, 10–100 eV (Barabash *et al.*, 1995).

The ENA signal associated with the pick-up oxygen (see Equation (6.20)) was recently investigated in detail by solving numerically the kinetic equations to obtain the global distribution of oxygen ions (Barabash *et al.*, 2002). The O^+ distribution was calculated using self-consistent kinetic simulations (Kallio and Janhunen, 2002), and found, as expected, to have a large asymmetry with respect to the direction of the interplanetary electric field, $\mathbf{E}_{SW} = -\mathbf{U}_{SW} \times \mathbf{B}_{SW}$. When this O^+ distribution was converted to the corresponding O-ENA flux it was found that the fluxes of the oxygen ENA's can reach 10^4 cm^{-2} s^{-1} sr^{-1} eV^{-1} and that it fully reflects the morphology of the oxygen ion population, demonstrating that O-ENA's provide a way to determine the instantaneous oxygen ion escape rate. Similarly, comparison

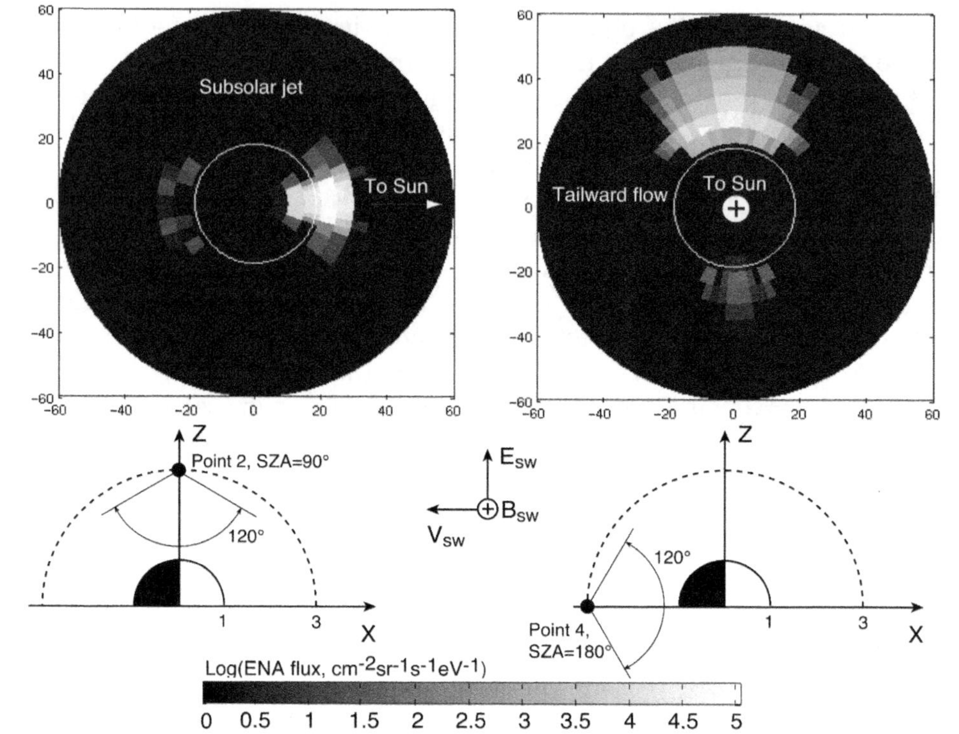

Figure 6.10. Oxygen ENA images for two vantage points: At the terminator plane (point No. 2, the left image) and at the midnight (point No. 4, the right image). The energy range is 0.1–1.65 keV. The position of the vantage points are shown in the inserts as well as the direction of the convective electric field ($\mathbf{E}_{sw}$) in the solar wind, and the interplanetary magnetic field ($\mathbf{B}_{sw}$). Both vantage points are on the *OXZ* plane. The solar wind parameters are $B_{IMF} = 2$ nT, $U_{sw} = 500$ km s^{-1}.

of oxygen ion fluxes measured by the SLED energetic particle telescope onboard Phobos-2 near Mars with predictions from a test particle calculations of pick-up O^+ ions created far from Mars puts limits on the neutral oxygen escape rate from that planet (Cravens *et al.*, 2002).

Figure 6.10 shows an example of the simulated O-ENA images for the energy range of 0.1–1.65 keV. The figure shows the vantage points and the corresponding ENA images in the fish-eye projection for two vantage points, No. 2 and No. 4. The projection is a polar one, with the polar axis pointing towards the center of Mars. The center of Mars is at the center of the figures. The distance from the center of the figures represent the value of the polar theta angle, that is, theta is 0° at the center of the figures. The polar angle is the angle to the solar direction in the plane perpendicular to the planetary center direction.

The energy range selected covers the main oxygen ion population near Mars, and it still makes the result relevant to modern ENA instrumentation. The angular resolution for the line-of-sight grid is 2.5° to 5°. The field of view is 120° in order to cover the entire interaction region. As expected, O-ENA emissions are concen-

Figure 6.11. A computer simulated image of Martian X-ray emissions from charge-exchange between solar wind ions and Mars' exosphere. The view direction is perpendicular to the Mars–Sun line, with the Sun to the left. Mars is seen as a white disc. The solar wind parameters are $n_{SW} = 2.5$ cm^{-3}, $U_{SW} = 400$ km s^{-1}.

trated close to the planetary surface, where the neutral density is highest reflecting also the strong asymmetry observed in the ion distribution. Note that Figure 6.10 (point No. 2) shows a strong jet of oxygen ENA coming out from the subsolar point, where the electric and magnetic field configuration effectively accelerates newly formed planetary ions. The tailward flow is also clearly reproduced (Figure 6.10, point 4). In contrast, when simulations for vantage points on the $+Y$ axis at the distance 3 R_M were made, only very weak O-ENA fluxes were obtained. Finally, simulations for a higher energy band of 1.65–5.0 keV showed that O-ENA image morphology is similar to the ones for the lower energy band, but that fluxes are two orders of magnitude lower.

Cravens (2000) suggested that significant X-ray fluxes, resulting from charge-exchange, may be present at Mars (Equation (6.21)). Figure 6.11 shows an example of a computer simulated X-ray image of the solar wind-Mars interaction (Holmström *et al.*, 2002). The maximum directional fluxes were found to be on the order of 10^6 eV cm^{-2} sterad^{-1} s^{-1} and depend on the solar wind conditions and view direction. The energies of the emitted X-rays are mostly below 1 keV.

Several other mechanisms can also generate X-ray emissions at Mars besides charge-exchange between solar wind ions and Mars' exosphere. In particular, fluorescence and elastic scattering of solar X-rays by Mars' atmosphere can also be significant sources. The estimated intensity of Martian X-rays from fluorescent and elastic scattering of solar X-rays somewhat exceeds the charge transfer produced X-rays (Cravens and Maurellis, 2001), as has just been confirmed by recent Chandra X-ray Observatory observations of Mars (Dennerl *et al.*, 2003). Future observations combined with more precise models for X-ray emissions should allow X-ray imaging to be a complement to ENA imaging in providing global information about the near-Mars space. X-ray emissions from Venus by the X-ray telescope Chandra have also been reported (Dennerl, 2001).

6.3. Ionosphere Models

As indicated in Section 5.1 the first detection of the ionosphere of Mars was by the Mariner 4 spacecraft in 1965 with the radio occultation experiment (Fjeldbo and Eshleman, 1968), which was soon followed by theoretical interpretations of these results (and that of Venus, as observed by Mariner 5, (Kliore *et al.*, 1967, Eshleman, 1970)). These models were based on photoionization of CO_2 and dissociative recombination of the resulting major ion, CO_2^+ (Herman *et al.*, 1970; McElroy, 1969; Whitten, 1970). However, difficulties arose with this type of model (cf., Bauer, 1973), which motivated the adoption of a more complex chemical scheme involving dissociation products of CO_2, and in particular, atomic oxygen (McElroy and McConnell, 1971; Stewart, 1972; Kumar and Hunten, 1974). The CO_2^+ and O^+ ions created by ionization of the major neutral species, CO_2, are converted to O_2^+ ions, as indicated by the following, simplified chemistry scheme:

$$CO_2 + h\nu \rightarrow CO_2^+ + e, \tag{6.22}$$

$$CO_2^+ + O \rightarrow O_2^+ + CO, \tag{6.23}$$

$$\rightarrow O^+ + CO_2, \tag{6.24}$$

$$O^+ + CO_2 \rightarrow O_2^+ + CO, \tag{6.25}$$

$$O_2^+ + e \rightarrow O + O. \tag{6.26}$$

The dissociative recombination of the O_2^+ ions (Equation (6.26)) produces the hot oxygen corona at both Venus and Mars. (cf., Schunk and Nagy, 2000).

The correctness of this scheme for Mars was confirmed by measurements made by instruments onboard the Viking 1 and 2 landers – the first *in situ* ionospheric measurements ever made for a planet other than Earth. A variety of theoretical models of the ionospheres of Venus and Mars have been developed over the past thirty years. The earliest models were photochemical, but models have become increasingly sophisticated, including both the energetics and magnetohydrodynamic (MHD) processes (cf., Cravens, 1991; Nagy and Cravens, 2002). A photochemical model is appropriate for explaining ion composition for altitudes below about 200 km where chemical time lifetimes are less than transport times. Hanson *et al.* (1977) used such a model to explain the RPA measurements and to deduce the neutral oxygen abundance (i.e., daytime O to CO_2 ratio of unity at around 200 km, in comparison to 150 km for Venus).

The next step in model sophistication for Mars was the model of Chen *et al.* (1978), who solved the coupled continuity, momentum, and energy equations in one-dimension. Among their findings was that ion transport only affects higher altitudes, although in order to obtain good agreement with measurements above about 250 km, an ion outflow (or escape) upper boundary condition was needed and was presumed to be somehow associated with the solar wind interaction processes. The same model indicated that energy inputs into the ion and electron gases were required at the upper boundary (again presumably associated with the solar wind interaction) in order to obtain agreement with measurements. Other work on the ionospheric energetics has also invoked either energy inputs and/or ionospheric thermal conductivities altered by ionospheric magnetic fields which are induced by the solar wind interaction (see Figure 5.3) (Rohrbaugh *et al.*, 1979; Choi *et al.*, 1998). For example, in the absence of magnetic field effects an input flux of 4×10^9 eV cm^{-2} s^{-1} is needed for the electrons (Choi *et al.*, 1998). Similar studies of the ion composition (Nagy *et al.*, 1980) and the energetics (Cravens *et al.*, 1980) of the Venus ionosphere were undertaken as data was returned from the Pioneer Venus Orbiter (PVO), starting in 1978.

The ion composition and the airglow of the Martian upper atmosphere and ionosphere were examined further by Fox and Dalgarno (1979), and Fox (1993, 1996). Figure 5.1, from the chemical/vertical transport model of Fox (1997), shows both solar minimum and maximum density profiles for a variety of species, including NO^+, N_2^+, CO^+, and C^+, in addition to the three major ions previously discussed. The dominance of O_2^+ as the major ion species independent of solar cycle is evident. By contrast, O^+ is the major ion species at Venus for altitudes above about 170 km (cf., Schunk and Nagy, 2000). To obtain agreement with the Viking data at higher altitudes, Fox (1997) had to assume ion outflow upper boundary conditions, just as Chen *et al.* (1978) did. The deduced flux of O atoms (in all forms) was 5.5×10^7 cm^{-2} s^{-1} for solar minimum and 1.8×10^8 cm^{-2} s^{-1} for solar maximum, values which are comparable to the direct escape rates of O from dissociative recombination of O_2^+ (Fox and Hac, 1997; Kim *et al.*, 1998; Hodges, 2000). Most of the dayside ion outflow will travel to the night, as pointed out by Fox (1997). Some of this trans-terminator flux, will flow down on the nightside and help to maintain the nightside ionosphere, while the rest will escape through the tail (Ma *et al.*, 2002; 2004).

Our understanding of the Martian ionosphere has greatly benefited from the Venus analogy. One-dimensional, multispecies, MHD models of the Venus ionosphere were able to explain the observed structure in both field and density (Shinagawa *et al.*, 1987; Shinagawa and Cravens, 1988). Ion loss due to horizontal day to night flow was included in these models, albeit not self-consistently. Shinagawa (1996) also constructed a two- dimensional MHD model of the Venus ionosphere in which the horizontal flow was handled self-consistently. Similar multispecies MHD models were constructed for the Martian ionosphere, with and without the inclusion of a small (50 nT or smaller) horizontal intrinsic magnetic field (Shinagawa and Cravens, 1989, 1992). Small intrinsic fields did not significantly affect the ionospheric density structure produced by the models, which was in agreement with the Viking measurements. On the other hand, for a case with an unmagnetized ionosphere (or for ionospheric magnetic fields with significant radial components), plus zero outflow upper boundary condition, Shinagawa and Cravens (1989), found a diffusive equilibrium topside density profile with a plasma scale height larger than the scale height evident in the Viking data. They also stated that this case 'is very similar to what would exist for Mars with a very large intrinsic field, which is completely shielded from the solar wind.' The latest version of the global, 3D, multi-species, MHD model of Ma *et al.* (2004) now has sufficient altitude resolution do be able to do a reasonably good job of estimating the relevant ionospheric parameters. The MGS observations (Acuña *et al.*, 1998; Connerney *et al.*, 1999) showed that: (1) the global-scale intrinsic field is negligibly small, and (2) the field can in places be dominated by large crustal magnetization. The presence of induced horizontal magnetic fields and/or strong crustal fields strongly impacts the ionospheric structure. Theory and modeling efforts have, so far, mainly addressed the Venus-like case of solar wind induced magnetic fields and the associated ionospheric behavior, because of the richness of data from the Pioneer Venus Orbiter. Theoretical interpretations (Krymskii *et al.*, 2002; Ness *et al.*, 2000; Brain *et al.*, 2003) and model calculations (Ma *et al.*, 2002; 2004) of the ionosphere in the presence of crustal magnetization have just begun to appear, as was briefly discussed earlier.

References

Acuña, M. H. *et al.*: 1992, 'Mars Observer Magnetic Fields Investigation', *J. Geophys. Res.* **97**, 7799.

Acuña, M. H. *et al.*: 1998, 'Magnetic Field and Plasma Observations at Mars: Initial Results of the Mars Global Surveyor Mission', *Science* **279**, 1676.

Acuña, M. H. *et al.*: 1999, 'Global Distribution of Crustal Magnetization Discovered by the Mars Global Surveyor MAG/ER Experiment', *Science* **284**, 790.

Acuña, M. H., Connerney, J. E. P., Ness, N. F., Lin, R. P., Mitchell, D., Carlson, C. W., McFadden, J., Anderson, K. A., Rème, H., Mazelle, C., Vignes, D., Wasilewski, P. and Cloutier, P.: 2001, 'The Magnetic Field of Mars : Summary of Results From the Aerobraking and Mapping Orbits', *J. Geophys. Res.* **106**, 23403.

Albee, A. L., Palluconi, F. D. and Arvidson, R. A.: 1998, 'Mars Global Surveyor mission: Overview and Status', *Science* **279**, 1671.

Alfvén, H.: 1957, 'On the Theory of Comet Tails', *Tellus* **9**, 92.

Alexander, C. J. and Russell C. T.: 1985, 'Solar Cycle Dependence of the Location of the Venus Bow Shock', *Geophys. Res. Lett.* **12**, 369.

Arkani-Hamed, J.: 2001, 'A 50-degree Spherical Harmonic Model of the Magnetic Field of Mars', *J. Geophys. Res.* **106**, 23197.

Axford W. I.: 1991, 'A Commentary on Our Present Understanding of the Martian Magnetosphere', *Planetary Space Sci.* **39**, 167.

Barabash, S. *et al.*: 1995, 'Diagnostic of Energetic Neutral Particles at Mars by the ASPERA-C Instrument For the Mars 96 Mission', *Adv. Space Res.* **16**, (4)81.

Barabash, S., Holmström, M., Lukyanov, A. and Kallio, E.: 2002, 'Energetic Neutral Atoms at Mars IV: Imaging of Planetary Oxygen, *J. Geophys. Res.* **107**(A10), 1280.

Bauer, S. J.: 1973, *Physics of Planetary Ionospheres*, Springer-Verlag, New York.

Bauer, S. J. and Hantsch, M. H.: 1989, 'Solar Cycle Variations of the Upper Atmosphere Temperatur of Mars', *Geophys. Res. Lett.* **16**, 373.

Bauer, S. J.: 1999, 'Mars Upper Atmosphere: Response to Solar Activity', *Anzeiger Abt. II* **136**, 19.

Baumgärtel K. and Sauer, K.: 1992, 'Interaction of a Magnetized Plasma Stream with an Immobile Ion Cloud', *Ann. Geophys.* **10**, 763–771.

Bertucci, C., Mazelle, C., Crider, D. et al., 2002, 'Magnetic Field Line Draping Enhancement across the Martian Magnetic Pileup Boundary', EGU Meeting, Nice, France.

Bertucci, C., Mazelle, C., Vignes, D., Crider, D. H., Acuña, M. H., Connerney, J. E. P., Mitchell, D. L., Lin, R. P., Rème, H., Cloutier, R. A., Ness, N. F. and Winterhalter, D.: 2003a, 'Magnetic Field Draping Enhancement at the Magnetic Pileup Boundary from Mars Global Surveyor', Observations, Geophys. Res. Lett. **30**(2), 1099, 10.1029/2002 GL015713.

Bertucci, C., Mazelle, C., Slavin, J. A., Russell, C. T. and Acuna, M. H., 2003b, Magnetic Field Draping Enhancement at Venus; Evidence of Magnetic Pileup Boundary, Geophys. Res. Lett. **30**(17), 1876, doi: 10.1029/2003 GL017271.

Bezrukikh, V. V., Verigin, M. I. and Shutte, N. M.: 1978, 'On the Disclosure of the Heavy Ions in the Region of the Solar Wind Interaction with Mars planet, Kosmicheskie Issledovaniya (Space Research)', **16**(4), 583–587, (in Russian).

Bougher, S. W., Engel, S., Hinson, D. P. and Forbes, J. M.: 2001, Mars Global Surveyor Radio Science Electron Density Profiles: Neutral Atmosphere Implications', *Geophys. Res. Lett.* **16**, 3091.

Brace, L. H. and Kliore, A. J.: 1991, 'The Structure of the Venus Ionosphere', *Space Sci. Rev.* **55**, 81.

Brain, D. A. *et al.*: 2003, 'Reconnection of Martian Crustal Magnetic Fields to the Solar Wind', *J. Geophys. Res.* **108** (A12), 1424, 10.1029/2002 JA009482.

Brannon, J. F., Fox, J. L. and Porter, H. S.: 1993, 'Evidence for Day-to Night Transport at Low Solar Activity in the Venus pre-dawn Ionosphere', *Geophys. Res. Lett.* **20**, 2739.

Brecht, S. H. and Thomas, V. A.: 1988, 'Multidimensional Simulations Using Hybrid Particle Codes', *Comp. Phys. Comm.* **48**, 135.

Brecht, S. H. and Ferrante, J. R.: 1991, 'Global Hybrid Simulation of Unmagnetized Planets: Comparison of Venus and Mars, *J. Geophys. Res.* **96**, 11209.

Brecht, S. H., Ferrante, J. R. and Luhmann, J. G.: 1993, 'Three-dimensional Simulations of the Solar Wind Interaction With Mars', *J. Geophys. Res.* **98**, 1345.

Brecht, S. H.: 1997a, 'Hybrid Simulations of the Magnetic Topology of Mars, *J. Geophys. Res.* **102**, 4743.

Brecht, S. H.: 1997b, 'Solar Wind Proton Deposition into the Martian Atmosphere', *J. Geophys. Res.* **102**, 11287.

Brecht, S. H.: 2002, 'Numerical Techniques Associated with Simulations of Solar Wind Interactions with Non-Magnetized Bodies', *Comparative Aeronomy in The Solar System*, eds. M. Mendillo, A. Nagy and H. Waite, American Geophysical Union, Washington.

Breus, T. K., Krymskii, A. M., Lundin, R., Dubinin, E. M., Luhmann, J. G., Yeroshenko, Ye. G., Barabash, S. V., Mitnitskii, V. Ya., Pissarenko, N. F. and Styashkin, V. A.: 1991, 'The Solar Wind Interaction with Mars: Consideration of Phobos 2 Mission Observations of an Ion Composition Boundary on the Dayside', *J. Geophys. Res.* **96**, 11165–11174.

Breus, T. K. and Krymskii, A. M.: 1992, 'Turbulent Pick-up of New-born Ions near Venus and Mars and Problems of Numerical Modelling of the Solar Wind Interaction with These Planets. –I. Features of the Solar Wind Interaction with Planets', *Planetary Space Sci.* **40**, 121.

Breus, T. K. *et al.*: 1992, 'Turbulent Pick-up of New-born Ions Near Venus and Mars and Problems of Numerical Modelling of the Solar Wind Interaction with These Planets. –II. Two-fluid HD-model', *Planetary Space Sci.* **40**, 131.

Breus, T. K. *et al.*: 1998, 'Conditions in the Martian Ionosphere/atmosphere from a Comparison of a Thermospheric Model with Radio Occultation Data', *Planetary Space Sci.* **46**, 367.

Bryant D., Krimigis, S. M. and Haerendel, G.: 1985, *IEEE Trans. Geosc. Remote Sensing* **GE-23**, 177.

Chen, R. H., Cravens, R. E. and Nagy, A. F.: 1978, 'The Martian Ionosphere in Light of the Viking Observations', *J. Geophys. Res.* **83**, 3871.

Chen, Y, Cloutier, P. A., Crider, D. H., Mazelle, C. and Rème, H.: 2001, 'On the Role of Charge Exchange in the Formation of the Martian Magnetic Pileup Boundary', *J. Geophys. Res.* **106**, 29387.

Choi, Y. W. *et al.*: 1998, 'Effect of the Magnetic Field on the Energetics of Mars's Ionosphere', *Geophys. Res. Lett.* **25**, 2753.

Cloutier P. A. and Daniell, R. E. and Butler: 1973, 'Ionospheric Currents Induced by Solar Wind Interaction with Planetary Atmospheres', *Planetary Space Sci.* **21**, 463.

Cloutier, P. A. and Daniell, R. E.: 1979, 'An Electrodynamic Model of the Solar Wind Interaction with the Ionospheres of Mars and Venus', *Planetary Space Sci.* **27**, 1111.

Cloutier, P. A. *et al.*: 1999, 'Venus-like Interaction of the Solar Wind with Mars', *Geophys. Res. Lett.* **26**, 2685.

Connerney, J. E. P. *et al.*: 1999, 'Magnetic Lineations in the Ancient Crust of Mars, *Science* **284**, 794.

Cravens, T. E., Gombosi, T. I., Kozyra, J. U., Nagy, A. F., Brace L. H. and Knudsen, W. C.: 1980, 'Model Calculations of the Dayside Ionosphere of Venus: Energetics, *J. Geophys. Res.* **85**, 7778.

Cravens, T. E.: 1991, Ionospheric models for Venus and Mars, p. 277 in *Venus and Mars: Atmospheres, Ionospheres, and Solar Wind Interactions*, Geophysical Monograph **66**, American Geophysical Union, Washington DC.

Cravens, T. E., Shinagawa, H. and Luhmann, J. G.: 1997, in S. Bougher, D. Hunten, and R. Phillips (eds.), Magnetohydrodynamic Processes: Magnetic Fields in the Ionosphere of Venus, *Venus II*, University of Arizona Press, Tucson, pp. 61–93.

Cravens, T. E.: 2000, 'X-ray Emission from Comets and Planets', *Adv. Space Res.* **26**, 1443.

Cravens, T. E. and Maurellis, A. N.: 2001, 'X-ray Emission from Scattering and Fluorescence of Solar X-rays at Venus and Mars'. *Geophys. Res. Lett.* **28**, 3043.

Cravens, T. E., Hoppe, A., Ledvina, S. A. and McKenna-Lawlor, S.: 2002, 'Pickup Ions near Mars Associated with Escaping Oxygen Atoms', *J. Geophys. Res.* **107**.

Crider, D. H. *et al.*: 2000, 'Evidence for Electron Impact Ionization in the Magnetic Pile-up Boundary of Mars', *Geophys. Res. Lett.* **27**, 45.

Crider, D. H., Acuña, M., Connerney, J. *et al.*: 2001, 'Magnetic Field Draping Around Mars: Mars Global Surveyor Results', *Adv. Space Res.* **27**(11), 1831.

Crider, D. H., Acuña, M. H., Connerney, J. E. P., Mitchell, D. L., Lin, R. P., Cloutier, P. A., Rème, H., Mazelle, C., Brain, D., Ness, N. F., and Bauer, S.: 2002, 'Observations of the Latitude Dependence of the Location of the Martian Magnetic Pileup Boundary', *Geophys. Res. Lett.* **29**(8), 10.1029/2001 GL013860.

Crider, D. H. *et al.*: 2002, 'Mars Global Surveyor Observations of Solar Wind Magnetic Field Draping Around Mars', *Space Sci. Rev.* **111**(1–2), 203–221.

Crider, D. H., Vignes, D., Krymskii, A. M., Breus, T. K., Ness, N. F., Mitchell, D. L., Slavin, J. A. and Acuña, M. H.: 2003, 'A Proxy for Determining Solar Wind Dynamic Pressure at Mars using Mars Global Surveyor Data', *J. Geophys. Res.* **108** (A12): 1461, 10.1029/2003 JA009875.

Dennerl, K.: 2001, 'Discovery of X-rays from Venus with Chandra. In 'High Energy Universe at Sharp Focus: Chandra Science'. Proceedings of a conference held in St. Paul, MN, 16–18 July 2001. *ASP Conference Series.*

Dennerl, K.: 2002, 'Discovery of X-rays from Mars with Chandra, *Astron. Astrophys.*, **394**, 1119 doi: 10.1051/0004-6361: 20021116.

Dobe, Z., Nagy, A. F. and Fox, J. L.: 1995, 'A Theoretical Study Concerning the Solar Cycle Dependence of the Nightside Ionosphere of Venus', *J. Geophys. Res.* **100**, 14507.

Dolginov Sh., Yeroshenko, Ye. G. and Zhuzgov, D. N.: 1976a, 'Magnetic Field of Mars According to Data Mars-3 and Mars-5 Satellites', *J. Geophys. Res.* **81**, 3353.

Dolginov, Sh., Yeroshenko, Ye. G., Zhuzgov, L. N. Sharova, V. A., Gringauz, K. I., Bezrukikh, V. V., Breus, T. K., Verigin, M. I. and Remizov, A. P.: 1976b, in N. F. Ness (ed.), 'Magnetic Field and Plasma Inside and Outside of the Martian Magnetosphere', *Solar Wind Interaction with the Planets Mercury, Venus, and Mars*, NASA SP-397, 1–20.

Dubinin, E. and Podgorny, I. M.: 1980, 'Combined magnetosphere', *Cosmic Research, Engl. Transl.* **18**, 470.

Dubinin, E. M., Lundin, R., Norberg, O. and Pissarenko, N.: 1993a, 'Ion Acceleration in the Martian Tail: The Phobos Observations, *J. Geophys. Res.* **98**, 3991.

Dubinin, E. M., Lundin, R., Koskinen, H. and Norberg, O.: 1993b, 'Cold Ions at the Bow Shock: Phobos Observations', *J. Geophys. Res.* **98**, 5617.

Dubinin E. M., Lundin, R. and Schwingenschuh, K.: 1994, 'Solar Wind Electrons as Tracers of the Martian Magnetotail Topology, *J. Geophys. Res.* **99**, 21233.

Dubinin E. M., Obod, D., Lundin, R., Schwngenschuh, K. and Grard, R.: 1995, 'Some Features of the Martian Bow Shock', *Adv. Space Res.* **15**(8/9), 423.

Dubinin E. M., Sauer, K., Lundin, R., Baumgartel, K. and Bogdanov, A.: 1996a, 'Structuring of the Transition Region (Plasma Mantle) of the Martian Magnetosphere', *Geophys. Res. Lett.* **23**, 785.

Dubinin, E. M. *et al.*: 1996b, 'Plasma Characteristics of the Boundary Layer in the Martian Magnetosphere', *J. Geophys. Res.* **101**, 27,061.

Dubinin E. M., Sauer, K., Baumgartel, K. and Srivastava, K.: 1998, 'Multiple Shocks Near Mars', *Earth Planets Space* **50**, 279.

Dubinin E. M. and Sauer, K.: 1999, 'The Martian Magnetosphere-A Laboratory for Bi-ion Plasma Investigation', *Astrophys Space Sci.* **264**, 273.

Dubinin E. M., Sauer, K., McKenzie, J. F. and Chanteur, G.: 2002, 'Nonlinear Waves and Solitons Propagating Perpendicular to the Magnetic Field in Bi-ion Plasma with Finite Plasma Pressure', *Nonlinear Processes Geophys.* **9**(2), 87.

Eastman, T. E., DeCoster, R. J. and Frank, L. A.: 1986, 'Velocity Distributions of Ion Beams in the Plasma Sheet Boundary Layer, in *Ion Acceleration in the Magnetosphere and Ionosphere, Geophysical Monograph 38*, AGU, pp. 117, Washington D.C..

Eastwood, J. W.: 1972, 'Consistency of Fields and Particle Motion in the 'Speiser' Model of the Current Sheet', *Planetary Space Sci.* **20**, 1555.

Elphic, R. C. *et al.*: 1980, 'Observations of the Dayside Ionopause and Ionosphere of Venus', *J. Geophys. Res.* **85**, 7679.

Eshleman, V. R.: 1970, 'Atmospheres of Mars and Venus: A Review of Mariner 4 and 5 and Venera 4 Experiments', *Radio Sci.* **5**, 325.

Fjeldbo, G. and Eshleman, V. R.: 1968, 'The Atmosphere of Mars Analyzed by Integral Inversion of the Mariner 4 Occulation Data', *Planetary Space Sci.* **16**, 1035.

Fox, J. L. and Dalgarno, A.: 1979, 'Ionization, Luminosity, and Heating of the Upper Atmosphere of Mars', *J. Geophys. Res.* **84**, 7315.

Fox, J. L.: 1993, 'The Production and Loss of Nitrogen Atoms on Mars', *J. Geophys. Res.* **98**, 3297.

Fox, J. L.: 1996, in G. W. F. Drake (ed.), 'Aeronomy', *Atomic, Molecular, and Optical Physics Handbook*, 940, Am. Institute Phys. Press, Woodbury, NY.

Fox, J. L.: 1997, 'Upper Limits to the Outflow of Ions at Mars: Implications for Atmospheric Evolution', *Geophys. Res. Lett.* **24**, 2901.

Fox, J. L. and Stewart, A. I. F.: 1991, 'The Venus Ultraviolet Aurora a Soft Electron Source', *J. Geophys. Res.* **96**, 9829.

Fox, J. L. and Hac, A.: 1997, 'Spectrum of Hot O at the Exobase of the Terrestrial Planets', *J. Geophys. Res.* **102**, 24005.

Gan, L., Cravens, T. E. and Horanyi, M.: 1990, 'Electrons in the Ionopause Boundary Layer of Venus', *J. Geophys. Res.* **95**, 19023.

Gombosi T. I. *et al.*: 1981, 'The Role of Charge-exchange in the Solar Wind Absorption by Venus4, *Geophys. Res. Lett.* **8**, 1265.

Grard, R., Pedersen, A., Klimov, S., Savin, S., Skalsky, A., Trotignon, J. G. and Kennel, C.: 1989, 'First Measurements of Plasma Waves near Mars', *Nature* **341**, 607.

Grard, R., Nairn, C., Pedersen, A., Klimov, S., Savin, S., Skalsky, A. and Trotignon, J. G.: 1991, 'Plasma and Waves Around Mars', *Planetary Space Sci.* **39**, 89.

Grard, R., Skalsky, S. and Trotignon, J. G.: 1993, in T.J. Combosi (ed.), 'Selected Wave and Plasma Features of the Martian Environment, *Plasma Environment of non-Magnetic Planets, COSPAR Colloq. Ser.* **4**, Pergamon, New York, pp. 321.

Gringauz K. I., Bezrukhikh, V. V., Verigin, M. I. and Remizov, A. P.: 1975, 'Studies of Solar Plasma Near Mars and Along the Earth-Mras Path, 3. Characteristics of Ion and Electron Components Measured on Satellite Mars-5', *Cosmic Res.* **13**, 107.

Gringauz, K. I., Bezrukikh, V. V., Verigin, M. I. and Remizov, A. P.: 1976, 'On Electron and Ion Component of Plasma in the Antisolar Part of Near-martian Space, *J. Geophys. Res.* **81**, 3349.

Gringauz, K. I.: 1976, 'Interaction of Solar Wind with Mars as Seen by Charged Particle Traps on Mars 2, 3 and 5 Satellites', *Rev. Geophys. Space Phys.* **14**, 391.

Gringauz, K. I.: 1981, 'A Comparison of the Magnetospheres of Mars, Venus and the Earth', *Adv. Space Res.* **1**(1), 5.

Haider, S. A., Kim, J., Nagy, A. F., Keller, C. N., Verigin, M. I., Gringauz, K. I., Shutte, N. M., Szego, K. and Kiraly, P.: 1992, 'Calculated Ionization Rates, Ion Densities, and Aiglow Emission Rates Due to Precipitating Electrons in the Nightside Ionosphere of Mars', *J. Geophys. Res.* **97**, 10637.

Hanson, W. B., Sanatani, S. and Succaro, D. R.: 1977, 'The Martian Ionosphere as Observed by the Viking Retarding Potential Analyzers', *J. Geophys. Res.* **82**, 4351.

Hanson, W. B. and Mantas, G. P.: 1988, 'Viking Electron Temperature Measurements: Evidence for a Magnetic Field in the Martian Ionosphere', *J. Geophys. Res.* **93**, 7538.

Harned, D. S.: 1982, 'Quasineutralhybrid Simulation of Macroscopic Plasma Phenomena', *J. Comp. Phys.* **47**, 452.

Harnett, E. M. and Winglee, R. M.: 2003, 'The Influence of a Mini-magnetopause on the Magnetic Pileup Boundary of Mars' *Geophys. Res. Lett.*, **30**, 20, 2074, doi: 1029/2003 GLO17852.

Herman, J. R., Hartle, R. E. and Bauer, S. J.: 1970, 'The Dayside Ionosphere of Venus', *Planetary Space Sci.* **19**, 443.

Hill, T. W.: 1975, 'Magnetic Merging in a Collisionless Plasma', *J. Geophys. Res.* **80**, 4689.

Hodges, R. R.: 2000, 'Distributions of Hot Oxygen for Venus and Mars', *J. Geophys. Res.* **105**, 6971.

Holmström, M., Barabash, S. and Kallio, E.: 2001, 'X-ray Imaging of the Solar Wind-Mars Interaction'. *Geophys. Res. Lett.* **28**, 1287.

Holmström, M., Barabash, S. and Kallio, E.: 2002, 'Energetic Neutral Atoms at Mars I: Imaging of Solar Wind Protons', *J. Geophys. Res.* **107**(A10), 1277.

Ip, W.-H.: 1992a, 'Ion Acceleration at the Current Sheet of the Martian Magnetosphere', *Geophys. Res. Lett.* **19**, 2095.

Ip W.-H.: 1992b, 'Neutral Particle Environment of Mars: The Exosphere-plasma Interaction Effects', *Adv. Space Res.* **12**(9), 205.

Ip W.-H., Breus, R. K. and Zarnowiecki, T.: 1994, 'Termination of the Solar Wind Flow Near Mars by Charge-exchange', *Planetary Space Sci.* **42**, 435.

Izakov, M. N. and Roste, O. Z.: 1996, 'Martian Upper Atmosphere Structure Variation', *Cosmich. Issled.* **34**, N3.

Israelevich P. L. *et al.*: 1994, 'The Induced Magnetosphere of Comet Halley: Interplanetary Magnetic Field During Giotto Encounter', *J. Geophys. Res.* **99**, 6575.

Kallio, E., Koskinen, H., Barabash, S., Lundin, R., Norberg, O. and Luhmann, J. G.: 1994, 'Proton Flow in the Martian Magnetosheath', *J. Geophys. Res.* **99**, 23547.

Kallio, E.: 1996, 'An Empirical Model of the Solar Wind Flow Around Mars' *J. Geophys. Res.* **101**, 11133.

Kallio, E., Luhmann, J. G. and Barabash, S.: 1997, 'Charge Exchange Near Mars: The Solar Wind Absorption and Energetic Neutral Atom Production', *J. Geophys. Res.* **102**, 22183.

Kallio, E. and Koskinen, H.: 1999, A Test Particle Simulation of Oxygen Ions and Solar Wind Protons Near Mars', *J. Geophys. Res.* **104**, 557.

Kallio, E. and Barabash, S.: 2000, 'On the Elastic and Inelastic Collisions Between the Precipitating Energetic Hydrogen Atoms and the Martian Atmospheric Neutrals', *J. Geophys. Res.* **105**, 24973.

Kallio, E. and Barabash, S.: 2001, 'Atmospheric Effects of Precipitating Energetic Hydrogen Atoms to the Martian Atmosphere', *J. Geophys. Res.* **106**, 165–177.

Kallio, E. and Janhunen, P.: 2001, 'Atmospheric Effects of Proton Precipitation in the Martian Atmosphere and its Connection to the Mars-Solar Wind Interaction', *J. Geophys. Res.* **106**, 5617.

Kallio, E. and Janhunen, P.: 2002, 'Ion Escape from Mars in a Quasi-neutral Hybrid Model', *J. Geophys. Res.* **107**(A3), 10.1029/2001 JA000090.

Kass, D. M. and Yung, Y. L.: 1995, 'Loss of Atmosphere from Mars due to Solar Wind Sputtering, *Science* **268**, 697.

Kass, D. M. and Yung, Y. L.: 1996, 'Response: The Loss of Atmosphere from Mars', *Science* **274**, 1932.

Keating G. M. *et al.*: 1998, 'The Structure of the Upper Atmosphere of Mars. In situ Accelerometer Measurements from Mars Global Surveyor', *Science* **279**, 1672.

Kim, J., Nagy, A. F., Fox, J. L. and Cravens, T. E.: 1998, 'Solar Cycle Variability of Hot Oxygen Atoms at Mars', *J. Geophys. Res.* **103**, 29339.

Kivelson, M. G., Khurana, K. K. and Volwerk, M.: 2002, 'The Permanent and Inductive Magnetic Moments of Ganymede.' *Icarus*, **157**, 2, 502.

Kliore, A. J. *et al.*: 1967, 'Atmosphere and Ionosphere of Venus from Mariner 5 S-band Radio Occultation Experiment', *Science* **158**, 1683.

Kliore A. J., Cain, D. C., Fjeldbo, G., Seidel, B. L. and Rasool, S. I.: 1972, 'Mariner-9 S-band Occultation Experiment: Initial Results on the Atmosphere and Topography of Mars', *Science* **175**, 313.

Kliore A. J., Fjeldbo, G., Seidel, B. L., Sykes, M. J. and Woiceshyn, P. M.: 1973, 'S-band Radio Occultation Measurements of the Atmosphere and Topography of Mars with Mariner 9 - Extended Mission Coverage of Polar and Intermediate Latitudes', *J. Geophys. Res.* **78**, 4331.

Kliore, A. J.: 1992, 'Radio Occultation Observations of the Ionospheres of Mars and Venus, *Venus and Mars: Atmospheres, Ionospheres and Solar Wind Interactions*', 265, *Geophys. Monograph* **66**, American Geophysical Union.

Kotova, G. A., Verigin, M. I., Shutte, N. M., Remizov, A. P., Rosenbauer, H., Riedler, W., Schwingenschuh, K., Zhang, T.-L., Szego, K. and Tatrallyay, M.: 1997a, 'Planetary Heavy Ions in the Magnetotail of Mars: Results of the TAUS and MAGMA Experiments Aboard PHOBOS 2', *Adv. Space Res.* **20**(2), 173.

Kotova, G., Verigin, M., Remizov, A., Shutte, N., Slavin, J., Szego, K., Tatrallyay, M., Rosenbauer, H., Livi, S., Richter, A., Schwingenschuh, K. and Zhang, T.-L.: 1997b, 'The study of the Solar Wind Deceleration Upstream of the Martian Terminator Bow Shock', *J. Geophys. Res.* **102**, 2165.

Kotova, G. A., Verigin, M. I., Remizov, A. P., Shutte, N. M., Rosenbauer, H. *et al.*: 2000a, 'Heavy Ions in the Magnetosphere of Mars: Phobos 2/TAUS Observations', *Phys. Chem. of the Earth (C)* **25**, 157.

Kotova G. A., Verigin, M. I., Remizov, A. P., Rosenbauer, H., Livi, S., Riedler, W., Schwingenschuh, K., Tatrallyay, M., Szego, K. and Apathy, I.: 2000b, 'On the Possibility of Identifying of Heavy Ion Acceleration Processes in the Magnetotail of Mars', *Earth Planets Space* **52**.

Krasnopolsky, V. A. and Gladstone, G. R.: 1996, 'Helium on Mars: EUVE and PHOBOS Data and Implications for Mars'Evolution', *J. Geophys. Res.* **101**, 15765.

Krymskii, A. M.: 1992, in J. G. Luhmann, M. Tatrallyay, and R. O. Pepin (eds.), 'An Interpretation of the Large Scale Ionospheric Magnetic Fields and the Altitude Distribution of the Ionosphere Plasma on the Dayside of Venus and Mars, *Venus and Mars, Atmospheres, Ionospheres, and Solar Wind Interactions, Geophys. Monograph* **66**, 289.

Krymskii, A. M., Breus, T. K., Ness, N. F. and Acuña, M. H.: 2000, 'The IMF Pile-up Regions near the Earth and Venus: Lessons for the Solar Wind – Mars Interaction', *Space Sci. Rev.* **92**, 535.

Krymskii, A. M., Breus, T. K., Ness, N. F., Acuña, M. H., Connerney, J. E. P., Crider, D. H., Mitchell, D. L. and Bauer, S. I.: 2002, 'Structure of the Magnetic Field Flux Connected with Crustal Magnetization and Top-side Ionosphere of Mars', *J. Geophys. Res.* **107** (A9), 1245, doi: 10.1029/2001 JAO000239.

Kumar, S. and Hunten, D. M.: 1974, 'An Ionospheric Model With an Exospheric Temperature of 350 K', *J. Geophys. Res.* **79**, 2529.

Leblanc, F. and Johnson, R. E.: 2002, 'Role of Molecular Species in Pickup Ion Sputtering of the Martian Atmosphere', *J. Geophys. Res.* **107**(E2), 10.1029/2000JE001473.

Lichtenegger, H., Schwingenschuh, K., Dubinin, E. and Lundin, R.: 1995, 'Particle Simulation in the Martian Magnetotail', *J. Geophys. Res.* 21659.

Lichtenegger, H., Dubinin, E. and Ip, W.-H.: 1997, 'The Depletion of the Solar Wind near Mars', *Adv. Space Res.* **20**(2), 143.

Lichtenegger, H. and Dubinin, E.: 1998, 'Model Calculations of the Planetary Ion Distribution in the Martian Tail', *Earth Planets Space* **50**, 445.

Lichtenegger, H. and Dubinin, E.: 1999, 'Charge-exchange in the Magnetosheath of Mars', *Adv. Space Res.*.

Lichtenegger, H., Dubinin, E., Schwingenschuh, K. and Riedler, W.: 2000, 'The Martian Plasma Environment: Model Calculations and Observations', *Adv. Space Res.* **26**, 1623.

Lichtenegger, H., Lammer, H. and Stumptner, W.: 2002, 'Energetic Neutral Atoms at Mars III: Flux and Energy Distribution of Planetary Energetic H atoms', *J. Geophys. Res.* **107**, (A10), 1279, doi: 10.1029/2001 JA000322.

Liu, Y., Nagy, A. F., Clinton, P. T., Groth, D. L., DeZeeuw, L. and Gombosi, T. I.: 1999, '3D multi-fluid MHD Studies of the Solar Wind Interaction with Mars', *Geophys. Res. Lett.* **26**, 2689.

Liu, Y, Nagy, A. F., Gombosi, T. I., DeZeeuw, D. L. and Powell, K. G.: 2001, 'The Solar Wind Interaction with Mars: Results of Three-dimensional Three-species MHD Studies', *Adv. Space Res.* **27**, 1837.

Luhmann, J. G.: 1986, 'The Solar-Wind Interaction with Venus', *Space Sci. Rev.* **44**(3–4, 241.

Luhmann, J. G. and Cravens, T. E.: 1991, 'Magnetic Fields in the Ionosphere of Venus', *Space Sci. Rev.* **55**, 201.

Luhmann, J. G. and Kozyra, J. U.: 1991, 'Dayside Pickup Oxygen Ion Precipitation at Venus and Mars: Spatial Distributions, Energy Deposition and Consequences', *J. Geophys. Res.* **96**, 5457.

Luhmann, J. G. *et al.*: 1987, 'Characteristics of the Mars-like Limit of the Venus – Solar Wind Interaction', *J. Geophys. Res.* **92**, 8545.

Luhmann, J. G., Russell, C. T., Schwingenschuh, K. and Yeroshenko, Ye.: 1991, 'A Comparison of Induced Magnetotails of Planetary Bodies: Venus, Mars, and Titan', *J. Geophys. Res.* **95**, 11199.

Luhmann, J. G. *et al.*: 1992, 'Evolutionary Impact of Sputtering of the Martian Atmosphere by O+ Pickup Ions', *Geophys. Res. Lett.* **19**, 2151.

Lundin, R., Zakharov, A., Pellinen, R., Borg, H., Hultqvist, B., Pissarenko, N., Dubinin, E. M., Barabash, S. V., Liede, I. and Koskinen, H.: 1989, 'First Measurements of the Oonospheric Plasma Escape from Mars', *Nature*, **341**, 6243, 609.

Lundin, R. *et al.*: 1990, 'Plasma Composition Measurements of Martian Magnetosphere Morphology', *Geophys. Res. Lett.* **17**, 877.

Lundin, R. and Dubinin, E. M.: 1992, 'Phobos-2 Results on the Ionospheric Plasma Escape from Mars', *Adv. Space Res.* **12**(9), 255.

Lundin, R. *et al.*: 1993, in T. I. Gombosi (ed.), 'ASPERA Observations of Martian Magnetospheric Boundaries', *Plasma Environments of Non-magnetic Planets* Pergamon Press, p. 311.

Ma, Y., Nagy, A. A. F., Hansen, K. C., DeZeeuw, D. L., Gombosi, T. I. and Powell, K. G.: 2002, '3D Multi-fluid MHD Studies of the Solar Wind Interaction with Mars in the Presence of Crustal Fields', *J. Geophys. Res.* **107**, (in press).

Ma, Y., Nagy, A. F. and Sokolov, I. V. and K. C. Hansen: 2004, 3D, Multispecies, High Spatial Presolution MHD Studies of the Solar Wind Interaction with Mars, *J. Geophys. Res.*, **109**, 2003 JAO10367.

Mahajan, K. K. and Mayr, H. G.: 1989, 'Venus Ionopause During Solar Minimum', *Geophys. Res. Lett.* **16**, 1477.

Mantas, G. P. and Hanson, W. B.: 1979, 'Photoelectron Fluxes in the Martian Ionosphere', *J. Geophys. Res.* **84**, 369.

Marubashi, K., Grebowsky, J. M., Taylor, H. A. *et al.*: 1985, 'Magnetic Field in the Wake of Venus and the Formation of Ionospheric Holes', *J. Geophys. Res.* **90**, 1385.

Mazelle, C., Rème, H., Sauvaud, J. A., d'Uston, C., Carlson, C. W., Anderson, K. A., Curtis, D. W., Lin, R. P., Korth, A., Mendis, D. A., Neubauer, F. M., Glassmeier, K. H. and Raeder, J.: 1989, 'Analysis of Suprathermal Electron Properties at the Magnetic Pile-up Boundary of Comet P/Halley', *Geophys. Res. Lett.* **16**(9), 1035.

Mazelle, C, Rème, H., Neubauer, F. M. and Glassmeier, K.-H.: 1995, 'Comparison of the Main Magnetic and Plasma Features in the Environments of Comets Grigg-skjellerup and Halley', *Adv. Space Res* **16**, (4)41–(4)45.

Mazelle C., Vignes, D., Rème, H., Sauvaud, J. A., d'Uston, C., Acuña, M. H., Connerney, J. E. P., Wasilewski, P., Lin, R. P., Mitchell, D. L., Anderson, K. A., Carlson, C. W., McFadden, J., Curtis, D. W., Cloutier, P. A., Crider, D. H., Law, C. C., Bauer, S. J., Ness, N. F. and Winterhalter, D.: 1998, 'Analysis of the Magnetic 'Pile-up' Boundary at Mars, *Eos Trans. AGU* Fall Meet. Suppl., Abstract P12A-09.

Mazelle, C., Bertucci, C., Rème, H., Mitchell, D. L., Lin, R. P., Vignes, D., Crider, D. H., Acuña, M. H., Connerney, J. E. P., Sauer, K., Chen, Y., Cloutier, P. A., Ness, N. F., and Winterhalter, D.: 2002, 'The Magnetic Pileup Boundary at Mars: A Comet-like Feature in the Interaction of the Planet Atmosphere with the Solar Wind', *in preparation.*

McComas D. J., Spence, H. E., Russell, C. T. *et al.*: 1986, 'The Average Magnetic-field Draping and Consistent Plasma Properties of the Venus Magnetotail'. *J Geophys Res.* **91**, 7939.

McElroy, M. B.: 1969, 'Structure of the Venus and Mars Atmospheres', *J. Geophys. Res.* **74**, 29.

McElroy, M. B. and McConnell, J. C.: 1971, 'Atomic Carbon in the Atmospheres of Mars and Venus', *J. Geophys. Res.* **76**, 6674.

McElroy, M. B. and McConnell, J. C.: 'Dissociation of CO_2 in the Martian Atmosphere', *J. Atmos. Sci.* **28**, 1437.

McKenzie J. F., Marsch, E., Baumgartel, K. and Sauer, K.: 1993, 'Wave and Stability Properties of Multi-ion Plasmas with Applications to Winds and Flows', *Annal. Geophysikae* **11**, 341.

McKenzie J. F., Sauer, K. and Dubinin, E.: 2001, 'Stationary Waves in a Bi-ion Plasma Transverse to the Magnetic Field', *J. Plasma Physics* **65**, 197.

Mellott, M. M. and Livesey, W. A.: 1987, 'Shock Overshoots Revisited', *J. Geophys. Res.* **92**, 13661.

Michel, F. C.: 1971, 'Solar-wind-induced Mass Loss from Magnetic Field-free Planets', *Planetary Space Sci.* **19**, 1580.

Mitchell, D. L. *et al.*: 2000, 'Oxygen Auger Electrons Observed in Mars' Ionosphere', *Geophys. Res. Lett.* **27**, 1827.

Mitchell, D. L., Lin, R. P., Mazelle, C., Rème, H., Cloutier, P. A., Connerney, J. E. P., Acuña, M. H. and Ness, N. F.: 2001, 'Probing Mars' Crustal Magnetic Field and Ionosphere with the MGS Electron Reflectometer', *J. Geophys. Res.* **106**, 23419.

Mitchell, D. L., Lin, R. P., Rème, H., Cloutier, P. A., Connerney, J. E. P., Acuña, M. H. and Ness, N. F.: 2002, 'Probing Mars' Crustal Magnetic Field and Ionosphere with the MGS Electron Reflectometer, Lunar and Planetary Science XXXIII.

Motschmann, U., Sauer, K., Roatsch, T. and McKenzie, J. F.: 1991, 'Subcritical Multiple-ion Shocks', *J. Geophys. Res.* **96**, 13841.

Möhlmann, D. *et al.*: 1991, 'The Question of an Internal Martian Magnetic Field', *Planetary Space Sci.* **39**, 83.

Möhlmann D.: 1992, 'The Question of a Martian Planetary Magnetic Field', *Adv. Spave. Res.* **12**(8), 213.

Mura, A., Milillo, A., Orsini, S., Kallio, E. and Barabash, S.: 2002, 'Energetic Neutral Atoms at Mars II: Energetic Neutral Atom Production near Phobos', *J. Geophys. Res.* **107**(A10), 1278.

Nagy, A. F., Cravens, T. E., Smith, S. G., Taylor, H. A. and Brinton, H. C.: 1980, 'Model Calculations of the Dayside Ionosphere of Venus: Ionic Composition', *J. Geophys. Res.* **85**, 7795.

Nagy, A. F., Gombosi, T. I., Szego, K., Sagdeev, R. Z., Shapiro, V. D. and Shevchenko, V. I.: 1990, 'Venus Mantle - Mars Planetosphere: What are the Similarities and Differences', *Geophys. Res. Lett.* **17**, 865.

Nagy, A. F. and Cravens, T. E.: 1997, in S. W. Bougher, D. M. Hunten and R. J. Phillips (Eds.), 'Ionosphere: Energetics', *Venus II* p. 189.

Nagy, A. F. and Cravens, T. E.: 2002, in M. Mendillo, A. Nagy and H. Waite (Eds.), 'Solar System Ionospheres', *Atmospheres in the Solar System: Comparative Aeronomy* p. 39, Geophys. Mon. 130, American Geophysical Union.

Ness, N. F., Acuña, M. H., Connerney, J., Wasilewski, P., Mazelle, C., Sauvaud, J., Vignes, D., d'Uston, C., Rème, H., Lin, R., Mitchell, D. I., McFadden, J., Curtis, D., Cloutier, P. and Bauer, S.: 1999, 'MGS Magnetic Fields and Electron Reflectometer Investigation: Discovery of Paleomagnetic Fields due to Crustal Remnance', *Adv. Space Res.* **23**(11), 1876.

Ness, N. F. *et al.*: 2000, 'Effects of Magnetic Anomalies Discovered at Mars on the Structure of the Martian Ionosphere and Solar Wind Interaction as Follows from Radio Occultation Experiment' *J. Geophys. Res.* **105**, 15991.

Neubauer, F. M.: 1987, 'Giotto Magnetic-field Results on the Boundary of the Pile-up Region and the Magnetic Cavity', *Astron. Astropys.* **187**, 73.

Neubauer, F. M., Marschall, H., Pohl, M., Glassmeier, K.-H., Musmann, G., Mariani, F., Acuña, . H., Burlaga, L. F., Ness, N. F., Wallis, M. K., Schmidt, H. U. and Ungstrup, E.: 1993, 'First Results from the Giotto Magnetometer Experiment During the P/Grigg-Skjellerup Encounter', *Astron. Astrophys.* **268**, L5–L8.

Nier, A. J. and McElroy, M. B.: 1977, 'Composition and Structure of the Mars' Upper Atmosphere: Results from the Neutral Mass Spectrometers on Viking 1 and 2', *J. Geophys. Res.* **82**, 4241.

Norberg, O., Barabash, S. and Lundin, R.: 1993, in T. Gombosi (ed.), 'Observations of Molecular Ions in the Martian Plasma Environment', *Plasma Environments of Non-magnetic Planets, COSPAR Colloquia Series*, **4**, p. 299.

Paxton, L. J. and Anderson, D. E.: 1991, 'Far Ultraviolet Remote Sensing of Venus and Mars', in *Venus and Mars: Atmospheres, Ionospheres, and Solar Wind Interactions*, 112, Geophysical Monograph 66, American Geophysical Union, Washington DC.

Pedersen A., Nairn, C., Grard, R. and Schwingenschuh, K.: 1991, 'Deviation of Electron Densities from Differential Potential Measurements Upstream and Downstream of the Bow Shock and in the Magnetosphere of Mars', *J. Geophys. Res.* **96**, 11243.

Phillips, J. L., Luhmann, J. G. and Russell, C. T.: 1984, 'Growth and Maintenance of Large-scale Magnetic Fields in the Dayside of Venus', *J. Geophys. Res.* **89**, 10676.

Powell *et al.*: 1999, 'A solution-adaptive Upwind Scheme for Ideal Magnetohydrodynamics', *J. Comp. Phys.* **154**, 284.

Rasool, S. I. and Stewart, R. W.: 1971, 'Results and Interpretation of S-band Occultation Experiments n Mars and Venus', *J. Atmos. Sci.* **28**, 869.

Rème, H., Mazelle, C., Sauvaud, J. A., d'Uston, C., Froment, F., Lin, R. P., Anderson, K. A., Carlson, C. W., Larson, D. E., Korth, A., Chaizy, P. and Mendis, D.A.: 1993, 'Electron Plasma Environment at Comet Grigg-Skjellerup: General Observations and Comparison with the Environment at Comet Halley', *J. Geophys. Res.* **98**, 20965.

Riedler, W,., Mohlmann, D., Oraevsky, V. N., Schwingenschuh, K., Eroshenko, Ye., Rustenbach, J., Aydogar, Oe., Berghofer, G., Lichtenegger, H., Delva, M., Schelch, G., Pirsch, K., Fremuth, G., Steller, M., Arnold, H., Raditsch, T., Ayster, U., Fornacon, K.-H., Schenk, H. J., Michaelis, H., Motschmann, U., Roatsch, T., Sauer, K., Schroter, R., Kurths, J., Lenners, D., Linthe, J., Kobzev, V., Styashkin, V., Achache, J., Slavin, J., Luhmann, J. G. and Russell, C. T.: 1989, 'Magnetic Fields Near Mars: First Results', *Nature* **341**, 604.

Roelof, E. C. and Skinner, A. J.: 2000, 'Extraction of Ion Distributions from Magnetospheric and EUV Images', *Space Sci. Rev.* **91**, 437.

Rohrbaugh, R. P., Nisbet, J. S., Blauler, E. and Hesman, J. R.: 1979, 'The Effect of Energetically Produced O2+ on the Ion Temperatures of the Martian Thermosphere', *J. Geophys. Res.* **84**, 3327, 1979.

Rosenbauer, H., N. Shutte, I. Apathy, A. Galeev, K. Gringauz, H. Gruenwaldt, P. Hemmerich, K. Jockers, P. Kiraly, G. Kotova, S. Livi, E. Marsh, A. Richter, W. Riedler, A. Remizov, R. Schwenn, K. Schwingenschuh, M. Steller, K. Szego, M. Verigin, and M. Witte, Ions of martian origin and plasma sheet in the martian magnetotail: Initial results of TAUS experiment, Nature, 341, 612, 1989.

Rosenbauer, H., M. Verigin, G. Kotova, S. Livi, A. Remizov, W. Riedler, K. Schwingenschuh, N. Shutte, J. Slavin, K. Szego, M. Talrallyay and T.-L. Zhang: 1994, On the Correlation of the Magnetic Field in the Martian Magnetotail to the Solar Wind Parameters, *J. Geophys. Res.* **99**, 17199.

Russell, C. T., Hoppe, M. M. and Livesay, W. A.: 1982, 'Overshoots in Planetary Bow Shocks', *Nature* **296**, 45.

Russell, C. T.: 1985, in B. Tsurutani and R. Stone (eds) *Planetary bow shocks, Collisionless Shocks in the Heliosphere: Reviews of Current Research, Geophysical Monograph* **35**, Washington DC, p. 109.

Russell, C. T. *et al.*: 1988, 'Solar and Interplanetary Control of the Location of the Venus Bow Shock', *J. Geophys. Res.* **93**, 5461.

Sagdeev, R. Z. and Zakharov, A. V.: 1989, 'Brief History of the Phobos Mission', *Nature* **341**, 585.

Sauer K., Baumgärtel, K., Axnäs, I. and Brenning, N.: 1990, 'A Fluid Simulation of the AMPTE Solar Wind Lithium Release', *Adv. Space Res.* **10**, 95.

Sauer, K., Roatsch, T., Motschmann, U., Schwingenschuh, K., Lundin, R., Rosenbauer, H. and Livi, S.: 1992, 'Observations of Plasma Boundaries and Phenomena around Mars with Phobos 2', *J. Geophys. Res.* **97**, 6227.

Sauer, K., Roatsch, T., Baumgartel, K. and McKenzie, J. F.: 1992, 'Critical Density Layer as Obstacle at Solar Wind-Exosphere Ion Interaction', *Geophys. Res. Lett.* **19**, 645.

Sauer, K., Bogdanov, A. and Baumgartel, K.: 1994, 'Evidence of an Ion Composition Boundary (Protonopause) in Bi-Ion Fluid Simulations of Solar Wind Mass Loading', *Geophys. Res. Lett.* **21**, 2255.

Sauer, K., Bogdanov, A. and Baumgartel, K.: 1995, 'The Protonopause–an Ion Composition Boundary in the Magnetosheath of Comets, Venus and Mars', *Adv. Space Res.* **16**(4), 153.

Sauer, K., Dubinin, E., Baumgartel, K. and Bogdanov, A.: 1996, 'Bow-Shock 'Splitting' in Bi-Ion Flows', *Geophys. Res. Lett.* **23**, 3643.

Sauer, K., Dubinin, E. and Baumgartel, K.: 1998, 'Nonlinear MHD Waves and Discontinuities in the Martian Magnetosheath. Observations and 2D bi-ion MHD Simulations', *Earth Planets Space* **50**, 793.

Sauer, K., McKenzie, E. J. F. and Dubinin, E.: 2000, in M. Verheest, M. Goosens, M. A. Hellberg and R. Bharuthram (eds), *Waves and nonlinear structures in bi-ion plasmas, in: Waves in Dusty,*

Solar and Space Plasmas Vol. 537 *AIP Conference Proceedings*, p. 327, American Institute of Ophysics, Melville, N.Y..
Sauer, K. and Dubinin, E.: 2000, 'The Nature of the Martian 'Obstacle Boundary', *Adv. Space Res.* **26**(10), 1633.
Savich, N. A. *et al.*: 1979, in N. F. Ness (ed) *The nighttime ionosphere of Mars from Mars 4 and 5 radio occultation dual-frequency measurements, Solar Wind Interaction with the Planets Mercury, Venus, and Mars*, NASA Special Publication 397, Washington, DC.
Schunk, R. W. and Nagy, A. F.: 2000, *Ionospheres*, Cambridge University Press.
Schwingenschuh, K., Riedler, W., Yeroshenko, Y. *et al.*: 1987, 'Magnetic-Field Draping in the Comet Halley Coma - Comparison of Vega Observations with Computer-Simulations', *Geophys. Res. Lett.* **14**, 640.
Schwingenschuh, K., Riedler, W., Lichtenegger, H., Yeroshenko, Ye., Sauer, K., Luhmann, J. G., Ong, M. and Russell, C. T.: 1990, 'Martian Bow Shock: Phobos Observations', *Geophys. Res. Lett.* **17**, 889.
Schwingenschuh, K. *et al.*: 1992a, 'The Martian Magnetic Field Environment: Induced or Dominated by an Intrincic Magnetic Field', *Adv. Space Res.* **12**(9), 213.
Schwingenschuh, K. a. W. R.: 1992b, in H. K. Biernat, G. A. Bachmaier, S. J. Bauer and R. P. Rijnbeek (eds), Solar wind interaction with nonmagnetic and weakly magnetized bodies in the solar system. *Proceedings of the Workshop 'The Solar Wind-Magnetosphere System'*, 247.
Shabanskiy, V. P.: 1972, *Phenomena in the Near Terrestrial Space*, Nauka, pp. 167–190, Moscow (in Russian).
Shinagawa, H., Cravens, T. E. and Nagy, A. F.: 1987, 'A One-Dimensional Time-Dependent Model of the Magnetized Ionosphere of Venus', *J. Geophys. Res.* **92**, 7317.
Shinagawa, H. and Cravens, T. E.: 1988, A One-Dimensional Multi-Species Magnetohydrodynamic Model of the Dayside Ionosphere of Venus', *J. Geophys. Res.* **93**, 11263.
Shinagawa, H. and Cravens, T. E.: 1989, 'A One-Dimensional Multispecies Magneto Hydodynamical Model of the Dayside Ionosphere of Mars', *J. Geophys. Res.* **94**, 6506.
Shinagawa, H. and Cravens, T. E.: 1992, 'The Ionospheric Effects of a Weak Intrinsic Magnetic Field at Mars', *J. Geophys. Res.* **97**, 1027.
Shinagawa, H.: 1996, 'A Two-Dimensionsl Model of the Venus Ionosphere, 1. Unmagnetized Ionosphere', *J. Geophys. Res.* **101**, 26911.
Shutte, N. M. *et al.*: 1989, 'Observations of Electron and Ion Fluxes in the Vicinity of Mars with the HARP Spectrometer', *Nature* **341**, 614.
Slavin, J. A., Elphic, R. C. and Russell, C. T.: 1979, 'A Comparison of Pioneer Venus and Venera Bow Shock Observations: Evidence for a Solar Cycle Variation', *Geophys. Res. Lett.* **6**, 905.
Slavin, J. A. and Holzer, R. E.: 1982, 'The Solar Wind Interaction with Mars Revisited', *J. Geophys. Res.* **87**, 10285.
Slavin, J. A., Holzer, R. E., Spreiter, J. R., Stahara, S. S. and Chaussee, D. S.: 1983, Solar Wind Flow about the Terrestrial Planets, 2. Comparisons with Gasdynamic Theory and Implications for Solar Planetary Interactions', *J. Geophys. Res.* **88**, 19.
Slavin, J. A. and Holzer, R. E.: 1983, 'Solar Wind Flow about the Terrestrial Planets 2, Comparison with Gasdynamic Theory and Implications for Solar-Planetary Interactions', *J. Geophys. Res.* **88**, 19.
Slavin, J. A., Smith, E. J. and Intriligator, D. S.: 1984, 'A Comparative Study of Distant Magnetotail Structure at Venus and Earth', *Geophys. Res. Lett.* **11**, 1074.
Slavin, J. A., Smith, E. J., Tsurutani, B. T. *et al.*: 1986, 'Giacobini-Zinner Magnetotail: ICE Magnetic Field Observations', *Geophys. Res. Lett.* **13**, 283.
Slavin, J. A., Schwingenschuh, K., Riedler, W. and Yeroshenko, Ye.: 1991, 'The Solar Wind Interaction with Mars: Mariner 4, Mars 2, Mars 3, Mars 5 and Phobos 2 Observations of Bow Shock Position and Shape', *J. Geophys. Res.* **96**, 11235.

Speiser, T. W.: 1965, 'Particle Trajectories in Model Current Sheets, 1.Analytical Solutions', *J. Geophys. Res.* **70**, 4219.

Spenner, K., Knudsen, W. C., Miller, K. L., Novak, V., Russell, K. L. and Elphic, R. C.: 1980, 'Observation of the Venus Mantle, the Boundary Region Between Solar Wind and Ionosphere', *J. Geophys. Res.* **85**, 7655.

Spreiter, J. R. and Briggs, B. R.: 1962, 'Theoretical Determination of the Form of the Boundary of the Solar Corpuscular Stream Produced by the Interaction with the Magnetic Dipole Field of the Earth', *J. Geophys. Res.* **67**, 37.

Spreiter, J. R. and Stahara, S. S.: 1980: 'A New Predictive Model for Determining Solar Wind – Terrestrial Planet Interactions', *J. Geophys. Res.* **85**(6), 769.

Stewart, A. I.: 1972, 'Mariner 6 and 7 Ultraviolet Ultraviolet Spectrometer Experiment: Implications of CO_2^+, CO, and O Airglow', *J. Geophys. Res.* **77**, 54.

Stewart, A. I. and Hanson, W. B.: 1982, in A. Kliore (ed) Mars Upper Atmosphere: Mean and Variations, in the *Mars Reference Atmosphere, Adv. Space Res.* **2**, 87.

Szegö, K., Glassmeier, K. H., Brinca, A., Bingham, R., Cravens, T., Fischer, C., Fisk, L., Gombosi, T., Harendel, G., Lee, M., Mazelle, C., Moebius, E., Motschmann, U., Isenberg, P., Sauer, K., Shapiro, V., Schwadron, N., Tsurutani, B. and Zank, G.: 2000, 'Physics of Mass Loaded Plasma', *Space Sci. Rev.* **94**, 429.

Szego, K.: 2001, 'Present Understanding of the Dayside Mantle of Venus and Mars', *Adv. Space Res.* **28**, 841.

Trotignon, J. G., Grard, R. and Slavin, S.: 1991, 'Plasma Wave System Measurements of the Martian Bow Shock from the Phobos 2 Spacecraft', *J. Geophys. Res.* **96**, 11253.

Trotignon, J. G., Grard, R. and Skalsky, A.: 1993, 'Position and Shape of the Martian Bow Shock: the Phobos 2 Plasma Wave System Observations', *Planetary Space Sci.* **41**, 189.

Trotignon, J. G., Dubinin, E., Grard, R., Barabash, S. and Lundin, R.: 1996, 'Martian Planetopause as Seen by the Plasma Wave System Onboard Phobos 2, *J. Geophys. Res.* **101**, 24965.

Vaisberg, O. L., Bogdanov, A. V., Smirnov, V. N. and Romanov, S. A.: 1975, 'Initial Results of Ion Flux Measurements by RIEP-2801M Instrument on Mars-4 and Mars-5', *Cosmic Res.* **13**, 112.

Vaisberg, O. L.: 1976, in D. J. Williams (ed) Mars - plasma environment, *Physics of Solar Planetary Environments*, Vol. 2. AGU, Boulder, Colorado, 854.

Vaisberg, O. L.: 1992, in J. G. Luhmann, M. Tatrallyay and R. O. Pepin (eds), The solar wind interaction with Mars: A review of results from early soviet missions to Mars, in *Venus and Mars: Atmospheres, Ionospheres, and Solar wind Interactions*, *Geophys. Monogr.* **66**, NW, Washington, p. 311.

Vasiliev, M. B., Vyshlov, A. S., Kolosov, M. A., Savich, A. I. *et al.*: 1975, 'Preliminary Results of the Two Frequency Radio Sounding of the Martian Ionosphere by Using the Mars Interplanetary Stations in 1974', *Kosmicheskie issledovanija* **13**, 48 (in Russian).

Verigin, M. I., Shutte, N. M., Galeev, A. A., Gringauz, K. I., Kotova, G. A., Remizov, A. P., Rosenbauer, H., Hemmerich, P., Livi, S., Richter, A. K., Apathy, I., Szego, K., Riedler, W., Schwingenschuh, K., Stellar, M. and Yeroshenko, Ye. G.: 1991a, 'Ions of Planetary Origin in the Martian Magnetosphere (Phobos 2/Taus Experiment)', *Planetary Space Sci.* **39**, 131.

Verigin, M. I., Gringauz, K. I., Shutte, N. M., Haider, S. A., Szgo, K., Kiraly, P., Nagy, A. F. and Gombosi, T. I.,: 19916, *J. Geophys. Res.*, **96**, 19307.

Verigin, M. I., Gringauz, K. I., Kotova, G. A., Remizov, A. P., Shutte, N. M., Richter, A., Riedler, W., Schwingenschuh, K., Szego, K., Apathy, I. and Tatrallyay, M.: 1993, 'The Dependence of the Martian Magnetopause and Bow Shock on Solar Wind Ram Pressure According to Phobos 2 TAUS Ion Spectrometer Measurements', *J. Gephys. Res.* **98**, 1303.

Verigin, M., Kotova, G., Shutte, N., Remizov, A., Szego, K., Tatrallyay, M., Apathy, I., Rosenbauer, H., Livi, S., Richter, A. K. Schwingenschuh, K., Zhang, T.-L., Slavin, J. and Lemaire, J.: 1997, 'Quantitative Model of the Martian Magnetopause Shape and its Variation with the Solar Wind Ram Pressure Based on Phobos 2 Observations', *J. Geophys. Res.* **102**, 2147.

Verigin, M. I., Kotova, G. A., Remizov, A. P., Styazhkin, V. A., Shutte, N. M., Zhang, T.-L., Riedler, W., Rosenbauer, H., Szego, K., Tatrallyay, M. and Schwingenschuh, K.: 1999, 'Shape and Location of Planetary Bow Shocks', *Cosmic Res.* **37**, 34.

Verigin, M. I., Kotova, G. A., Remizov, A. P., Szegö, K., Tátrallyay, M., Slavin, J., Rosenbauer, H., Livi, S., Riedle, W., Schwingenschuh, K. and Zhang, T.-L.: 2001, 'Evidence of the Influence of Equatorial Martian Crustal Magnetization on the Position of the Planetary Magnetotail Boundary by Phobos 2 Data', *Adv. Space Res.* **28**(6), 885.

Vignes, D. *et al.*: 2000, 'The Solar Wind interaction with Mars: Locations and Shapes of the Bow Shock and the Magnetic Pile-up Boundary from the Observations of the MAG/ER Experiment Onboard Mars Global Surveyor', *Geophys. Res. Lett.* **27**, 49.

Vignes, D., Acuña, M. H., Connerney, J. E. P., Crider, D. H., Mazelle, C. and Rème, H.: 2002, 'Factors Controlling the Location of the Bow Shock at Mars', *Geophys. Res. Lett.* **29**, 9, 42, 1328 doi: 10.029/2001 GL014513.

Wallis, M. K.: 1978, 'Exospheric Density and Escape Fluxes of Atomic Isotopes on Venus and Mars', *Planetary Space Sci.* **26**, 949.

Whitten, R. C.: 1970, 'The Daytime Upper Atmosphere of Venus', *J. Geophys. Res.* **75**, 3707.

Yeroshenko, Ye., Riedler, W., Schwingenschuh, K., Luhmann, J. G., Ong, M. and Russell, C. T.: 1990, 'The Magnetotail of Mars: Phobos 2 Observations', *Geophys. Res. Lett.* **17**(6), 885–888.

Zakharov, A. V.: 1992, in J. G. Luhmann, M. Tatrallyay and R. Pepin (eds), The plasma environment of Mars: Phobos mission results, *Venus and Mars: Atmospheres, Ionospheres, and Solar Wind Interactions*, p. 327, *Geophysical Monograph* **66**, Washington, D.C..

Zhang, T. L., Luhmann, J. G. and Russell, C. T.: 1990a, 'The Solar Cycle Dependence of the Location and Shape of the Venus Bow Shock', *J. Geophys. Res.* **95**, 14961.

Zhang, M. H. G., Luhmann, J. G. and Kliore, A. J.: 1990b, 'An Observational Study of the Nightside Ionospheres of Venus and Mars with Radio Occultation Methods', *J. Geophys. Res.* **95**, 17095.

Zhang, M. H. G., Luhmann, J. G. and Kliore, A. J.: 1990c, 'A Post-Pioneer Venus Reassessment of the Martian Dayside Ionosphere as Observed by Radio Occultation Methods', *J. Geophys. Res.* **95**, 14829.

Zhang, T. L., Luhmann, J. G. and Russell, C. T.: 1991, 'The Magnetic Barrier at Venus', *J. Geophys. Res.* **96**, 11145.

Zhang, M. H. G., Luhmann, J. G., Nagy, A. F., Spreiter, J. R. and Stahara, S. S.: 1993a, 'Oxygen Ionization Rates at Mars and Venus: Relative Contributions of Impact Ionization and Charge Exchange', *J. Geophys. Res.* **98**, 3311.

Zhang, M. H. G., Luhmann, J. G., Bougher, S. W. and Nagy, A. F.: 1993b, 'The Ancient Oxygen Exosphere of Mars: Implication for Atmospheric Evolution', *J. Geophys. Res.* **98**, 10915.

Zhang, T. L., Schwingenschuh, K., Russell, C. T., Luhmann, J. G., Rosenbauer, H., Verigin, M. I. and Kotova, K.: 1994, 'The Flaring of the Martian Magnetotail Observed by the Phobos 2 Spacecraft', *Geophys. Res. Lett.* **21**, 1121.

BOW SHOCK AND UPSTREAM PHENOMENA AT MARS

C. MAZELLE[1], D. WINTERHALTER[2], K. SAUER[3], J.G. TROTIGNON[4],
M.H. ACUÑA[5], K. BAUMGÄRTEL[6], C. BERTUCCI[1], D.A. BRAIN[7], S.H. BRECHT[8],
M. DELVA[9], E. DUBININ[3], M. ØIEROSET[10] and J. SLAVIN[5]

[1] *C.E.S.R., CNRS/UPS, 9-11 Avenue du Colonel Roche, 31400 Toulouse, France*
[2] *Jet Propulsion Laboratory, California Institute of Technology, Pasadena, CA, U.S.A.*
[3] *Max-Planck-Institut für Aeronomie, Katlenburg-Lindau, Germany*
[4] *L.P.C.E., CNRS, Orléans, France*
[5] *NASA Goddard Space Flight Center, Greenbelt, MD 20771, U.S.A.*
[6] *Astrophysikalisches Institut Potsdam, Potsdam, Germany*
[7] *Department of Astrophysical and Planetary Sciences, University of Colorado, Boulder, CO, U.S.A.*
[8] *Berkeley Research Associates, Berkeley, CA, U.S.A.*
[9] *Institut für Weltraumforschung, Graz, Austria*
[10] *Space Science Laboratory, University of California, Berkeley, CA, U.S.A.*

(Accepted in final form 14 March 2003)

Abstract. Mars Global Surveyor is the sixth spacecraft to return measurements of the Martian bow shock. The earlier missions were Mariner 4 (1964), Mars 2 and 3 (1972), Mars 5 (1975) and Phobos 2 (1989) (see reviews by Gringauz, 1981; Slavin and Holzer, 1982; Russell, 1985; Vaisberg, 1992a,b; Zakharov, 1992). Previous investigations of planetary bow shocks have established that their position, shape and jump conditions are functions of the upstream flow parameters and the nature of the solar wind – planet interaction (Spreiter and Stahara, 1980; Slavin *et al.*, 1983; Russell, 1985). At Mars, however, the exact nature of the solar wind interaction was elusive due to the lack of low altitude plasma and magnetic field measurements (e.g., Axford, 1991). In fact our knowledge of the nature of the interaction of Mars with the solar wind was incomplete until the arrival of MGS and the acquisition of close-in magnetic field data (Acuña *et al.*, 1998). As detailed by a series of review papers in this monograph, the Mars Global Surveyor (MGS) mission has now shown that the Mars environment is very complex with strong, highly structured crustal magnetic remnants in the southern hemisphere, while the northern hemisphere experiences the direct impingement of solar wind plasma. This review paper first presents a survey of the observations on the Martian bow shock and the upstream phenomena in the light of results from all the missions to date. It also discusses the kinetic properties of the Martian bow shock compared to the predictions of simulations studies. Then it examines the current status of understanding of these phenomena, including the possible sources of upstream low-frequency waves and the interpretations of localized disturbances in the upstream solar wind around Mars. Finally, it briefly discusses the open issues and questions that require further study.

1. Introduction

The study of planetary bow shocks and their upstream region provides important insight both into the behavior of astrophysical collisionless shocks and into the nature of the planetary obstacle to the solar wind flow. Since the flow velocity of

Space Science Reviews **111:** 115–181, 2004.

the solar wind at each of the planets exceeds the velocity of compressional waves in the solar wind plasma, i.e., exceeds the fast magnetosonic speed, a shock wave must form in front of each planet if it does not absorb the incoming flow. The bow shock serves to slow down the flowing solar wind from supersonic to subsonic and to heat and deflect it around the planetary obstacle. The Mach number (magnetosonic, as well as sonic and Alfvénic) indicates the amount of deceleration and heating that occurs at the shock (the strength of the shock). The size of the solar wind obstacle largely determines the size of the shock relative to the planet. However, the physical size of the shock is also relevant since it dictates length scales for physical processes operating near the shock as well as the amount of space the upstream solar wind has available to be slowed and deflected around the obstacle.

Previous investigations of planetary bow shocks have established that their position, shape and jump conditions are functions of the upstream flow parameters and the nature of the solar wind – planet interaction (e.g., Spreiter and Stahara, 1980; Slavin *et al.*, 1983; Russell, 1985). The first observations of the Martian bow shock were taken by Mariner 4 (Smith *et al.*, 1965), from which the basic parameters of the shock were established. Other missions to Mars, such as Mars 2 and 3 (1972), Mars 5 (1975) and Phobos 2 (1989), continued the investigations (see reviews by Gringauz, 1981; Slavin and Holzer, 1982; Russell, 1985; Vaisberg, 1992a, b; Zakharov, 1992). The Phobos-2 spacecraft in particular provided a wealth of information about the plasma environment at relatively high altitude (> 850 km) above Mars.

However, the exact nature of the solar wind interaction was elusive due to the lack of low altitude plasma and magnetic field measurements (e.g., Axford, 1991). It was not clear whether Mars, as an obstacle to the solar wind, was Earth-like, i.e., if Mars possesses an appreciable internal magnetic field, or if it was Venus-like or comet-like. The measured and/or inferred shock parameters such as standoff distance, flare angles, and jump conditions, provided much room for speculation about the nature of the Martian obstacle.

All this has changed with the arrival of Mars Global Surveyor (MGS) at Mars, and its acquisition of close-in magnetic field data (Acuña *et al.*, 1998). MGS has now conclusively established that Mars is lacking, at present, an active dynamo. Thus the solar wind interaction is primarily Venus/comet-like. MGS has also shown the existence of magnetic remnants in localized, near-surface regions around the planet. These remnants, while complicating some detailed considerations at lower altitude, do not appear to influence the global structure of Mars's bow shock and upstream regions.

In comparison to Venus, Mars is located at a greater heliocentric distance and is also much smaller. Thus the solar wind density and interplanetary magnetic field (IMF) strength are smaller at Mars than at Venus, and therefore the Mach numbers are different. Further, the small planet size coupled with the low IMF strength at Mars makes the gyroradius of solar wind protons comparable to the size of the shock. At Mars, therefore, kinetic effects are potentially important. Finally, the

Martian exosphere extends beyond the bow shock due to the much weaker Martian gravity, with significantly higher fluxes than at Venus. This has major implications for the actual solar wind/Mars interaction process, and therefore for the shock characteristics and upstream dynamics.

Below we review the current status of understanding for the phenomena that occur at the Martian bow shock and in the interplanetary medium upstream from this shock. We also discuss their implications for the nature of the global Mars – solar wind interaction processes. Section 2 is devoted to the observations on the Martian bow shock in the light of results from all the missions to date and to the simulation studies, including the discussion on the kinetic properties. In Section 3 we review the observations on upstream phenomena at Mars. Section 4 is devoted to the possible sources of low-frequency waves upstream from the bow shock. Section 5 discusses the interpretations of localized disturbances in the upstream solar wind around Mars. Finally, in Section 6, we briefly comment the open issues and questions that require further study.

2. The Bow Shock

2.1. Bow Shock Structure

The plasma and magnetic field measurements performed by the Phobos-2 spacecraft have confirmed previous observations made on Mars-2, 3, 5 and Mariner-4 that the bow shock exists near Mars (Riedler *et al.*, 1989; Schwingenschuh *et al.*, 1990). The main shock substructures, a foot of reflected protons, overshoot, electron and ion foreshock were also clearly identified (Barabash and Lundin, 1993; Delva and Dubinin, 1998; Skalsky *et al.*, 1992, 1993; Tatrallyay *et al.*, 1997; Trotignon *et al.*, 1992).

Phobos 2 provided sufficient magnetic field and plasma data for performing the first statistical analysis of the structure of the Martian bow shock by Tatrallyay *et al.* (1997). They used 82 crossings from the circular orbits of Phobos 2 in the terminator region ($80° < SZA < 110°$) to investigate the characteristics of the transition layer at the Martian bow shock. They were interested in the investigation of the shock overshoot magnitude and thickness. They computed the overshoot magnitude as $(B_M - B_2)/B_2$, where B_M is the maximum field value in the overshoot and B_2 is the theoretical downstream value computed from Spreiter and Stahara's (1980) gasdynamic model, using measured upstream field values. They found that the overshoot amplitude scales with the fast magnetosonic Mach number M_f as typical for a fast-mode type shock. The thickness of the overshoot was taken as the distance along the shock normal between the beginning of the ion foot and the minimum of the first undershoot. The thickness of the Martian shock overshoots was in most of the cases between 0.5 and 2.5 r_L (r_L denotes the proton gyroradius calculated from the upstream solar wind speed and magnetic

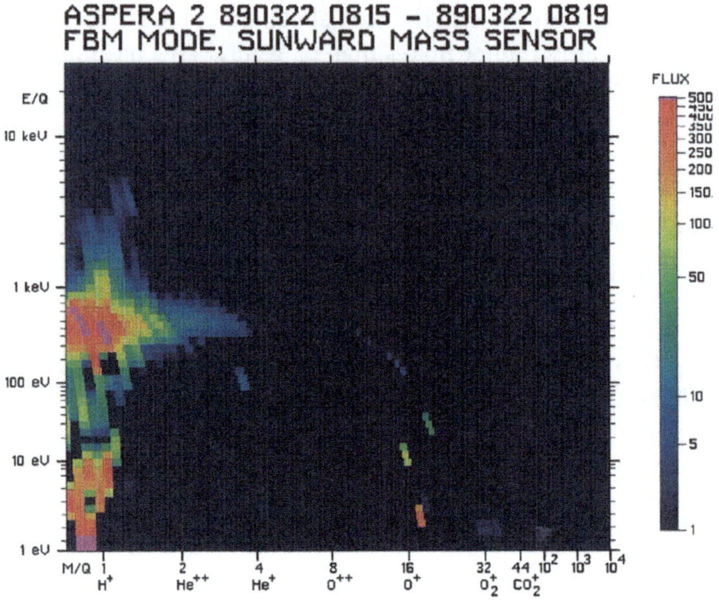

Figure 1. $E/q - m/q$ matrix collected by the ASPERA ion mass-spectrometer onboard Phobos-2 at the bow shock crossing near the terminator. Cold ions H^+, O^+ (and traces of molecular ions with $m/q = 32$) are of planetary origin.

field strength $r_L = V_1/\omega_{ci1}$). In ion inertial length, the typical thickness was in the range $2 < c/\omega_{pi} < 8$.

On the other hand, some new features of the Martian bow shock were found. Dubinin *et al.* (1993) observed a sudden appearance of the low-energetic protons (and sometimes oxygen ions) at the bow shock. The energy of these ions was ≤ 10 eV q^{-1}. Figure 1 shows the $E/q - M/q$ matrix measured by the ASPERA instrument during the 4 min interval corresponding to the crossing of the terminator bow shock. Three main proton populations are observed, shocked solar wind protons, more energetic ions that probably come from the reflected and subsequently accelerated protons and the protons with low energy. Spikes of the low-energy oxygen ions are also seen. Dubinin *et al.* (1993) attributed these low-energy particles to the ions originated from the extended hydrogen exosphere and hot oxygen corona. The authors confirmed the unique feature of the Martian bow shock associated with a small size of the obstacle to the solar wind. Moses *et al.* (1988) noted that the subsolar distance between the bow shock and the obstacle is about of the proton gyroradius based on its kinetic energy and therefore there is not enough space for the ion thermalization. Indeed, Dubinin *et al.* (1993) observed that a thermalization of the solar wind and exospheric protons occupies not only the whole magnetosheath (whose width near the terminator is much larger than in the subsolar region) but extends to the so-called 'plasma mantle'.

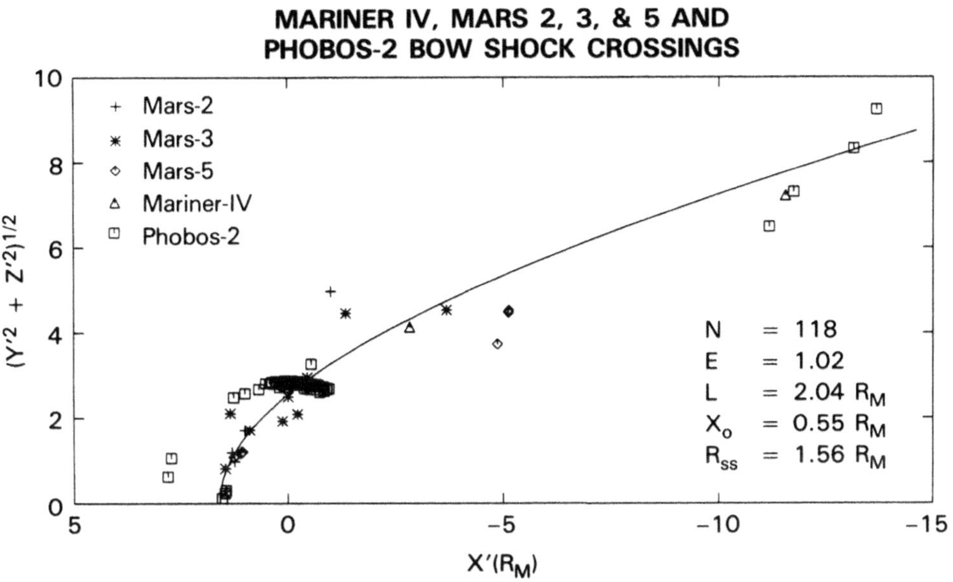

Figure 2. (Position and shape of the Mars bow shock based upon boundary encounters by Mariner IV, Mars 2, 3, and 5 and Phobos 2 along with a least-square, axisymmetric, second order fit in aberrated Mars Solar Orbital (MSO) coordinates (Slavin *et al.*, 1991).

Since the number density of the exospheric ions was, at least, comparable with the number density of the shocked solar wind protons, their influence on the ion dynamics at the bow shock and magnetosheath can be important. The authors raised also an interesting question how to explain the high fluxes of planetary ions at such large distances from the planet ($R \sim 2.8\ R_M$).

The problem of unusual properties of the Martian bow shock was also addressed by Brecht (1997) in hybrid simulations of the solar wind-Mars interaction. Brecht (1997) argued that proton dissipation caused by ion reflection does not exist around Mars and concluded the difference of the 'kinetic wave-like behavior of the Martian bow shock and standard hydrodynamic paradigm'. However, Skalsky *et al.* (1998) in their analysis of two typical bow shock crossings by Phobos-2 found that the observed changes in the bulk parameters are comparable with values expected for the MHD shocks and mechanisms responsible for the energy dissipation are similar to those operating at the Earth's bow shock. Further detailed measurements are necessary to resolve these discrepancies.

2.2. The Average Location

The particles and fields measurements collected by the various missions have clearly established that Mars possesses a fast mode bow shock whose structure closely resembles that of the other planets (e.g., Russell, 1985). The locations of the bow shock crossings returned by the pre-MGS missions (i.e., 2 from Mariner 4, 22 from

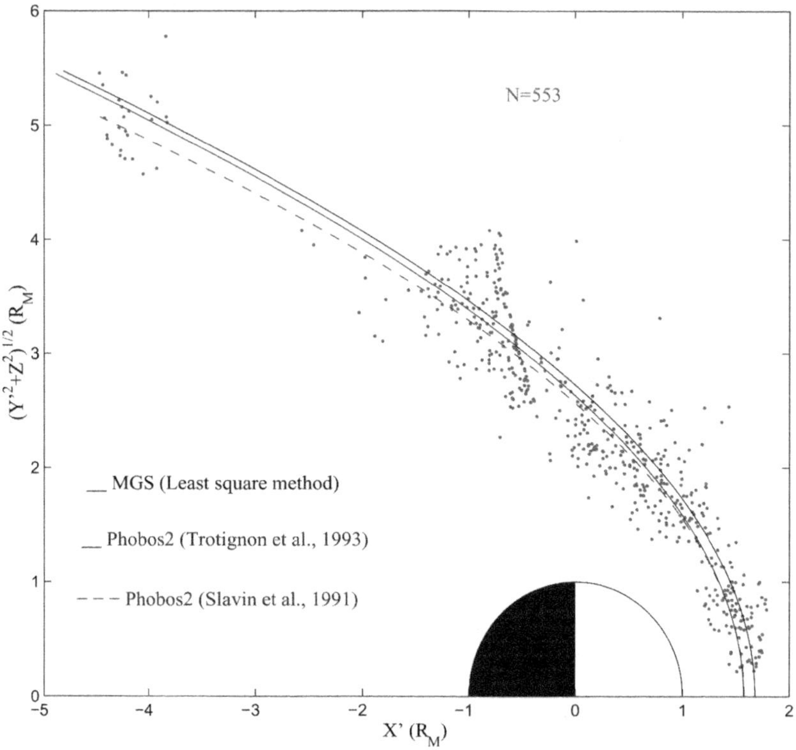

Figure 3. Position and shape of the Mars bow shock based upon boundary encounter by Mars Global Surveyor along with an axisymmetric second order fit in aberrated MSO coordinates (Vignes *et al.*, 2000) compared to previous fits from Phobos 2 data (Slavin *et al.*, 1991; Trotignon *et al.*, 1993).

Mars 2, 3, and 5, and 94 from Phobos 2) are displayed in Figure 2 along with a solar wind aberrated, axisymmetric, second order least-square fit (Slavin *et al.*, 1981, 1991). As shown, they span the region from the subsolar point to nearly 15 Mars radii downstream. These shock encounters nearly all occurred at low to moderate latitudes. Slavin *et al.* (1991) derived a planetocentric distance to the nose of the shock of 1.56 R_M and a Mach cone half-angle of 11.4 deg. This Mach cone implies an average magnetosonic Mach number of 5.1 which is close to the expected mean value of $\sim$ 6 at Mars (Slavin and Holzer, 1982). Independent models of the bow shock by the different Phobos 2 instrument teams (Schwingenschuh *et al.*, 1990; Verigin *et al.*, 1991; Trotignon *et al.*, 1991b) have produced very similar shock surfaces with an average subsolar shock distance $\sim$ 1.5 R_M.

During the first year of the MGS mission, 553 shock crossings were identified from a total of 363 orbits and the mean shape and location of the shock determined (Vignes *et al.*, 2000) using data from the magnetometer and electron reflectometer (MAG/ER). The MAG/ER package consisted of two magnetometers, which measure local magnetic field, and an electron reflectometer, which measures electrons fluxes in the energy range between 10 eV–20 keV (Acuña *et al.*, 1992). In Figure 3 a solar wind aberrated, axisymmetric fit of all these bow shock crossings using a conic section is displayed and compared with the earlier models of Slavin *et al.* (1991) and Trotignon *et al.* (1993). Comparing Figure 2 and Figure 3, MGS had many crossings on the dayside, while Phobos-2 and the other missions provided a small but complementary number of crossings down the tail at greater distance than 10 R_M. Overall, the agreement between the three surfaces based upon the MGS data alone, Phobos 2 alone, and the aggregate of the pre-MGS missions is very good.

Phobos 2 also investigated the variation in Martian bow shock location upon upstream parameters. The location of the shock was found to depend very weakly on the solar wind ram pressure $\sim (\rho V^2)^{0.02}$ (Verigin *et al.*, 1993). This implies that the height of the subsolar Martian obstacle is highly stable. It was also found that the Martian bow shock moves from the planet as the sonic and Alfvénic Mach numbers decrease, in a manner usual for other planetary shocks (Verigin *et al.*, 1997, 1999).

2.3. Solar Cycle Variations

The position of the Mars bow shock is highly variable on a day-to-day basis, but the comparison between surfaces taken at different points in the solar cycle (see Figure 3) does not show anything like the $\sim$35% increase in the planetocentric distance to the shock in the terminator plane from solar minimum to maximum which was observed at Venus (Slavin *et al.*, 1979). MGS and Phobos 2 observations give nearly the same mean bow shock fits for sunspot numbers of $\sim$30–90 and $\sim$140–180, respectively, suggesting that the mean surface is independent of the phase of the solar cycle (Vignes *et al.*, 2000). In fact, the solar minimum shock surface from MGS is located at a slightly greater altitude than the solar maximum results from Phobos. However, this may be due to the high inclination of the MGS orbit as opposed to the very low inclination Phobos mission. It is known from studies of the bow shock at the Earth and other planets that the higher magnetosonic wave speed perpendicular, as opposed to along the local magnetic field, creates a small asymmetry in the cross section of the shock in the terminator plane with it being located at slightly greater altitudes over the poles as compared with the equator (e.g., Russell, 1985).

Slavin *et al.* (1979, 1983) proposed that the underlying cause of the solar cycle variation at Venus was the enhanced absorption of the incident solar wind via charge exchange at solar minimum as the ionopause recedes to lower altitudes and exposes more of the neutral atmosphere to the solar wind in the magnetosheath (Cravens *et al.*, 1997). Conversely, Alexander and Russell (1985) and Russell *et al.* (1988) argued that the variation in bow shock location at Venus was due to the shock moving outward at solar maximum, because of enhanced pre-shock mass loading caused by the enhanced solar EUV flux. Zhang *et al.* (1990) found evidence in support of both mechanisms – subsolar absorption at solar minimum and enhanced mass loading at maximum. Finally, solar wind deceleration upstream of the Martian bow shock was detected by Phobos 2 orbiter and has been at least partially attributed to the effect of mass-loading operating at this planet (Verigin *et al.*, 1991; Kotova *et al.*, 1997). However, Dubinin *et al.* (2000a) proposed another interpretation from a multi-instrument data analysis of the observations in the upstream region which is detailed in Section 3.1. They showed that transient upstream wave phenomena interfere mass-loading signatures and can dominate over them.

2.4. Effects of Pickup Ions

In order to study the variability of the BS position, Vignes *et al.* (2002) extrapolate each crossing to the terminator plane. Conic section fits were used to project them to the terminator plane using the same eccentricity and focus location than for the mean fit. This is helpful to examine the asymmetric dependence of the BS position relatively to the direction of locally upward electric field of convection. This dependence is related to the mass loading of the sheath by picked-up oxygen: the additional mass added to the flow in the North slows more the flow and creates a bigger effective obstacle there.

In the case of the solar wind interaction with Venus, mass loading of the solar wind results when atomics oxygen, the most important neutral species in the exosphere, are ionized. Atomics hydrogen, also present in Venus exosphere, are picked-up. Mass loading is thought to increase the distance of the bow shock from the planet and also to enhance the magnetic flux contained in the induced magnetotail.

In the case of Venus, the magnetic field direction was found to control the shock location. The pickup effect is at its maximum value when the solar wind flow and the IMF are perpendicular. This particular angle, the angle between the interplanetary magnetic field and the direction of the solar wind, is called the cone angle. The Venus BS was found to be sensitive to the cone angle (Alexander *et al.*, 1986). Thus, to show the effect of the pick-up ions, we also need to look on the hemisphere of locally upward electric field of convection ($\mathbf{E} = -\mathbf{v} \times \mathbf{B}$). The relative clock angle correspond to the angle between the projection of the radial vector at the shock crossing location in the terminator plane and the local direction of the magnetic field also projected in this plane. Then, bow shock crossing in the northern hemisphere, by respect to the direction of the interplanetary magnetic field, occurs for clock angle between 0° and 180°.

As the solar wind interactions with Mars and Venus are thought to be similar Vignes *et al.* (2002) look for evidence of bow shock asymmetry related to pick-up ions. The MGS observations obtained during periods when the solar panels are illuminated are estimated to be accurate to ~ 1 nT (Acuña *et al.*, 2001). As the value of the interplanetary magnetic field at Mars is typically a few nanotesla, this study of the mass loading effect on the Martian bow shock is made exclusively with cases where the IMF amplitude is strong enough to have an uncertainty on the direction of the IMF that is less than 20 deg, assuming this ~ 1 nT error. This data

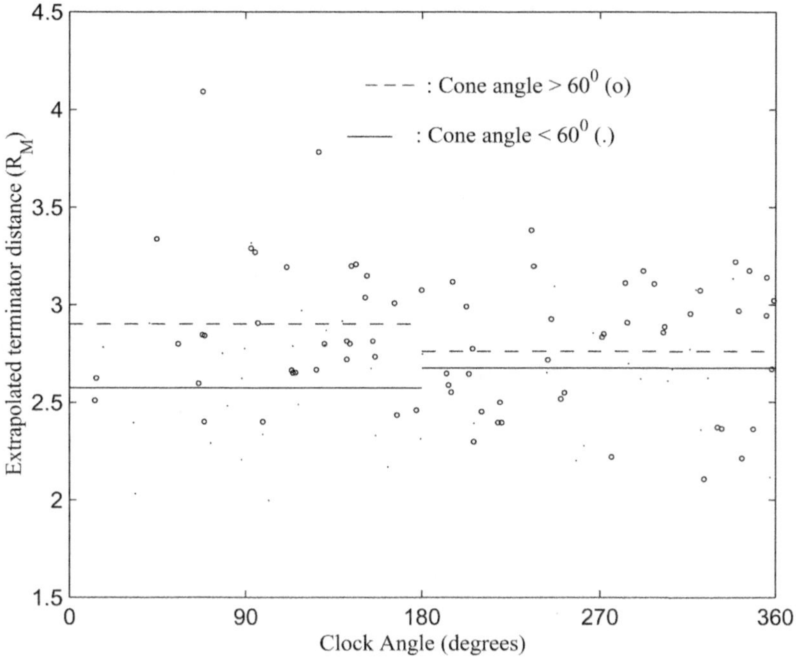

Figure 4. Location of Mars BS extrapolated to the terminator plane versus Clock Angle. Large and small cone angle are distinguished. Medians of the shock position are given depending on the hemisphere and depending on large or small cone angle (from Vignes *et al.*, 2002).

restriction enables this study to look at cases of favorable pick-up effect: strong IMF, large cone angle and north clock angle.

Figure 4 shows the extrapolated terminator distance of Mars bow shock crossings as function of the clock angle. Vignes *et al.* (2002) distinguish cases of large and small cone angle in order to show the cumulative effect of large cone angle in the hemisphere of locally upward electric field. In the northern hemisphere, corresponding to the hemisphere of locally upward electric field, the mean value of the terminator distance for large cone angle (2.9 R_M) is 13% greater than for small cone angle in the same hemisphere. While, in the southern hemisphere we found a small difference (3%) between cases of large and small cone angle. Thus, the shock appears significantly farther from the planet in the hemisphere of locally upward electric field when the angle between the solar wind direction and the interplanetary magnetic field is large. This North-South asymmetry is consistent with the idea of asymmetric mass loading of the sheath by picked-up oxygen (Vignes *et al.*, 2002). The additional mass added in the northern hemisphere slows more the flow and creates a bigger obstacle. Thus mass loading appears as playing some role on controlling the BS location. This role is expected to be more important during solar maximum since EUV fluxes increase and thus photo-ionization.

2.5. Influence of Crustal Sources

During the Phobos 2 circular orbit phase, it was found that the magnetic field measurements contain spectral peaks at 12 and 24 hours. This periodicity was attributed the perturbation of the magnetic tail by a planetary magnetic field as Mars rotates about its spin axis (Mohlmann *et al.*, 1991). MGS observations made from many periapsis passes over the northern hemisphere and at different longitudes only indicated the existence of weak fields (≤ 5 nT) at low altitudes (down to ~ 100 km). This leads to an upper limit on a 'planet centered' intrinsic magnetic moment of $\sim 2 \times 10^{17}$ A m^2 (or 2×10^{20} G cm^3) for the magnitude of a putative present-day Mars dipole equivalent to 0.5 nT equatorial field (Acuña *et al.*, 2001). However, MGS has found that while Mars has little or no low order intrinsic magnetic field, there are strong localized crustal magnetic fields mainly in the southern hemisphere (Acuña *et al.*, 1999). The most intense magnetic crustal sources detected by MGS-MAG/ER lie in the Terra Sirenum region (150° E to 240° E; 30° S to 85° S). Connerney *et al.* (1999) present analytical models of the sources in this region.

The total equivalent surface magnetic moment of this region, $\sim 1.3 \times 10^{17}$ A m^2 (or 1.3×10^{20} G cm^3), is sufficiently high to shield portions of Mars from the solar wind. Using the usual pressure balance expressions (e.g., see Slavin and Holzer, 1982) the Terra Sirenum magnetic fields appear sufficient to stand-off the typical solar wind ram pressure at ~ 1000 km. However, in reality, some of the Terra Sirenum magnetic fields will close at lower altitudes via fields emanating from other anomalies. Furthermore, the net effect of the normal stress exerted by solar wind and the tangential stress due to reconnection between the interplanetary magnetic field and the Terra Sirenum magnetic fields will be to limit the altitude reached by these crustal fields and to spread them over a broader range of aerographic latitude and longitude (Brain *et al.*, 2003). Theoretical model calculations (Ma *et al.*, 2002), discussed in Section 6, have provided some good insight into how the crustal fields impact the interaction processes and create mini-magnetocylinders (Mitchell *et al.*, 2001).

Figure 5 shows the extrapolated MGS terminator bow shock positions versus the longitude of the subsolar point at the time of the crossing in order to study the bow shock location dependence on crustal magnetic sources. If the variability of the bow shock position at all latitudes and longitudes were caused by high field anomalies embedded in the crust rotating with Mars, then a distinct maximum would be present over the Terra Sirenum region. However, the mean position of the shock averaged over all latitudes appears to be independent of the longitude of the subsolar point. This result is consistent with the Phobos analyses by Schwingenschuh *et al.* (1990) and Slavin *et al.* (1991) who also have found that the distance to the bow shock is independent of subsolar longitude in the Phobos 2 observations.

However, the high inclination of the MGS orbit relative to Phobos 2 also provides an opportunity to examine bow shock location as a function of latitude. This is significant, because the vast majority of the strong magnetic anomalies detected

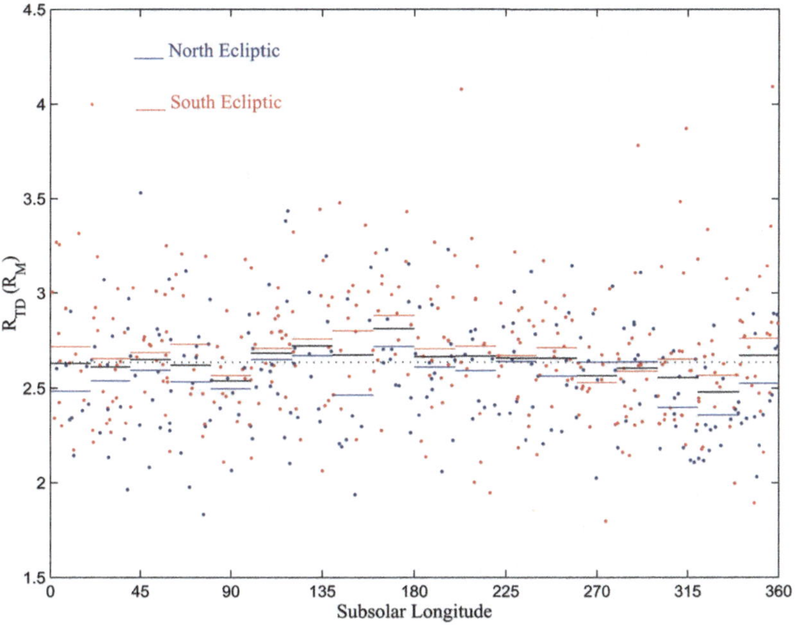

Figure 5. MGS bow shock crossings mapped into the terminator plane and plotted against the longitude of the subsolar point at the time of each event. Crossings over the northern and southern hemispheres are plotted in blue and red, respectively, and medians are indicated as horizontal line segments.

by MGS are located south of the Martian equator. Figure 5 plots the shock crossings over the northern and southern hemispheres in blue and red, respectively, and indicates the corresponding medians as horizontal line segments. The results do show a clear tendency for the Martian bow shock to be found at slightly greater heights over the southern hemisphere, i.e., the red median bars are higher than the blue medians in 16 of 18 longitude sectors. If verified by further analysis of the MGS measurements, the asymmetric distribution of bow shock crossings suggests that the effect of the crustal magnetic anomalies on the solar wind flow about Mars are felt most strongly above and just downstream of the southern hemisphere.

2.6. Bow Shock Simulations and Related Problems

Simulations should help to find out what are the reasons for the different signatures seen in the observations of the Martian bow shock. The fact that the bow shock at Mars is located close to the planet that there is not enough space for the solar wind protons to fulfill complete gyro motions was the motivation for kinetic studies using hybrid simulations (Moore *et al.*, 1991; Brecht, 1990, 1997; Brecht *et al.*, 1993). Another point of interest is the different variability of the bow shock location at Venus and Mars with the solar cycle. Models should give an answer why it happens. At Mars a specific situation arises from the existence of an extended

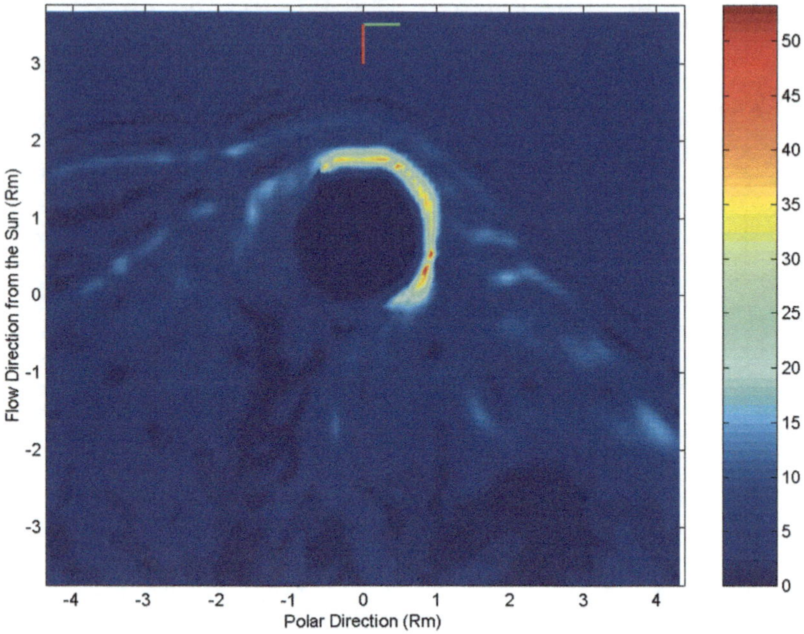

Figure 6. This figure shows the tendency for the magnetic field to convect over the Northern magnetic Pole defined by the direction of the convection electric field $-\mathbf{V} \times \mathbf{B}$. One can see the tendency for the field in the barrier region to coalesce. Further the asymmetry of the shock and barrier structure over the poles is seen (Brecht, 2002).

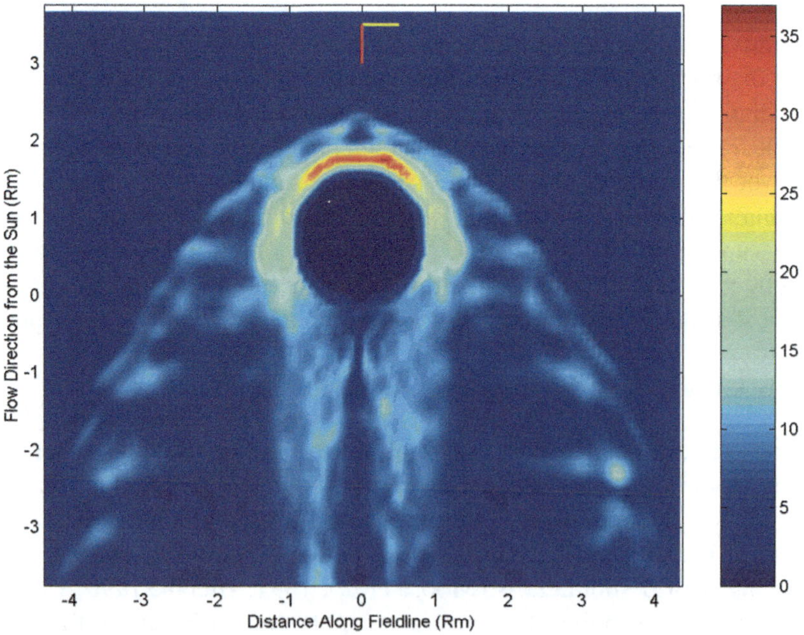

Figure 7. This figure shows the same simulation but the opposite plane. In this plane the magnetic field and the solar wind exist. This is nominally the ecliptic plane being plotted. The extreme symmetry is remarkable since this simulation is fully kinetic and one is seeing only a slice through the three dimensional simulation (Brecht, 2002).

hydrogen/oxygen exosphere which may already lead to the interaction of the solar wind with planetary ions in the vicinity of the bow shock. The question is how this process modifies the shock structure. Finally one has to solve the problem about what are the dominant processes in shock formation. After the MGS mission there are no doubts that the shock at Mars is caused by braking of the solar wind due to its interaction with the exospheric/ionospheric plasma which acts as an obstacle. But, the nature of the 'obstacle boundary' is still a subject of ongoing discussions which concern the interpretation of the so-called 'magnetic pile-up boundary' (MPB) seen in MGS data; see the companion review paper by Nagy *et al.* (this issue). In this context one can also hope to find a final answer what processes determine the shape of the bow shock, which means to clarify whether it is symmetric or asymmetric with respect to the direction of the interplanetary magnetic field and the convection electric field. From different models different answers have been obtained.

In first three-dimensional kinetic simulations (kinetic ions and mass-less fluid electrons) of the bow shock of Venus by Moore *et al.* (1991), and Mars by Brecht (1990, 1997) and Brecht *et al.* (1993) the planet was considered as a obstacle with an absorbing boundary. The aim of these studies was to find out the effect of finite Larmor radius on the structure of the bow shock. They showed that the large Larmor radius of the solar wind protons prevents the formation of a traditional collisionless shock. An asymmetry was found in the direction of the convection electric field as shown in Figure 6 (Brecht, 2002). One interesting note is that the asymmetry is a function of the Parker spiral angle, because the angle determines the relative strength of the solar wind convection electric field. In the ecliptic plane both the data and kinetic simulations show a very symmetric result, as shown in Figure 7 (Brecht, 2002). In simulations by Shimazu (1999) using a reflecting boundary similar results have been obtained. The observed asymmetry of the magnetic field configuration is interpreted as result of differences in the ion acceleration due to the convection electric field.

Another feature of the Martian bow shock as revealed by kinetic simulation is that because of the closeness between the bow shock and the planet, there is barely room for the number of gyroradii necessary for thermalization. Indeed, while simulations of Venus show that the plasma is thermalized when passing through the shock, the simulations of Mars show that the plasma is not completely thermalized (Brecht *et al.*, 1993). This is coupled with the strong wave activity generated by the high Alfvénic Mach number shock created at Mars. For the normal solar wind parameters, the Alfvén Mach number is above 10. A collisionless shock with a Alfvén Mach number of this size or greater is very electromagnetically active and will generate a substantial suite of electromagnetic waves with frequencies ranging from the whistler down to the electromagnetic modes with frequencies near or even below the ion gyrofrequency, T_{ci} (Thomas and Brecht, 1987)

Shimazu (2001) has modeled more realistic conditions with respect to the situation at Mars by treating the planet as an ionized gaseous body. In this model planetary ions are distributed uniformly in a sphere of a radius $R = D/8$ which

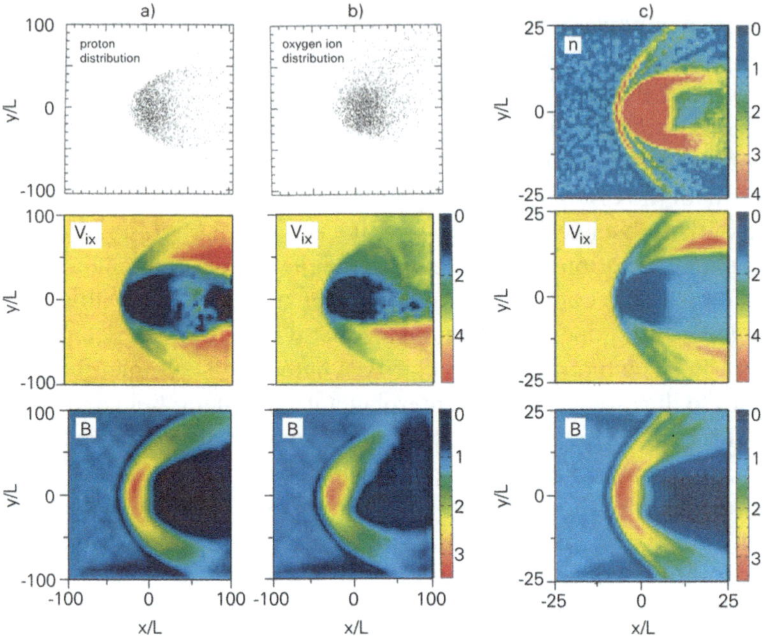

Figure 8. 3D hybrid code simulations of solar wind – Mars interaction, after Shimazu (2001). The ionosphere of Mars is modeled by a source of planetary ions, distributed uniformly in a sphere of radius $R = d/8$ (d: box size); (a) proton source, $d = 200L$ (L: proton skin length); (b) source of oxygen ions, the same parameters as in (a); (c) proton source, $d = 50L$. The Mach number is $M_A = 4$.

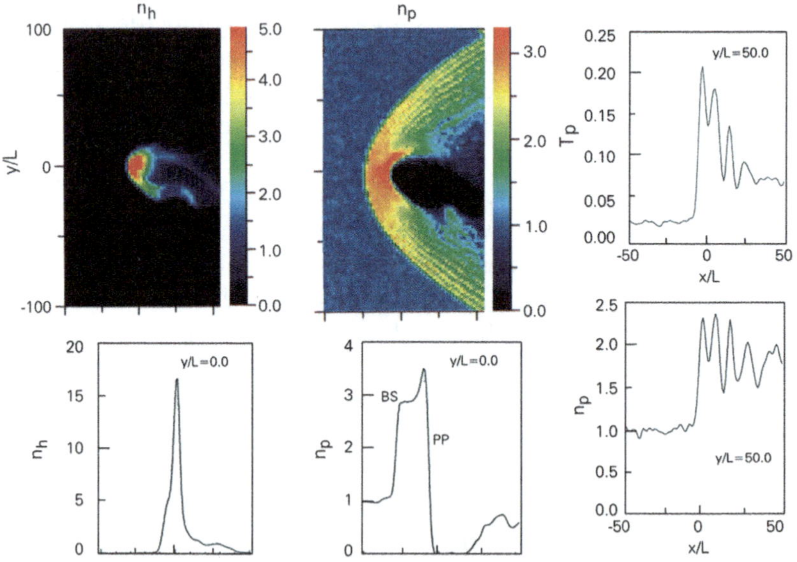

Figure 9. 2D hybrid code simulations of solar wind – (cometary) exosphere interaction, after Sauer and Dubinin (2003). The color plots on top show the heavy ion density n_h (left) and the proton density n_p (right). The most remarkable signature is the formation of a proton cavity. The cuts along x trough the center of the box (middle two panels) clearly show that the proton density jumps down to about zero where the heavy ion density strongly increases (change of ion composition). The lower panels show that the bow shock is structured; T_p is the proton temperature.

is placed at the center of the simulation box with the size D. The supply rate q of ionospheric ions was assumed to be uniform and constant, and with a rate $q = 0.256\,\omega_p$ (ω_p: proton plasma frequency) the supply and escape to the tail were almost in balance. This rate corresponds to an ion production rate of $\approx 3 \times 10^{24}$ s^{-1} which is about 5 times smaller than the estimated loss rate of Mars from Phobos-2 measurements.

Results of the simulations are summarized in Figure 8. The solar wind Mach number is $M_A = 4.0$. Three cases are shown. In panel a) and b) the radius of the sphere is $R = 6.4\lambda$ where $\lambda = M_A L$ (L =proton skin length) using protons (a) and oxygen ions (b) as planetary ions. Panel c) shows results for a small planet ($R = 1.6\lambda$) with proton production. The magnetic field is directed out of the plane (z axis), that means, the convection electric field E shows in positive y-direction.

For oxygen ion production (Figure 8(b)) a clear asymmetry in the direction of E can be seen. This is caused by the oxygen ions escaping from the side of the planet to which the electric field is pointing. The calculated asymmetry in the magnetic barrier intensity is consistent with the observations. The asymmetry in the shock size is the same as calculated by Brecht and Ferrante (1991) and Shimazu (1999), however, do not match the asymmetry observed near either Venus (Alexander *et al.*, 1986; Russell *et al.*, 1988) or Mars (Zhang *et al.*, 1991). This might be an effect of mass loading far away from the planet which is not described by the present ionospheric model with the abrupt onset of planetary ion production. In the case of a reduced planet size a multiple-shock structure is observed in the simulations. This is clearly visible as ray structure in the upper panel of Figure 8(c) where the density distribution is shown. Although an explanation is given due to finite Larmor radius effects, one cannot exclude that a beam-plasma configuration by reflected protons might cause the periodic density structure. This conclusion is supported by the fact that similar phenomena also appear for larger ‘obstacles’, as shown later.

A further interesting aspect of the Shimazu (2001) paper concerns the discussion about the behavior of the solar wind if it interacts with the plasma of planetary origin, even if this is not the main focus of this paper. It is stated that the two kinds of ions (solar wind and ionosphere) do not mix very much and that a reflecting or absorbing boundary as taken in several earlier models (Brecht, 1990, 1997; Moore *et al.*, 1991; Brecht *et al.*, 1993; Shimazu, 1999) appears to be an extreme and unrealistic assumption.

In a recent work by Sauer and Dubinin (2003) 2D hybrid code simulations which were originally done to study how the solar wind interacts with weakly outgassing comets (Lipatov *et al.*, 1997) have been extended to cometary-type objects with larger production rates. The results shown in Figure 9 are relevant for the conditions at Mars where extended hydrogen and oxygen exospheres seem to have essential influence on the solar wind before it interacts with the denser ionosphere closer to the planet. Figure 9 shows results for a cometary-type source with a production rate of $Q = 10^{27}$ s^{-1} (photo-ionization rate 10^{-6} s^{-1}) for an incoming solar wind with $M_A = 5.0$. The simulation box was 200×200 in units of the proton skin length L. The center of the source is in $x = y = 0$. More details are given in Lipatov *et al.* (1997). The color plots on top of Figure 9 show the heavy ion distribution (left) and the solar wind density distribution (right). The shock is nearly symmetric, although the motion of the main heavy ion population opposite to the convection electric field direction causes an asymmetric picture. This motion results from the momentum conservation in y-direction: the momentum of the low-dense high-velocity ions escaping along the convection electric field is balanced by the momentum of the dense ions moving slowly in the opposite direction. Further, there are two remarkable effects to mention. One is the clear bow shock structuring with wave lengths of about 5 L (visible in the cuts shown in the lower panels of Figure 9) which are probably caused by steepened magneto-sonic waves as discussed in earlier papers (Hopcroft and Chapman, 2001; Lipatov *et al.*, 2002) and in a recent publication by Bagdonat and Motschmann (2002).

The other important result is related to the crucial problem of 'obstacle' formation at solar wind interaction with a source of cometary/planetary ions, as it happens at comets, Mars and Venus. The color plots of the density distributions in Figure 9 and the cuts below clearly show that a boundary is formed where the solar wind proton density sharply decreases and the heavy ion density jumps up to values of about the (upstream) peak proton density. So, a 'replacement' of protons by heavy ions takes place. An accompanying effect (not shown here) is an abrupt pile up of the magnetic field. The same characteristic features of the 'obstacle boundary' have been described before with the help of two-ion fluid simulations (Sauer *et al.*, 1994, 1995a; Sauer and Dubinin, 2000) demonstrating that the macroscopic momentum coupling and not single-particle effects are important in forming the boundary. The Phobos and MGS missions to Mars observed such new type of plasma boundary. It has been shown recently from the analysis of the magnetic field that this boundary is also present at Venus (Bertucci *et al.*, 2003a,b) and coincides with the proposed 'mantle boundary' both at Venus and Mars (Vaisberg and Zelenyi, 1984; Zelenyi and Vaisberg, 1985). It is now called 'magnetic pile-up boundary' and will be described in more details in Nagy *et al.* (this issue).

Another class of simulations, not directly related to the Mars bow shock, which generally help to understand the pick-up dynamics near bodies with an ionized environment have been done using both hybrid codes (Lipatov *et al.*, 1997; Hopcroft and Chapman, 2001; Bagdonat and Motschmann, 2002) and two-ion MHD models (Bogdanov *et al.*, 1996; Sauer and Dubinin, 2000; Szegö *et al.*, 2000). The aim of these studies was to see how the character of solar wind interaction with a source of cometary/planetary ions changes if the production rate varies from low to high values and what are the signatures of the related plasma structures. In all models, cited above, a neutral gas source of $n_0 = Q/(4\pi r^2 v_0)$ was taken, where Q is the production rate in molecules per second, r the distant from the center of the source and v_0 the thermal velocity of the ejected neutrals. The difference to the conditions at planets is that, instead of the $1/r^2$-profile for comets a source with exponential dependence ($\sim e^{-r/h}$) must be taken. In this sense the following results give hints how the existence of planetary exospheres, which in some cases (as at Mars) extends far upstream the bow shock, may influence the global interaction region.

As shown first in a paper by Bogdanov *et al.* (1996), the Mach number and the production rate are essential parameters which determine the character of the interaction. If the Mach number is fixed at $M_A = 5$, for example, a transition from a very asymmetric type of interaction in form of Mach cones to nearly symmetric bow shock structures takes place if the production rate is enhanced from $Q \sim 10^{25}$ s^{-1} to $Q \sim 10^{27}$ s^{-1}. First, in Figure 10(a), plasma structures are shown for a low production rate, comparing kinetic hybrid code results (Bagdonat and Motschmann, 2002) with the outcome of two-ion fluid MHD simulations (Sauer and Dubinin, 2000; Szegö *et al.*, 2000). As a somewhat surprising fact the overall pattern, showing the heavy-ion cycloid in the convection electric field direction and the related strongly asymmetric Mach cone, coincides very well for both models. This demonstrates that the electromagnetic coupling of the proton and heavy-ion fluid dynamics, which is obviously well described in the two-ion fluid approach, primarily causes the observed plasma behavior. The reason of the strong asymmetry in the magnetic field is due to the proton deflection opposite to the cycloidal motion of the heavy ions by the Lorentz force which arises from the differential motion of high-speed protons and slowly moving heavies near the center of the source. So, even for relatively weak interaction processes test particle description and one-fluid models will fail.

With increasing production rate the interaction pattern becomes more symmetric. This is illustrated in Figure 10(b) where results of two-ion fluid simulations are shown for two production rates, $Q \approx 10^{26}$ s^{-1} (left panels) and $Q \approx 10^{27}$ s^{-1}. For the lower production rate there is a clear asymmetry that arises from the strongly different motion of the two types of ions, as discussed before. If Q further increases, the cycloidal motion is suppressed and the heavy ions begin to form ray-like structures which is accompanied by a relatively symmetric magnetic field pattern which is confirmed by hybrid code simulations, e.g., Lipatov *et al.* (1997). More discussion about these effects and their relevance for the Martian bow shock one can find in papers by Dubinin *et al.* (1998), Sauer *et al.* (1998b) and Dubinin and Sauer (1999).

In summary one can state that progress was made over the last years in simulating the bow shock of Mars using different approaches and models. It seems that the existence of an extended exosphere is an essential signature at Mars, which makes it necessary to take into account the effect of pick-up processes on the bow shock structure. Up to now, this was insufficiently done and, hopefully, in next future better models for describing the particular conditions of the solar wind interaction with the exosphere/ionosphere of Mars will improve the present situation.

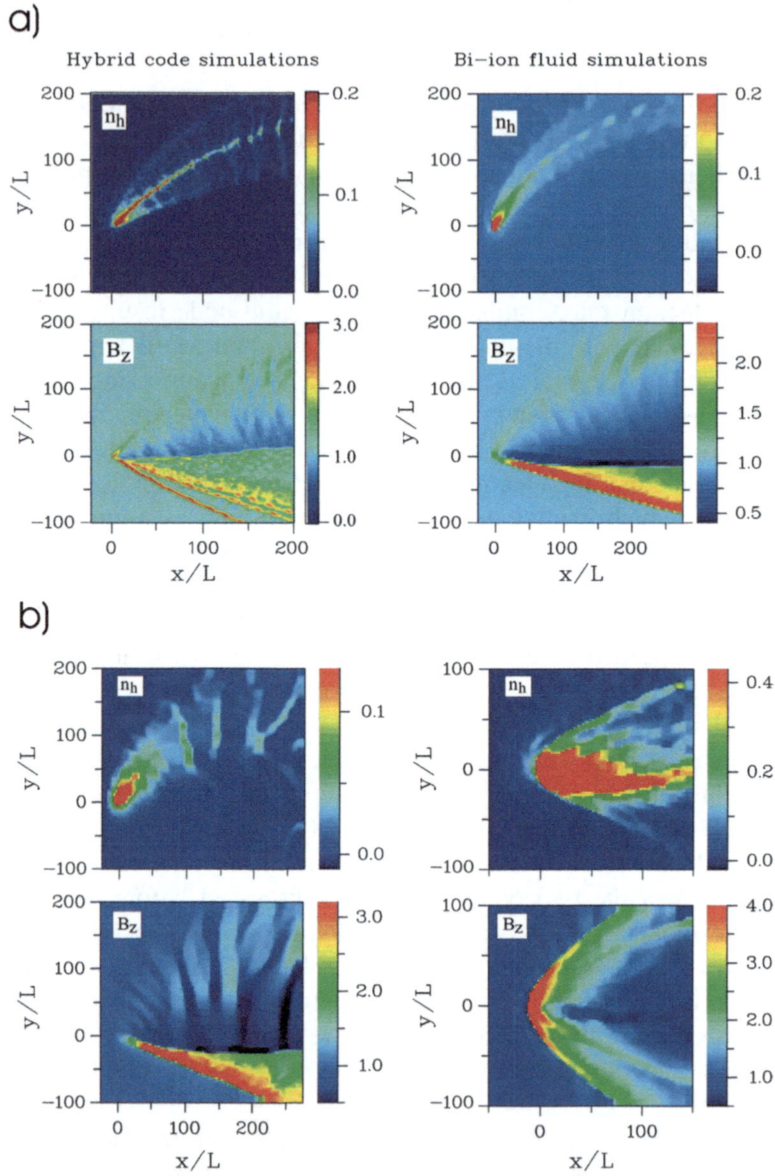

Figure 10. Influence of mass loading on the bow shock structure: Interaction pattern for different production rates. (a) Comparison between hybrid code results (left panels), after Bagdonat and Motschmann (2002) with bi-ion fluid simulations (right panels), after Sauer and Dubinin (1999); $M_A = 10$, $Q \approx 10^{26}$ s^{-1}. The good agreement indicates the crucial role of momentum exchange between both ion fluids, which is obviously well described in the fluid approach. (b) Interaction pattern from bi-ion fluid simulations, after Sauer and Dubinin (1999), for two production rates: $Q \approx 10^{25}$ s^{-1} (left panels) and $Q \approx 10^{26}$ s^{-1} (right panels), $M_A = 5$.

3. Upstream Phenomena

3.1. Mass Loading and Foreshock Phenomena

Due to the lack of the shielding global intrinsic magnetic field the solar wind exposes an extended exosphere/atmosphere of Mars far from the bow shock. Photoionization, electron impact and charge exchange lead to the creation of newly born ions which may affect the incoming solar wind. At large distances the most abundant neutral components are hydrogen atoms which could reach very high altitudes as the consequence of a weak gravitational attraction of Mars. Fast oxygen atoms originated in the process of the dissociative recombination of molecular oxygen O_2^+ which dominates at the ionospheric heights may additionally contribute to the neutral exosphere. If these neutral gaseous shells are not very rare mass-loading effects similar to be observed near the comets could be expected at Mars.

Indeed, already the first measurements made by the Phobos 2 spacecraft were interpreted as the evidence of strong mass loading. Figure 11 shows the variations in the potential difference between a conductive spacecraft at floating potential and the electric field probe, negatively biased in relation to the spacecraft along four successive elliptical orbits as a function of time centered around times of periapsis (Pedersen *et al.*, 1991). A gradual decrease of the potential difference, which is approximately inversely proportional to the electron number density, indicates an increase in the plasma density with decreasing distance to the planet. Under some assumptions, Pedersen *et al.* (1991) derived the electron number density and found that a distinct increase in n_e starts far upstream from the Martian bow shock. The electron number density increased by 1–2 cm^{-3} between the undisturbed solar wind and the outer limit of the shock foot.

Verigin *et al.* (1991) have reported the essential solar wind deceleration upstream the bow shock along the elliptical orbits and attributed these observations to mass loading on the hot oxygen corona. In front of the bow shock the decrement in the bulk speed reaches ~ 100 km s^{-1}. Assuming an instantaneous assimilation of planetary ions to the solar wind, Verigin *et al.* (1991) suggested that the oxygen corona near Mars is five times denser than that predicted by Ip (1988). Since the model by Ip (1988) overestimated the number density in the hot oxygen corona for solar maximum conditions at a factor of 2–3 (Ip, 1990) the value obtained by Verigin *et al.* (1991) far exceeds even the extreme number density of oxygen atoms. Later Kotova *et al.* (1997) have analyzed the TAUS observations on the circular orbits and concluded that deceleration of the solar wind occurs on oxygen corona with the number density at ~ 3 times higher than the values predicted by the extreme model. Recent reexamination of the altitude distribution of neutrals in the hot oxygen corona at Mars for maximum solar activity made by Kim *et al.* (1998) gives the oxygen density nearly one order of magnitude less than the values estimated by Kotova *et al.* (1997). There is also another serious problem in the mass-loading scenario. Unlike the comet case, there is not enough room near Mars

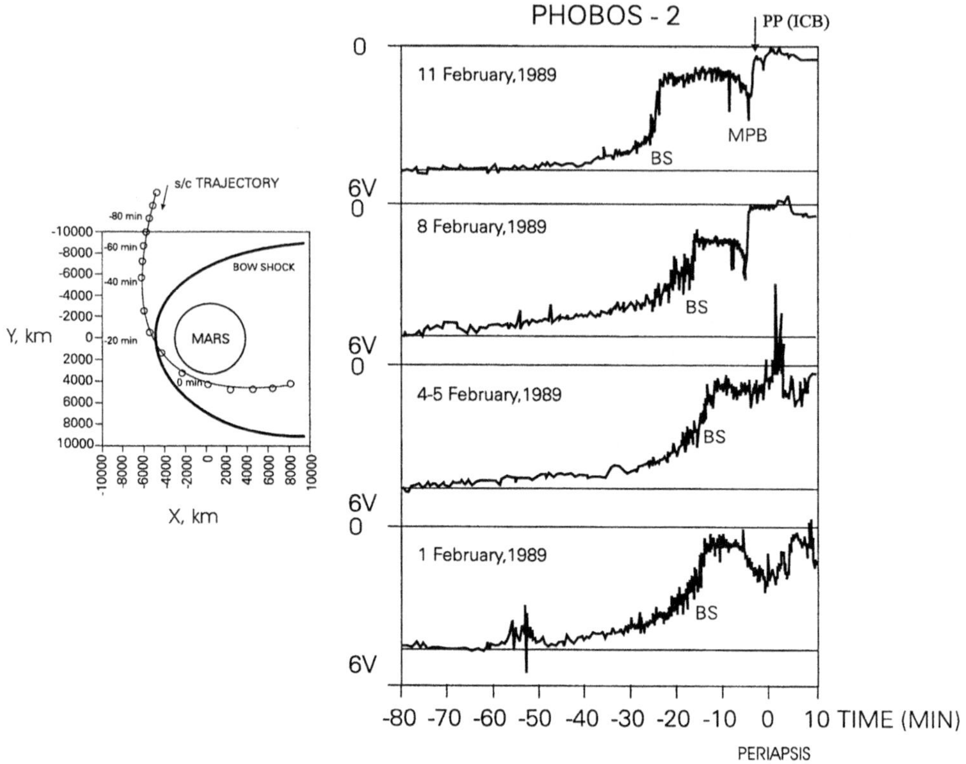

Figure 11. The potential difference between a conductive spacecraft at floating potential and the electric field probe, negatively biased in relations to the satellite body, is approximately inversely proportional to the electron number density. The data for the four successive elliptical orbits around Mars centered in time relative to the periapsis passage are presented (adapted from Pedersen *et al.*, 1991). The orbit of Phobos-2 in XY ecliptic plane is shown too. When Phobos-2 was approaching Mars within the ion foreshock (1, 4 and 8 February 1989) the onset of increase in electron number density took place far from the planet. When the spacecraft was traveling outside the foreshock (11 February) signatures of increase in *Ne* were observed only after the crossing of the foreshock boundary close to the bow shock.

for assimilation of pickup ions. Gyroradius of pickup oxygen ions is much larger than the deceleration distance and the assumption of ion assimilation into the solar wind is not valid.

Dubinin *et al.* (1994) have shown that a decrease in the difference potential measured by the PWS instrument was observed only when the spacecraft was approaching Mars within the foreshock. On February 1, 4 and 8 Phobos 2 was moving in the foreshock only with short excursions into the solar wind. On February 11 the spacecraft was out the foreshock most of the time and crossed the foreshock boundary only downstream from the tangent point close to the bow shock. The foreshock geometry and the position of the spacecraft with respect to the foreshock were reliably determined by different methods (Dubinin *et al.*, 2000a). It was suggested

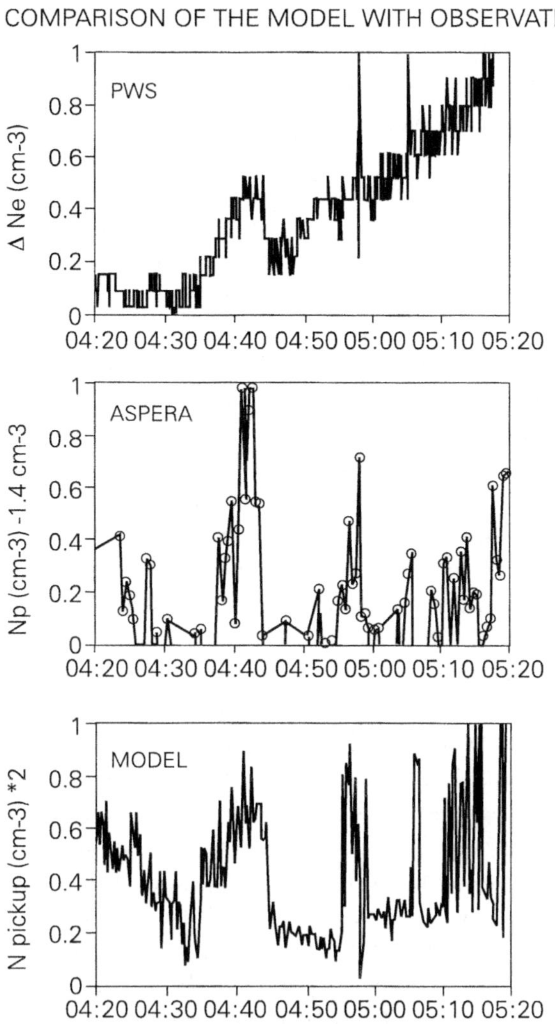

Figure 12. Comparison of observed and simulated variations in plasma density along the orbit of Phobos-2 on February 8, 1989. From top to the bottom are the variations in the electron number density derived from the measurements of the floating potential, variations in the proton number density measured by the ASPERA instrument, modeled values of the density variations (adapted from Dubinin *et al.*, 1994).

that such asymmetry is caused by reflection at the bow shock of pickup H^+ ions originating from the extended atomic hydrogen atmosphere (Dubinin *et al.*, 1994, 1995). The importance of pickup ion reflection is motivated by the significant fraction of pickup protons with small normal velocity components at the bow shock that prevent their overcoming through electrostatic potential barrier.

Dubinin *et al.* (1994, 1995) have simulated the dynamics of pickup protons and have shown that their altitude distribution becomes strongly asymmetrical due to reflection at the bow shock. The model qualitatively well explains not only the global asymmetry in the floating potential measurements but also a fine structure of the potential variations. Figure 12 compares changes in the total number density as compared to the unperturbed solar wind: Δn_e derived from the difference potential measurements (the bottom panel), Δn_p (ASPERA) (middle) and $\Delta n_{\mathrm{pickup}}$ derived from the model (the top) for one of the elliptical orbits. A fraction of reflected ions is mass-dependent $\sim (m_p/m_{pickup})^{1/2}$ where m_p and m_{pickup} are masses of protons and pickup ions, respectively and therefore heavy ion species (He^+, O^+) less effectively participate in this process. It has been shown that the flux of back-streaming particles in the Martian foreshock may be anomalously high ($> 10\%$) and cause a slowdown of the solar wind (Ip, 1992; Dubinin *et al.*, 2000b). Note that the deceleration of the solar wind is also observed in the Earth's foreshock where the flux of backstreaming ions is much less (Bonifazi *et al.*, 1980; Zhang *et al.*, 1995, 1997). Particles reflected at the bow shock may turn back by the Lorentz force and execute several reflections. Strong magnetic and electric field turbulence observed near the bow shock (Grard *et al.*, 1991) may significantly increase the number of ion oscillations and provide the formation of a peculiar trap for pickup protons that in their turn may lead to a strong solar wind deceleration in front of the bow shock observed by the TAUS instrument.

Multi-instrument data analysis of the observations in the upstream region made by Dubinin *et al.* (2000a) showed that transient phenomena interfere mass-loading signatures and dominate over them. It was shown that variations in the solar wind speed at $R \geq 6000$ km are mainly related with propagating large-amplitude Alfvén waves. The flow and magnetic field perturbations are well correlated and follow Walén's relation $v = \pm b/(\rho^{1/2}) = \pm K_{\mathrm{A}} b$, where ρ is the plasma mass density and $\pm$ sign is taken as the sign of the product $-(k \cdot B_0)$ and varies in dependence on the wave propagation vector k with respect to the ambient magnetic field B_0. Figure 13 show typical examples of relationship between the v_x and b_x components measured upstream of the bow shock. A linear dependence between the perturbations $v_x = K\ b_x$ with the factor K close to the Alfvén wave factor K_A strongly suggests the wave origin of the observed perturbations. The in-phase relationship between the velocity and field perturbations in all these cases suggests the antisunward direction of wave propagation.

It is interesting to note that propagation of such waves explains decrease of the solar wind speed in the foreshock as compared to the velocity outside the foreshock. Figure 14 shows a scheme illustrating the effect of waves propagating antiparallel to the IMF in the antisunward direction. If the perturbation $b_x > 0$, the tangent line (TL) turns from its nominal position shown by the thick arrow to the position (1). Then the observer that was in the solar wind close to the tangent line may occur in the foreshock and records a decrease of the speed ($|V_0 + v| < |V_0|$, $V_0 < 0$, $v > 0$). For the perturbations with $b_x < 0$ the tangent line goes to the position (2) and the observer occurs farther from the tangent line where the velocity increases. It was found that the observed variations in the flow can be provided

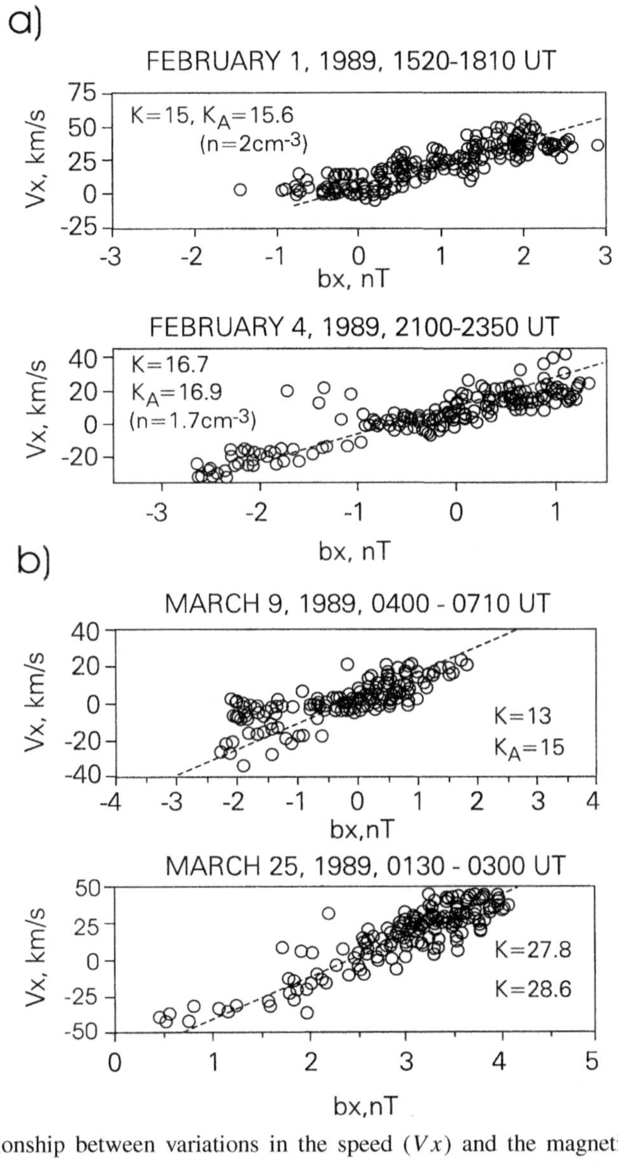

Figure 13. Relationship between variations in the speed (Vx) and the magnetic field (Bx) well follows the Walén relation for Alfvén waves propagating in the solar wind. At close distances to the bow shock perturbations in the flow and the magnetic field do not follow the Walén relation (not shown) (adapted from Dubinin *et al.*, 2000a).

either by waves of the solar wind origin or by waves excited in the foreshock or near its boundary.

In all cases the foreshock boundary is not just a topological boundary separating the regions either magnetically connected or disconnected with the bow shock. Transition across the tangent line is often accompanied with a sharp rotation of the

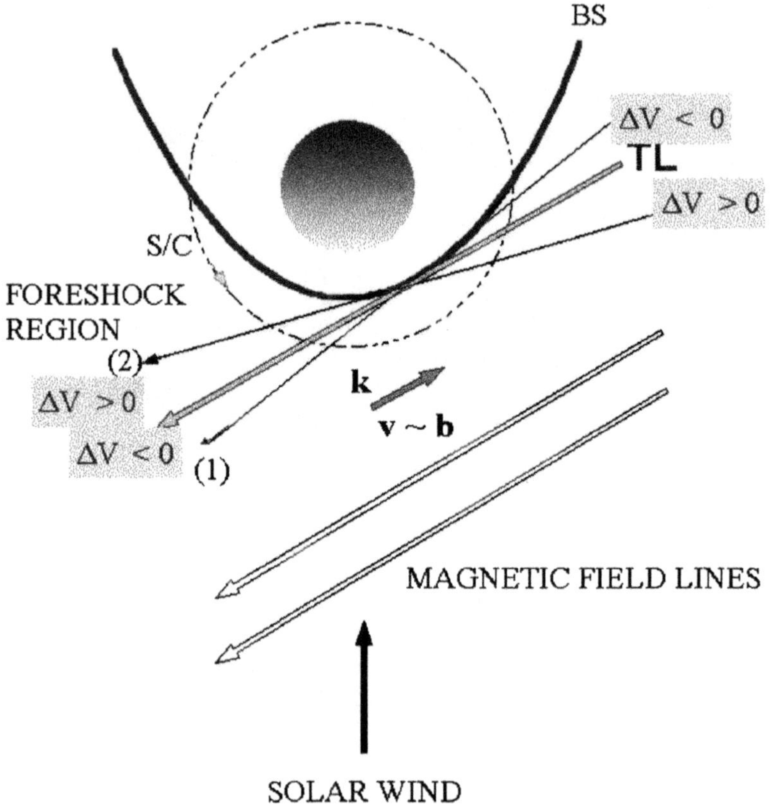

Figure 14. Scheme illustrating the effect of Alfvén waves propagating in antisunward direction. For sunward propagation, the structure of the solar wind/foreshock interface is reversed in dependence on the direction of wave propagation (adapted from Dubinin *et al.*, 2000a).

magnetic field, decrease in the field value and increase in the proton number density and temperature (Figure 15). Although variations of the solar wind speed at $R \geq 6000$ km are mainly controlled by Alfvén waves their generation or amplification in the foreshock may be closely related with large fluxes of backstreaming ions.

At closer distances to the planet where the solar wind slows down at ~ 100 km s^{-1} and the perturbations of the field and velocity do not follow the Walén relation, deceleration may be due to mass-loading on oxygen atmosphere or related with foot phenomena, e.g. with an enhancement of the number density of pickup ions in front of the bow shock due to multiple reflections of pickup ions.

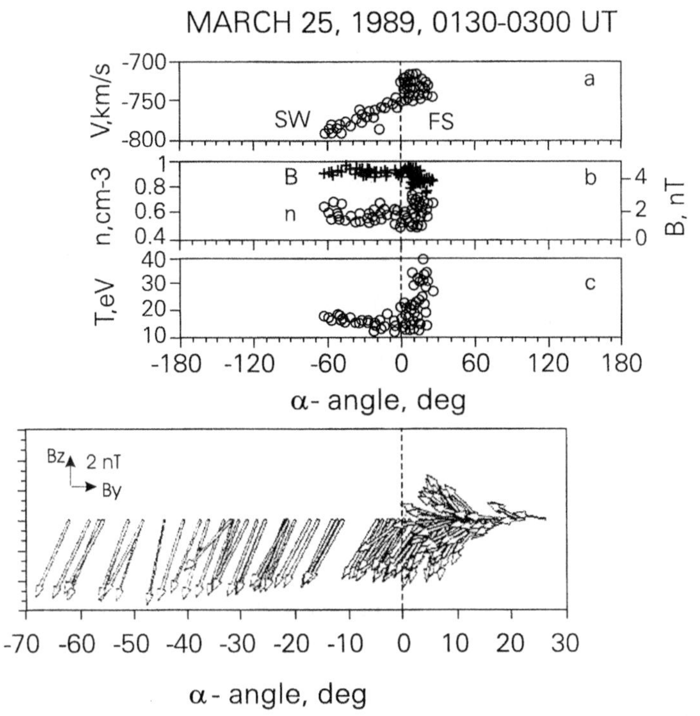

Figure 15. Variations in the bulk speed, the proton number density and proton temperature across the foreshock boundary along the circular orbit at $R =\sim 2.8\ R_M$. Position of the spacecraft with respect the foreshock boundary is given in angle between the tangent line and the direction to Phobos-2 from the tangent point. Positive (negative) values correspond to the location of the spacecraft in the foreshock (solar wind). The bottom panel depicts projections of the magnetic field onto the YZ plane (the MSO reference frame) (adapted from Dubinin *et al.*, 2000a).

3.2. Characteristics of the Upstream Foreshock Region

3.2.1. *General Aspects of Foreshock Geometry*

The three-dimensional geometry of the foreshock is determined through the main motion of particles back-streaming from the bow shock. The foreshock region is magnetically connected to the bow shock and limited upstream by the magnetic field line tangent to the model bow shock. Particles back-streaming from the bow shock are guided by the magnetic field lines (limited by the tangent field line). Close to the tangent field line, only back-streaming electrons are observed (owing to their higher velocities) and form the electron foreshock, the (slower) back-streaming ions are found further downstream from the tangent field line, generating VLF and ULF waves. An overview of the wave observations in the Earth's foreshock is shown in Figure 16 (from Greenstadt *et al.*, 1995).

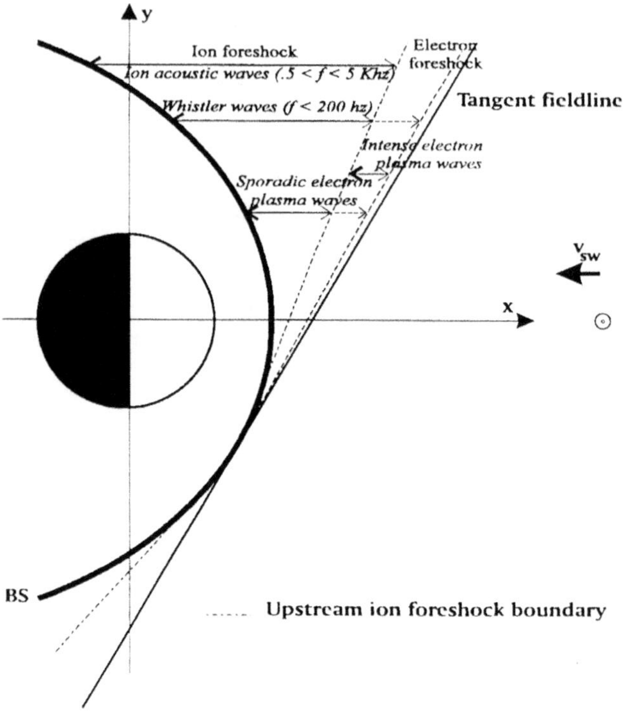

Figure 16. Wave modes, typical for the electron and ion foreshock at Earth (from Greenstadt *et al.*, 1995).

3.2.2. *Mars Foreshock Results*

From the Phobos 2 data, the electron foreshock was observed through the appearance of 5.8–28 kHz emissions (Skalsky *et al.*, 1993). Simultaneous electron measurements revealed enhanced fluxes of particles, back-streaming from the bow shock along the magnetic field. From the Phobos 2 data of multiple instruments (Dubinin *et al.*, 2000a) the foreshock geometry was also well traced by a change of the antisunward/sunward anisotropy of suprathermal electron fluxes. More details about electric field waves will be given below.

The existence of the ion foreshock at Mars was stated from the Phobos 2 observations by Russell *et al.* (1990), based on the magnetometer data, and by Grard *et al.* (1991), based on the broadband electric noise upstream of the bow shock. This noise (750 Hz–4 kHz) was interpreted as ion acoustic waves, with typically weak amplitudes and bursty in nature. Weak waves at the proton gyro frequency were observed upstream of the bow shock, associated with pick-up of protons from the Mars hydrogen exosphere (Russell *et al.*, 1990).

Statistics from the circular orbits of the Phobos 2 S/C (March 1989) showed the following characteristics (Delva and Dubinin, 1998; Skalsky *et al.*, 1998; Dubinin *et al.*, 2000a):

- Waves in the range of the electron plasma frequency are most intense close to the upstream boundary of the electron foreshock.
- ULF waves (0.2–10 Hz) gradually increase at larger distances from the tangent line, i.e., with the depth of the S/C location in the foreshock region.
- Deceleration of the solar wind starts closely after the crossing of the foreshock boundary, with significant deceleration closer to the bow shock, accompanied by a strong increase of the ULF turbulence.
- An enhanced level of electric field oscillations in the range 50–200 Hz is typically detected inside the foreshock.
- Intensities of the broadband electric emissions at frequencies of 0.5–4 kHz are weak, but the most intense bursts occur deeper in the foreshock, compared to the higher frequency waves.
- For about 30% of the Phobos 2 S/C passes through the foreshock region, the onset of the waves with frequencies 50–200 Hz occurred upstream of the foreshock boundary.
- ULF waves are also observed upstream of the tangent field line; they might be of solar wind origin or generated by pickup protons and oxygen ions, originating from the extended hydrogen atmosphere and hot oxygen corona.

As was already discussed in the previous section, the multi-instrument observations from Phobos 2 showed that the foreshock boundary is not simply a topological boundary separating the regions magnetically connected to the bow shock (Dubinin *et al.*, 2000a). Large amplitude Alfvén waves essentially contribute to the perturbations of the plasma and the IMF near the tangent line. Even for waves of solar wind origin, the transition across the tangent magnetic field line is often accompanied by a sharp change of the clock angle of the magnetic field, a decrease of the magnetic field value and an increase of the proton number density and temperature. Downstream of the upstream foreshock boundary, ULF waves are generated by back-streaming ions and ions of planetary origin interacting with the on-streaming solar wind.

From the Mars Global Surveyor S/C data, it is not always possible to distinguish if the S/C is located in the ion foreshock or not, due to a lower accuracy of the magnetometer data of the MGS MAG sensors, when the S/C is in the solar wind and the solar panels are illuminated (Acuña *et al.*, 2001). The obtained accuracy of ~ 1 nT prevents from directly using the MAG data to study the connection of the local field line to the bow shock when the ambient IMF magnitude is small, which is often the case at Mars. For orbits for which the IMF is larger than 4–5 nT, typically, it is possible to determine the tangent field line with high enough accuracy and check whether the spacecraft is in the foreshock or not. Since there was no wave instrument on MGS, only low frequency electromagnetic waves were accessible to observations.

Low frequency waves at the proton cyclotron frequency were frequently observed upstream of the bow shock (Brain *et al.*, 2002) and often appear as large amplitude highly coherent waves (Mazelle *et al.*, 2001). On MGS orbits for which

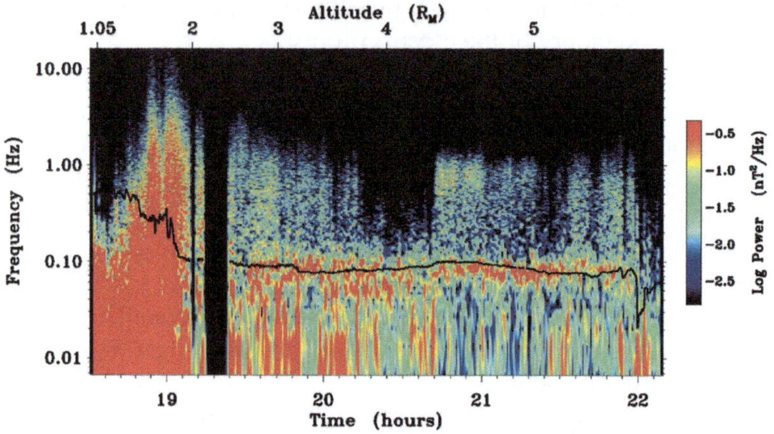

Figure 17. Time-frequency spectrogram of one vector field component of MGS MAG data from April 24, 1998 (Brain *et al.*, 2002). Two wave features are evident – one at the local proton gyro-frequency (plotted over the spectrogram in black), and one near 1 Hz. A bow shock crossing occurs at $\sim$ 19.1 hours, after which MGS is outside of the shock. The data were rotated into mean-field coordinates (with the mean field pointing in the $+z$ direction) before the spectrogram was applied.

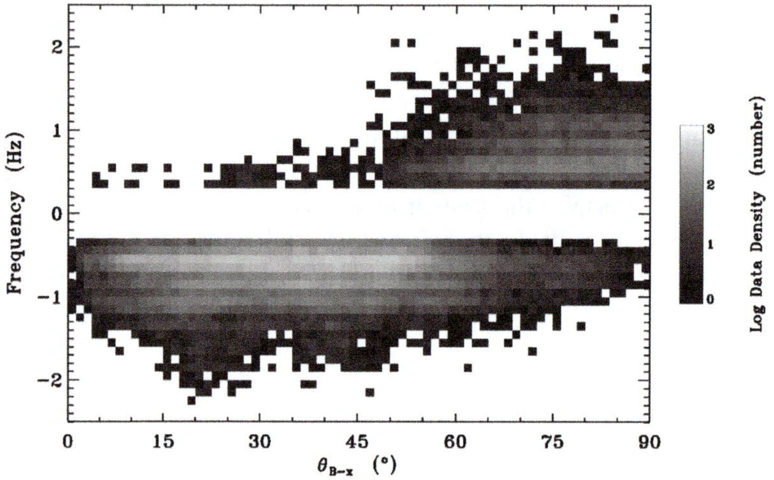

Figure 18. Frequency of whistler waves observed by MGS MAG as a function of the angle between the magnetic field and the solar wind velocity (Brain *et al.*, 2002). The solar wind velocity is taken in the x direction. Negative frequencies indicate left hand polarization in the rest frame of the spacecraft.

the foreshock boundary can be determined from the tangent field line, such waves are observed with the same characteristics for a large range of radial distances and even when the spacecraft is not magnetically connected to the bow shock (Mazelle *et al.*, 2001). This indicates that they are not ion foreshock-related, i.e., not associated to back-streaming ions. Moreover, the orbits for which coherent monochromatic upstream waves at the proton cyclotron frequency are observed, never show a sharp interruption of the waves in the time series of the magnetic field as the S/C leaves the foreshock region, although this is characteristic for observations at the Earth, when a S/C has crossed the ion foreshock boundary. These wave observations will be detailed below.

Higher frequency (~ 1 Hz) whistler waves were also detected in the MGS magnetometer data upstream of the Martian bow shock, which were not reported from the Phobos 2 data. They occurred mainly while MGS was magnetically connected to the bow shock, i.e., they appear to be foreshock phenomena (Brain *et al.*, 2002).

3.3. Observations of Low Frequency Waves

The magnetometer/electron reflectometer (MAG/ER) on board the Mars Global Surveyor (MGS) has recorded a variety of plasma wave oscillations in the magnetic field throughout the Martian plasma environment (Mazelle *et al.*, 2000, 2001; Brain *et al.*, 2002; Cloutier *et al.*, 1999; Crider, 1999). Two distinct wave frequencies are often observed upstream from the shock (Figure 17). The elliptical orbit used in the pre-mapping portion of the mission allowed substantial altitude and solar zenith angle coverage of the region upstream from the bow shock, ranging from ~ 30–150 in SZA and out to ~ 17 R_M for some orbits. Data is returned at 32, 16, or 8 samples per second, corresponding to Nyquist frequencies of 16 Hz, 8 Hz, or 4 Hz. A lower time resolution version of the data is available (with Nyquist frequencies of 0.17–0.67 Hz), and has the advantage of having magnetic field contributions from the spacecraft and solar panels removed (lack of high time resolution engineering data from the spacecraft make calibration impractical for the high time resolution MAG data). The calibrated data are accurate to ~ 1 nT in sunlight, and ~ 0.5 nT in shadow (Acuña *et al.*, 2001). Fortunately, the remaining spacecraft contributions to the magnetic field are due to the thermal response of the solar panels, which occurs at very low frequencies and can be filtered out of the data when searching for plasma waves. However, the direction of the ambient magnetic field is uncertain to within the accuracy of the calibration.

A recurring wave feature in the frequency range 0.4–2.3 Hz is evident in high time resolution data upstream from the shock (Brain *et al.*, 2002). This feature has many characteristics that are consistent with whistler waves observed at other solar system bodies, including Earth, Mercury, Venus, and Saturn (Orlowski *et al.*, 1990, 1992). First, the waves are associated with the foreshock. Second, the polarization sense for the waves is controlled by the angle between the ambient magnetic field and the solar wind flow (Figure 18). Third, the amplitude of the wave decreases with distance to the shock (along the magnetic field line). Fourth, the wave characteristics of Martian whistlers follow trends with heliocentric distance observed for whistlers at other solar system bodies. These trends include a decrease in frequency and amplitude with heliocentric distance, and an increase in eccentricity and the angle between the wave propagation vector and background magnetic field (θ_{kB}) (Orlowski and Russell, 1995).

It is believed that the whistlers are generated at or near the shock, and propagate upstream with a group velocity greater than the solar wind velocity of ~ 400 km s^{-1} (Fairfield, 1974). The direction of the ambient magnetic field determines the direction of propagation of the wave, but the component of the wave's group velocity in the direction of the solar wind flow will be Doppler shifted when the waves are observed in the rest frame of the spacecraft or planet. The whistlers are right handed in the plasma rest frame, but are observed as left hand waves when the angle between the background field and solar wind flow is low because of this Doppler shift. There are many open questions concerning upstream whistler waves (see Brain *et al.*, 2002, and references), some of which may be answered by future spacecraft to Mars. A wide variety of generation mechanisms have been proposed, and it is still not clear whether ions can be responsible for some of the observed waves, and how the waves are damped.

Left-hand circularly polarized waves (PCWs) at the proton cyclotron frequency were first observed by Russell *et al.* (1990), using magnetometer data from the Phobos spacecraft. These waves were observed 2–3 R_m from Mars on three spacecraft orbits, had low amplitude, and propagated at small angle to the background magnetic field. ASPERA observations of proton ring distributions in the hydrogen corona (Barabash et al. 1991), and lack of similar wave observations at Venus (which does not have an exosphere that extends beyond the shock) led Russell *et al.* (1990) to conclude that the waves were formed by solar wind pickup of newly-ionized exospheric hydrogen. The highly eccentric orbit of MGS enabled confirmation of the Phobos observations, and examination of the spatial distribution of PCWs outside of the Martian shock.

Over 500 orbits of pre-mapping MAG data were analyzed for the presence of PCWs (Brain *et al.*, 2002). The wave characteristics agreed with Phobos results in the observed frequency, eccentricity, and propagation angle, while the amplitudes recorded for MAG data were a factor of 2–3 higher than observed by Phobos. There is no correlation of wave observations with the foreshock, and the wave features are often long-lived in the data. The spatial distribution of power at the local proton gyrofrequency in MGS MAG data is shown in Figure 19. Wave power generally decreases with increasing altitude, and there are two spatial locations where PCWs are common. Waves near the subsolar point have lower amplitudes, lower eccentricities, and smaller values of θ_{kB} than waves observed near the flanks of the solar wind interaction. Wave power near the flanks decreases rapidly with altitude relative to waves near the subsolar point.

The analysis of MGS data by Brain *et al.* (2002) was based on a search for waves in specific frequency ranges. There is no reason that all observed whistler waves should be caused by a single mechanism, and the same is true for PCWs. It is even possible that the two different wave frequencies are related to each other for some of the observations. We note however, that this is not likely to be true for the majority of the observed upstream waves. First, given the constraints of the calibration to the data, the observed whistler waves are overwhelmingly associated

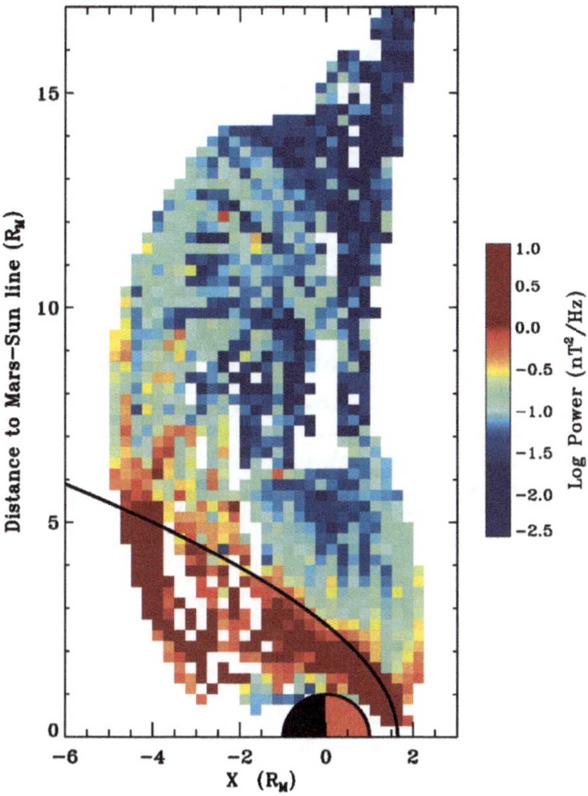

Figure 19. Spatial distribution of power at the local proton gyrofrequency in MGS MAG data (Brain *et al.*, 2002). Average power is shown as a function of location for the x component of magnetic field in mean-field coordinates. The best fit bow shock (Vignes *et al.*, 2000) is indicated in black.

(88%) with the foreshock, while the PCWs are not. Second, the occurrence of the two wave signatures is statistically independent of each other: whistlers are observed 18% of the time, PCWs are observed 26% of the time, and the two are observed concurrently 6% ($\approx 0.26 \times 0.18$) of the time.

Finally, large amplitude, highly coherent upstream waves at the proton cyclotron frequency f_c were reported by Mazelle *et al.* (2001). These waves are observed for a large range of radial distances and even when the spacecraft is not magnetically connected to the bow shock; therefore, they are not ion foreshock-related, i.e., not associated to back-streaming ions. The observed periods (spacecraft frame) always closely fit the local proton cyclotron periods. The waves are left-handed circularly polarized in the spacecraft frame. Their wave vectors lay at moderate angles from the ambient magnetic field. Thus, at first glance, such waves would be as usual interpreted as generated by a resonant helical beam instability fed by the planetary pick-up protons, which can lead to derive constraints on the hydrogen density in the Martian exosphere.

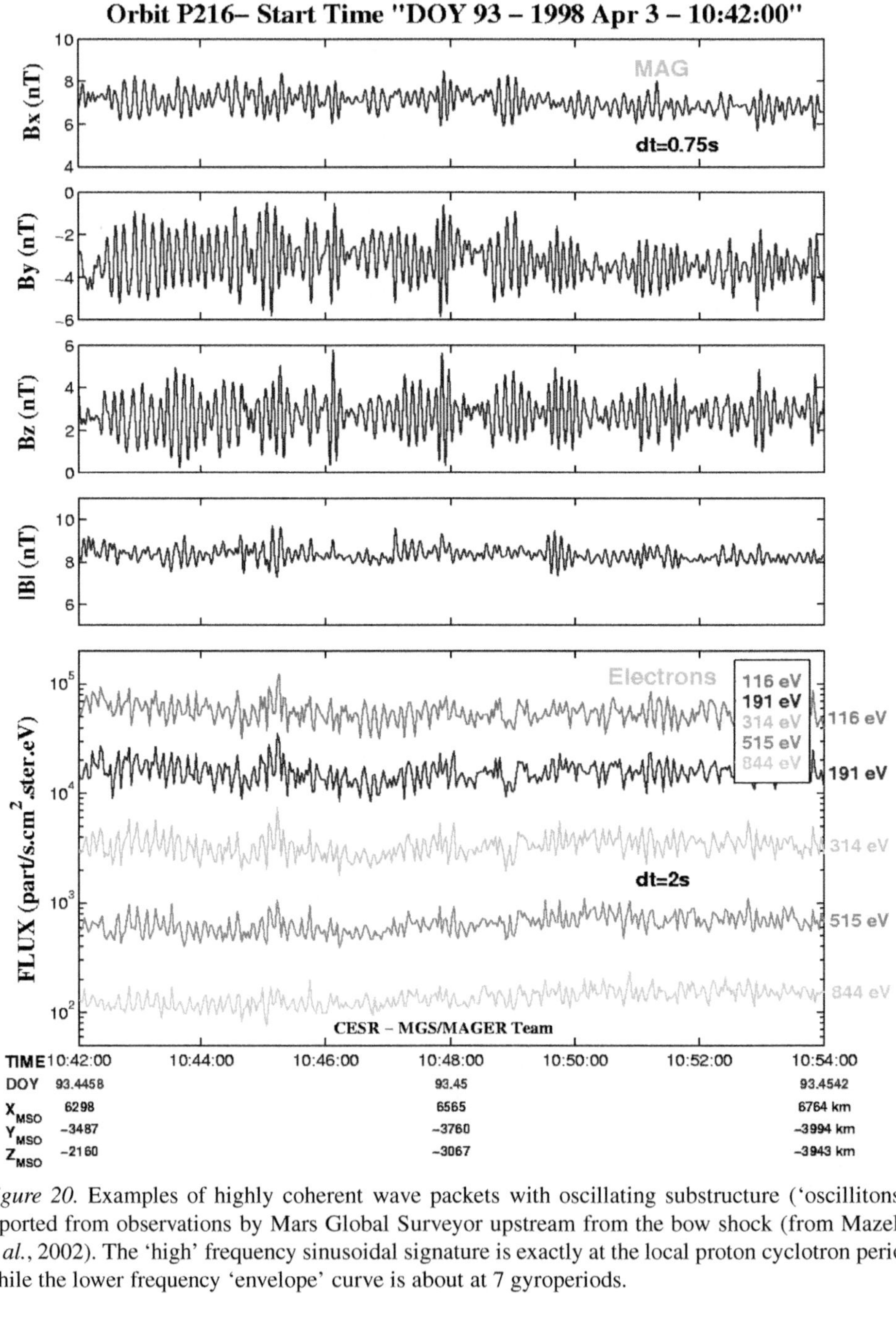

Figure 20. Examples of highly coherent wave packets with oscillating substructure ('oscillitons') reported from observations by Mars Global Surveyor upstream from the bow shock (from Mazelle *et al.*, 2002). The 'high' frequency sinusoidal signature is exactly at the local proton cyclotron period while the lower frequency 'envelope' curve is about at 7 gyroperiods.

However, these 'proton cyclotron waves' are frequently revealed as monochromatic wave packets embedded inside a regular lower frequency 'wave envelope' at about $f_c/10$ (Figure 20). They can have large amplitude (up to 5 nT peak-to-peak) even at large distance from the planet (more than 6 R_M). Their amplitude is too large to be consistent e.g., with the saturation value for the picked-up proton ring instability using realistic neutral densities from theoretical models of the Martian exosphere (e.g., Kim *et al.*, 1998). Moreover, a satisfactory mechanism is also needed for both the lower frequency 'envelope' signature and the high coherence of the ion gyrofrequency signature. Classical nonlinear processes occurring for finite amplitude circularly polarized Alfvén waves such as the decay or the modulational instability can be excluded (e.g., Spangler *et al.*, 1988, and references therein). The oscillations of the electron fluxes for all energy channels (and thus the density) are in close correlation with the magnetic field wave oscillations whereas ponderomotive effects for a circularly polarized Alfvén wave would make the density perturbation to be proportional to the square of the wave amplitude and thus be observed at half the wave period (Spangler *et al.*, 1988). These waves can be interpreted as oblique bi-ion stationary waves ('oscillitons') (Sauer *et al.*, 2001, 2002a). This will be detailed in Section 4.6. They can be produced (in the flow regimes in which they are permitted) by the existence of two ion populations (solar wind + planetary) with different densities, velocities and/or temperatures. This implies that direct connection between wave amplitude at the cyclotron frequency and neutral densities in the planetary exosphere is not always ensured.

3.4. Observations of High Frequency Plasma Waves

The bow shock is usually considered to be crossed when the ramp in the magnetic field is encountered. As this shock transition region separates the upstream region, where charged particles energized and reflected at the shock cause a variety of instabilities and attendant waves, from the downstream region, in which the wave activity fades away after reaching a maximum at the ramp, bow shock crossings are also clearly identified from plasma and wave data. This point is illustrated in Figure 21, which displays some of the parameters measured with the plasma wave system on February 11, 1989, along the fourth elliptical orbit of Phobos 2 around Mars: from top to bottom, (1) the r.m.s. fluctuation of the current collected with the Langmuir probe, (2) the high frequency variation of it, and (3) the electric field intensity measured by the 1.45 m long double sphere antenna as a function of time and frequency, from 5 Hz to 150 kHz. For a full description of the plasma wave system, the reader is referred to Grard *et al.* (1989). The color scaling shown on the right side of Figure 21 depends on the strength of the electric field, which is expressed in decibels above 1 μV m^{-1} Hz$^{-1/2}$, so that 60 dB correspond to a signal level of 1 mV m^{-1} Hz$^{-1/2}$. A vertical dashed line marks the shock ramp, which was crossed at 11:03 UT. The abrupt burst of the electric field level, from a few Hz up to 10 kHz, the jump in the electron density, which is seen as a jump

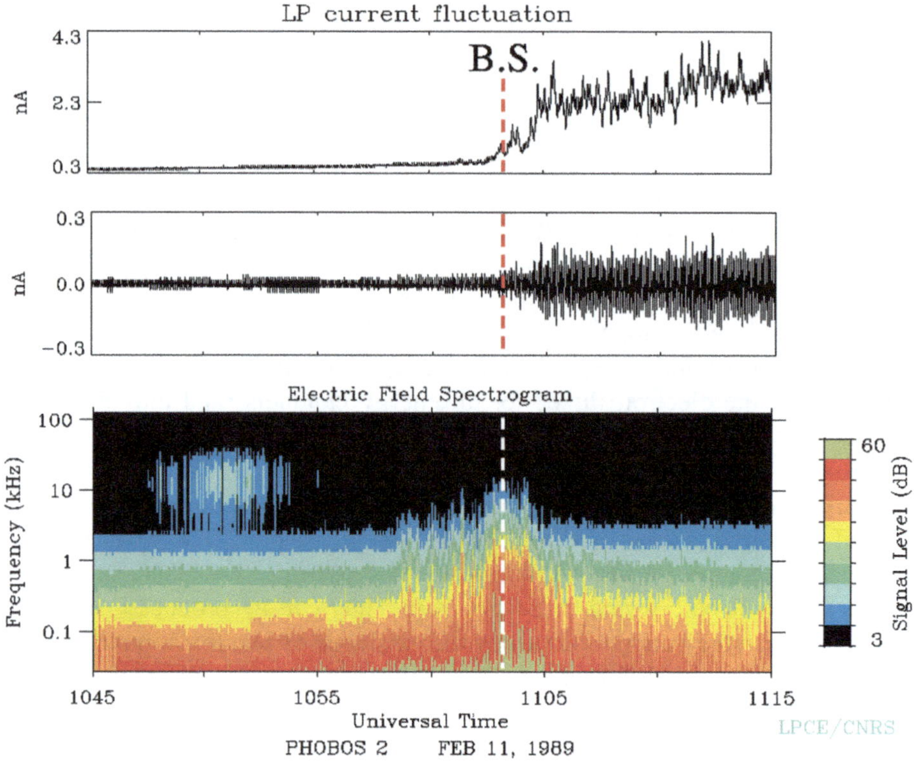

Figure 21. From top to bottom: root-mean-square of the current collected with the Langmuir probe, high frequency variation of this signal, and dynamic electric field spectrogram measured by the plasma wave system onboard Phobos 2. The Martian bow shock was crossed at 11:03 UT on February 11, 1989 (vertical dashed line). Electron plasma oscillations occurred at about 13 kHz in front of the shock (Trotignon *et al.*, 1993.

in the Langmuir probe current fluctuation, as well as the increase of the plasma turbulence just behind the shock transition region allow a reliable and accurate location of the shock front to be determined.

The electric field spectrum recorded by PWS in the Martian shock ramp at 11:02:42 UT on February 11, 1989, is plotted as crosses in Figure 22. In addition, several plasma characteristic frequencies, the electron cyclotron frequency (F_{ce}), the ion plasma frequency (F_{pi}), the Buneman frequency (F_B), the upper frequency cut-off of Doppler-shifted ion-acoustic waves ($V_{SW}/2\pi\lambda_D$, where V_{SW} is the solar wind velocity and λ_D the plasma Debye length), and the local plasma frequency (F_{pe}) are pointed out by arrows. Similar to the Earth's bow shock (Grard *et al.*, 1989, 1993; Trotignon *et al.*, 1991a) there is evidence that two well-defined wave components are present in the shock transition region. The 10:02:42 UT electric field spectrum shown as crosses in Figure 22 indeed consists of a low frequency component below F_{ce} and a high frequency one with an apparent frequency cut-off at $V_{SW}/2\pi\lambda_D$. The low-frequency noise is attributed to the electric component

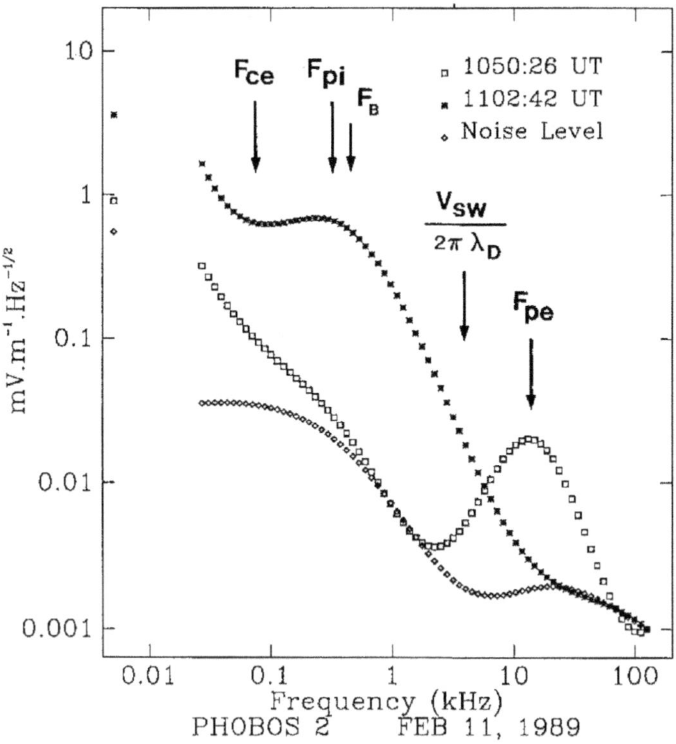

Figure 22. Electric field spectra recorded by the plasma wave system onboard Phobos 2 in the free solar wind (small diamonds), in the Mars' electron foreshock (squares) where electron plasma oscillations were observed, and in the shock ramp (crosses). Plasma characteristic frequencies are indicated by vertical arrows (Trotignon *et al.*, 1991c).

of the electromagnetic whistler mode emission, while Doppler-shifted ion-acoustic waves, which are electrostatic in nature, are thought to be responsible for the high frequency spectral component (Sagdeev *et al.*, 1990; Grard *et al.*, 1991; Trotignon *et al.*, 1991c). Consequently, the whole frequency range explored by the PWS has been split into three consecutive bandwidths: (1) the 25 Hz–100 Hz low-frequency bandwidth for whistler mode emissions, (2) the 100 Hz–6 kHz middle frequency bandwidth to cover the ion-acoustic Doppler-shifted frequency range, and (3) the 6 kHz–130 kHz very high frequency bandwidth for the upstream electron plasma waves, which will be introduced ahead. The result of the splitting process applied to the Figure 21 time interval is presented in Figure 23. A peak is observed in the wave intensity, in particular in the middle frequency bandwidth, which covers the ion-acoustic Doppler-shifted frequency domain.

The cluster of intense noise bursts shown in the top panel of Figure 23 is centered on 13 kHz. These wave bursts, also present in the color spectrogram of Figure 21 are usually observed in front of the Martian shock whenever the interplanetary magnetic field line that passes through the spacecraft position intersects

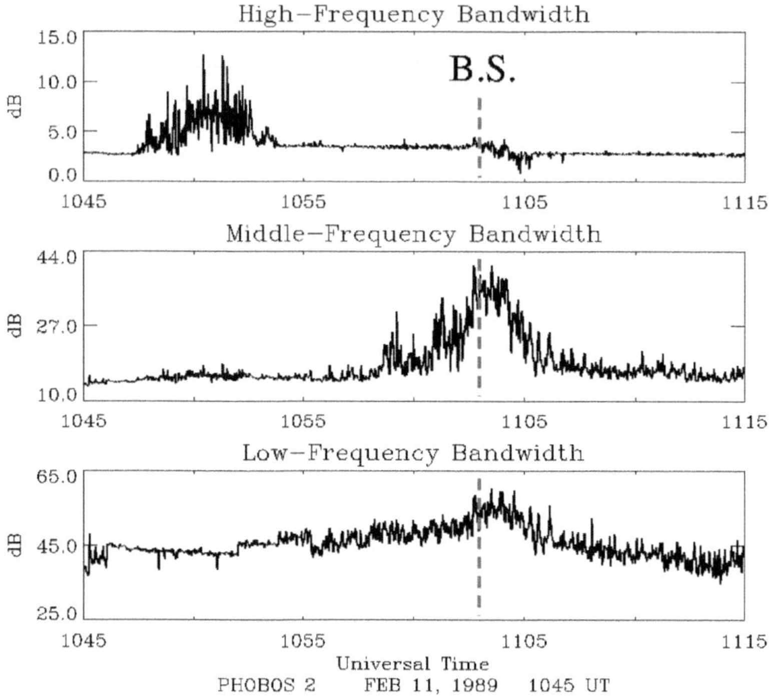

Figure 23. Intensity of the electric field averaged over three frequency bandwidths: 25–100 Hz (bottom), 100–6 kHz (middle), and 6–130 kHz (top). The bow shock has been identified from these Phobos 2 plasma wave system measurements at 11:03 UT, on February 11, 1989. Electron plasma waves are seen as an electric intensity increase in the high frequency bandwidth (Trotignon *et al.*, 1993).

the shock front. Such waves have commonly been reported upstream of most of the solar system planets and even some comets. These electrostatic waves, also called electron plasma oscillations, are known to be generated by a two-stream plasma instability in the electron foreshock region, where solar wind electrons that are energized and accelerated at the shock front stream back into the solar wind along the interplanetary magnetic line connected to the shock.

A typical spectrum of these electron plasma oscillations is plotted in Figure 22. This spectrum (squares) detected at 10:50:26 UT on February 11, 1989, exhibits a bump at or close to the solar wind plasma frequency, F_{pe}, from which the solar wind density may directly be deduced (Trotignon *et al.*, 1992): at this time F_{pe} and N_e were 14 kHz and 2.4 cm^{-3}, respectively. As can be seen in Figure 22, the electron plasma oscillation level is more than one order of magnitude higher than the noise level recorded by the PWS in the unperturbed solar wind (small diamonds). Skalsky *et al.* (1992) reported that the observed electron plasma oscillations were polarized along the magnetic field, i.e., longitudinal, as expected for such upstream electron plasma wave in the electron foreshock. Large electron flux anisotropies were detected by the ASPERA experiment in association with the PWS observations (Barabash *et al.*, 1993). Finally, the electric field emissions turned out to be associated with flux enhancements of electrons with energies from 100 eV to 530 eV that were backstreaming into the solar wind (Skalsky *et al.*, 1993).

The electron plasma oscillation occurrence and characteristics depend on connection parameters calculated in the so-called $\mathbf{B} - \mathbf{v}$ plane, where $\mathbf{B}$ and $\mathbf{v}$ are the interplanetary magnetic field and solar wind speed vectors, respectively. Some of them are displayed in the top panels of Figure 24, between 22:35 UT and 23:35 UT, on February 4, 1989, when Phobos 2 crossed the electron foreshock region along its second elliptical orbit around Mars. The shock subsolar distance is the distance from the center of Mars to the nose of the shock surface section. The connection depth is the distance from the spacecraft to the interplanetary magnetic field line that is tangent to the shock surface section. This distance measured parallel to $\mathbf{v}$ is positive whenever the spacecraft is located downstream of the tangent field line. The time of flight distance is the distance between the point of tangency and the spacecraft position, i.e., the distance covered by the supra-thermal electrons that are reflected at the tangent point. The distance covered by the electrons that stream back along the magnetic field line that intersects the spacecraft position is called the connection distance. Finally, the angle between the normal to the shock front and the spacecraft field line is the θ_{Bn} angle.

The top panel of Figure 24 thus shows that the shock subsolar distance increases regularly, i.e., the shock surface section appears to move regularly from the anti-sunward direction to the sunward direction until the mean shock subsolar distance of 1.57 R_M obtained by Trotignon *et al.* (1993) is reached (solid horizontal line). In the meantime, the interplanetary magnetic field line that passes through the spacecraft position is sometimes connected to the shock, sometimes not. When the field line is connected, the connection distance and θ_{Bn} may be calculated. Conversely, when it is not, the two parameters cannot be derived. It becomes consequently easy to see in Figure 24 that the electron plasma oscillations shown in the color spectrogram only occur when the connection takes place, i.e., when the connection distance and θ_{Bn} angle are plotted.

Assuming that the interplanetary magnetic field line is at a 57° angle to the subsolar direction, i.e., the nominal Parker interplanetary magnetic field orientation, a statistical image of the Mars' electron foreshock region may be constructed. In this image, the electric field intensity averaged over the 6 kHz–130 kHz frequency bandwidth is plotted as a function of the equatorial $\mathbf{B} - \mathbf{v}$ plane coordinates, i.e., the plane that is parallel to the $\mathbf{B} - \mathbf{v}$ plane and passes through the planet center. As can be seen in Figure 25 borrowed from Trotignon *et al.* (2000), the highest electric field levels, which correspond to electron plasma waves in the electron foreshock of Mars, are observed in a limited region behind the tangent field line. Figure 25 also shows that no instabilities and therefore electron plasma oscillations arise beyond 6 R_M from the point of tangency. The small size of the Martian shock

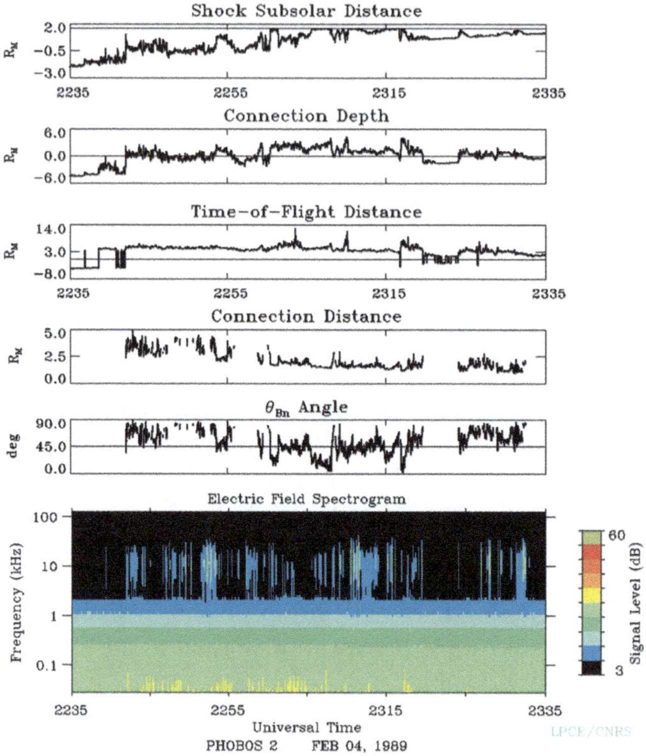

Figure 24. Electric field spectrogram measured by the plasma wave system in the electron foreshock of Mars along the second elliptical orbit of Phobos 2 around the planet (bottom) and connection parameters calculated in the **B** – **v** plane, where **B** and **v** are the interplanetary magnetic field and solar wind velocity vectors, respectively (Trotignon *et al.*, 2000).

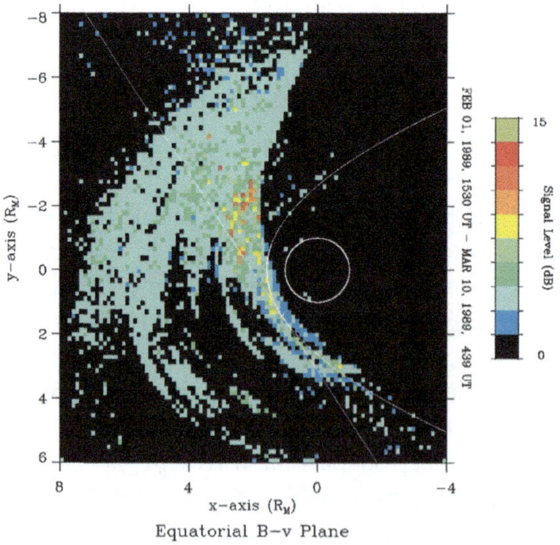

Figure 25. Statistical image of the electric field intensity integrated in the 6 kHz–130 kHz frequency bandwidth measured by the Phobos 2 plasma wave system. (X,Y) are the nominal equatorial **B** – **v** plane coordinates. The Martian bow shock section and the nominal Parker interplanetary magnetic field line tangent to the shock are shown as white conic curve and straight line, respectively. The highest intensities are seen in the electron foreshock region, as expected (Trotignon *et al.*, 2000).

is thought to be responsible for this limited extension along the tangent field line by limiting the electron energization processes (see for example, Strangeway and Crawford, 1995).

4. Sources of Low-Frequency Waves Upstream from the Bow Shock

Low-frequency electromagnetic wave phenomena observed by Phobos 2 and MGS outside of the Martian bow shock include waves near the local proton gyrofrequency (PCWs), waves in the range of whistler frequencies ('upstream whistlers') and strongly localized wave-like solar wind perturbations (e.g., 'Phobos-Deimos events'). The presence of these different wave types support the suggestion that the Martian interaction with the solar wind is unique among the other planets. Whereas upstream whistlers have been observed at many other solar system bodies (Mercury, Venus, Earth, Saturn), the presence of proton gyrofrequency waves, although observed under several circumstances in space plasmas, seems to be a special, permanent feature of the Martian environment. There is general consensus that the PCW's can be attributed to the pick-up process of protons from the extended hydrogen corona of Mars. The significant localized wave-like solar wind perturbations, detected in the neighborhood of Mars, are somewhat exceptional and originate likely from the two Martian moons Phobos and Deimos rather than from Mars itself. The generation mechanism of any of the above three wave phenomena is still under debate and several scenarios have been introduced in the discussion.

4.1. Upstream Whistlers

Upstream whistlers (or '1-Hz-waves') have long been known from Earth (Russell *et al.*, 1971; Fairfield, 1974) and other planets (Orlowski and Russell, 1995) but were first detected at Mars by MGS (Brain *et al.*, 2002). They are characterized by a propagation direction oblique to the magnetic field and by distinct polarization properties which tend to change from left-hand polarization to right-hand polarization with increasing propagation angle. The same tendency has been found with increasing heliocentric distances as a consequence of the varying spiral angle of the IMF (always left-handed at Mercury, about half the time left-handed at Earth, always right-handed at Saturn). The characteristics of upstream whistlers at Mars (frequency range 0.4–2.3 Hz) are consistent with observations at other planets: their frequency and amplitude decreases with increasing angle between propagation direction and solar wind flow direction and their polarization properties are as expected for the heliocentric distance of Mars. Although comprehensive observational material has been accumulated on upstream whistlers, their generation mechanism is still not well understood. It is commonly accepted that the shock or the shock ramp is the location of generation and that the waves propagate upstream due to their large group velocity, guided by the magnetic field. The similarities between the waves observed at different planets suggest that the source mechanism is not sensitive to details of size and shape of the shock and that similar plasma processes are at work to produce the waves.

4.2. Waves at the Proton Gyrofrequency

Waves at the proton gyrofrequency (PCWs) have been previously observed at comets (Tsurutani *et al.*, 1989; Mazelle and Neubauer, 1993), in the AMPTE experiment (Gleaves *et al.*, 1987) and in the high-altitude cusp (Le *et al.*, 2001), sometimes as monochromatic wave packets. At Mars PCWs were detected by both the Phobos 2 (Russell *et al.*, 1990) and the MGS spacecraft (Brain *et al.*, 2002). The MAG data from MGS, based on hundreds of orbits, provide comprehensive material to analyze the characteristics of the waves and to determine their spatial distribution with respect to Mars. The main features are: PCWs around Mars are always left-hand elliptically polarized and propagate at small to moderate angles to the ambient magnetic field ($\theta_{kB} \approx 20°$), in agreement with Phobos 2 observations, but have 2–3 times higher amplitude than the very low one (≈ 0.15 nT) observed by Phobos. The wave intensity varies spatially, is concentrated near the subsolar point and the flanks of the solar wind interaction, and decreases with altitude. In the following we discuss several potential generation mechanisms of PCWs.

4.2.1. *Ring-Beam Instabilities*

The PCWs observed at Mars are generally believed to result from the pick-up process of protons from the hydrogen corona of Mars which extends beyond the Martian bow shock (Russell *et al.*, 1990; Barabash *et al.*, 1991). Pick-up processes have also been favored to explain PCWs in the cometary environment (Mazelle and Neubauer, 1993). At a first look it seems likely that the same mechanisms are at work that have been extensively studied in the context of solar wind interaction with cometary atmospheres (Brinca, 1991; Tsurutani, 1991) to explain the origin of waves at the water group pick-up ion gyrofrequency. It is known from these studies that ring-beam distributions, established by heavy pick-up ions, can drive several solar wind plasma modes unstable, preferentially low-frequency MHD waves at the gyrofrequency of the pick-up ions.

According to the standard resonance condition $\omega - kv_{\parallel} \pm n\Omega_i = 0$ (ω wave frequency in the solar wind frame, $v_{\parallel}$ pick up ion drift velocity along the ambient field, Ω_i pick-up ion gyrofrequency) and in case of a super-Alfvénic drift velocity $v_{\parallel}$, two low-frequency MHD modes can resonantly interact with the pick-up population to produce amplification at Ω_i ($n = 1$). These are a right-hand polarized co-propagating (in direction of $v_{\parallel}$) fast mode and a counter-propagating left-hand polarized Alfvén wave (Price *et al.*, 1988). In both cases the pick-up ions sense the unstable wave electric field in their own frame as nearly coincident with their gyration motion, ensuring an intense wave-particle interaction. Due to an anomalous Doppler shift, the polarization of the right-hand fast mode is reversed to left-handed in the spacecraft ($\approx$ pick-up ion) frame. The fast mode resonant instability has largely been considered as the origin of low-frequency waves at the water group ion gyrofrequency observed around comets. In case of a sub-Alfvénic drift velocity $v_{\parallel}$ with a dominating ring character of the pick-up ion distribution, the left-hand polarized Alfvén/proton cyclotron mode turns to become the most interesting resonant mode. The above resonance mechanism, in spite of its attractiveness, is not the most likely candidate to explain the PCWs at Mars because analytical and numerical studies have in common shown that the presence of newly created ions in the solar wind support preferentially the amplification of parallel propagating MHD waves under most conditions of interest, inconsistent with the clearly observed obliquity of PCW propagation at Mars ($\theta_{kB} \approx 20°$). Thus, another possibilities to interpret the observations cannot be ruled out a priory.

4.2.2. *Ion-Whistler Instabilities*

It should be mentioned that electromagnetic ring beam and/or pure ion beam instabilities can also excite waves well above the proton cyclotron frequency in the whistler range of frequencies which may exhibit peak wave growth for oblique propagation. This mechanism may be relevant for PCW generation since the unstable whistler frequency in the solar wind frame is significantly down shifted in the spacecraft ($\approx$ pick-up ion) frame to a frequency near to the proton gyrofrequency. Ion-whistler instabilities have first pointed out in the early study of solar wind-comet interaction by Wu and Davidson (1972), were later analyzed by Goldstein and Wong (1987), and have been studied to explain upstream whistlers at Earth (Wong and Goldstein, 1988). More recently, the ion-whistler instability has been applied to interpret the so-called Phobos events (Baumgärtel *et al.*, 1998) and the generation of PCWs at the AMPTE Ba release (Sauer *et al.*, 1999). Since the generated waves are shorter wavelength whistlers, electron magnetization is essential rather than the gyration motion of the ion species involved, thus sometimes they are denoted as electron/ion whistler instabilities (Akimoto and Gary, 1987). These instabilities are believed to contribute little to the momentum coupling between pick-up ions and the solar wind but they may dominate over the relatively long wavelength resonant MHD modes in situations where the interaction region is spatially confined (smaller than the gyroradius of the pick-up ions), as, for example, for weak comets, in the AMPTE experiment or in the solar wind interaction with the Phobos neutral gas torus. The mechanism of the ion-whistler instability becomes apparent in an idealized picture in which one realizes that newborn ions, immediately after creation, tend to behave like a cold ion beam which is not in general aligned with the ambient magnetic field. The beam has super-Alfvénic velocity in general and thus can resonantly interact with the whistler branch of the solar wind plasma. In the approximation of an unmagnetized ion beam, the resonance can only be an '$n = 0$ resonance' $\omega - \mathbf{k}\mathbf{v}_b \approx 0$ in which the component of the beam velocity along the propagation direction meets the whistler phase velocity. In a dispersion diagram the instability thus arises near to the cross point of the whistler mode phase velocity with the beam velocity component along **k** where interchanges between whistler type modes and beam modes take place. This is illustrated in Figure 26(a) which depicts a solution of the fully kinetic electrodynamic dispersion relation for the solar wind plasma penetrated by a thin proton beam, demonstrating the very narrow unstable wave number region. Figure 26(b) displays in addition the growth rate, maximized over k, and other related quantities as function of the propagation angle θ relative to the ambient magnetic field. As seen from this figure, the relevant mode is stable for field-aligned propagation and acquires the highest growth rates for finite obliquity. In the observer frame (beam frame $\approx$ spacecraft frame) the waves are left-hand polarized and the frequency at peak growth is seen to vary around Ω_p, which appears typical for fairly thin and relatively cold beams. Finite electron temperature and a small but finite spread in the beam velocity combine to quench the unstable k's to a very narrow region resulting in the generation of an almost monochromatic wave. The generated waves thus appear to address all the constraints coming from the PCW observations. Along this way, even the oxygen exosphere of Mars may contribute to the generation of PCWs. Since their group velocity in the beam frame is almost zero, the waves do not tend to escape from the generation region and significant amplification may occur even in a spatially limited area. This property favors the above instability especially to account for the presence of Ω_p-fluctuations at the Phobos–Deimos events. One may argue against the ion-whistler instability as source of PCWs at Mars because of the assumption of an unmagnetized ion beam. This may be justified for heavy pick-up ions but is not a good approximation for pick-up protons since their gyroradius is not large compared to the spatial scale of the interaction region, which results in an increase of the velocity spread of the beam.

4.2.3. Ω_p-*Oscillations Seen in Bi-Ion Fluid Simulations*

An appropriate alternative mechanism for explaining PCWs at Mars has been found in bi-ion fluid simulations of mass-loading (Szegö *et al.*, 2000). The bi-ion fluid approach (Sauer *et al.*, 1994) uses a macroscopic concept which incorporates essential effects of the electromagnetic coupling between two ion fluids and the electrons through the Lorentz force and the condition of charge neutrality. The magnetic field is frozen into the electrons, so that Hall-current effects are included and differential streaming between the ion species automatically arises. The model anticipates the same local bulk velocity for all ions of each of the species which is not automatically satisfied in case of mass loading of a plasma flow, where, in general, freshly generated ions (at rest) coexist in the same region with pick-up ions which were born upstream at earlier time. However, in the two limit cases of a large or a small spatial extent of the mass-loading region, compared to the gyration radius of the secondary ions, the model can be expected to work satisfactorily. As an example, Figure 27 illustrates the results of one of the simulation runs. A proton-electron plasma (solar wind) with an Alfvén Mach number $M_{\mathrm{A}} = 5$ enters the simulation box in which a homogeneously distributed source injects protons at a

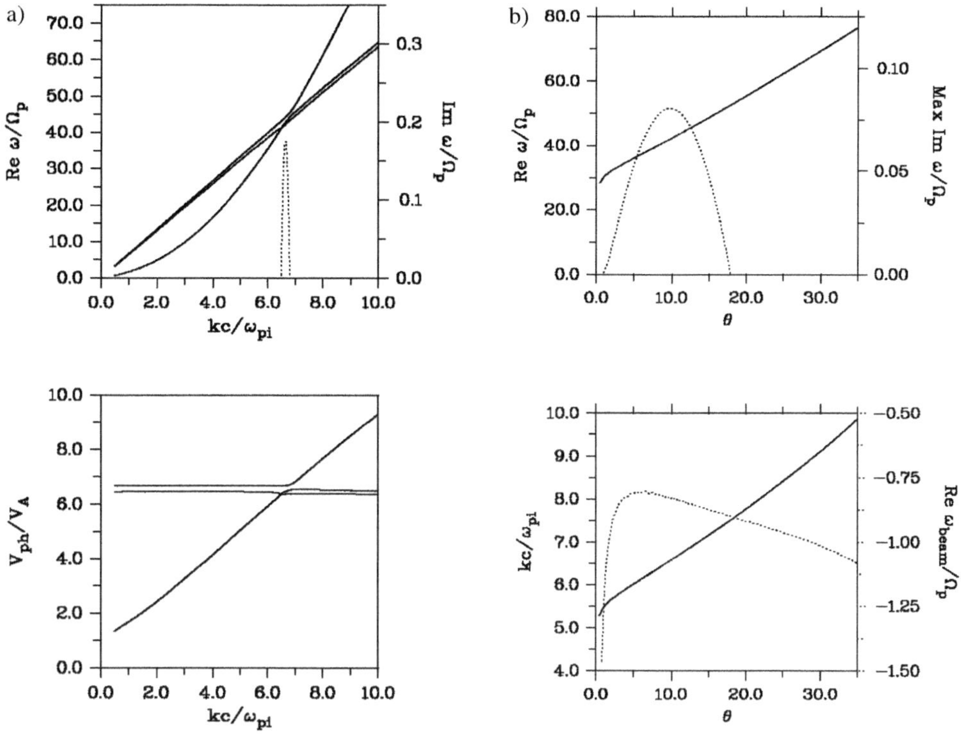

Figure 26. Warm plasma dispersion diagram for electromagnetic waves in the whistler range of frequencies in the solar wind plasma coexisting with a thin proton beam whose velocity vector $\mathbf{v}_b$ is 45° inclined to the IMF and whose density is 0.125% of the ambient density. The wave vector $\mathbf{k}$ is in the $v_b - B_0$-plane and makes an angle θ to the magnetic field direction. (a) Top: real part (solid) and imaginary part (dotted) of the frequency normalized to Ω_p versus wave number. Bottom: phase velocity normalized to the v_A versus wavenumber. The beam frame frequency varies within the unstable k region between $0.4\Omega_p$ and $1.1\Omega_p$. Parameters: $\theta = 10°$, $v_b = 8v_A$, $\beta_e = 0.5$, $\beta_i = 1$, $\beta_b = 0.01$. (b) Top: growth rate, maximized over k (dotted), and corresponding real frequency (solid) versus propagation angle θ relative to the magnetic field. Bottom: wave number and beam frame frequency for peak growth versus θ. Same parameters as in (a), except $\beta_b = 0.02$ (Baumgärtel *et al.*, 1998).

certain rate. After running the simulation a sufficiently long time, a quasi-stationary state is achieved which is shown in Figure 27. The most noticeable feature is seen in the spatial profiles of the magnetic field components B_y and B_z which indicate the presence of coherent waves. Their frequency can easily be calculated from the $x - t$-variation of B_z shown on top of Figure 27 and is found to be slightly below the proton cyclotron frequency ($\approx 0.9\Omega_p$) in the observer system. The relative velocity between the core protons and the pick-up protons is gradually reduced along the simulation box and adjusted to a value $\Delta_v \approx 2v_A$ in a self-consistent way by the electromagnetic coupling between the two ion species. It is suggested that the presence of the rather coherent Ω_p oscillations seen in the simulations can be traced back to peculiarities in the dispersion characteristics of bi-ion plasma. As

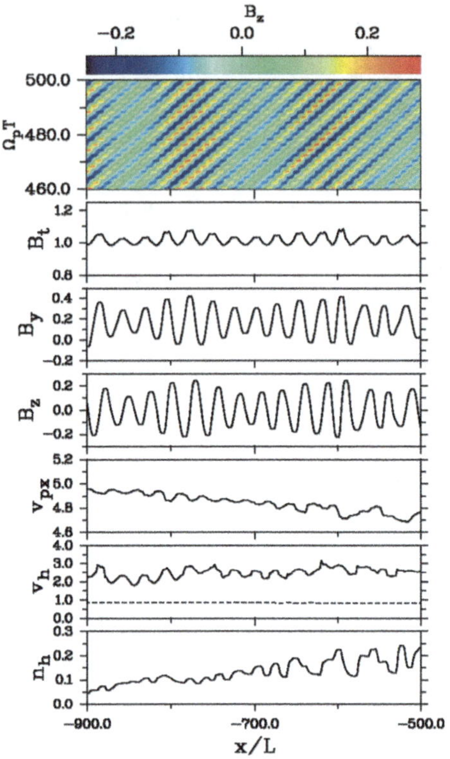

Figure 27. Bi-ion fluid simulation of solar wind mass loading by protons from an external source. The magnetic field is inclined to the x-axis by 10°. In the six panels (from bottom to top) the density and velocity of pick-up protons, the solar wind velocity, both components of the magnetic field (B_z, B_y) and the magnitude B_t are shown at $\Omega_p t = 500$. The diagram on top displays the $x - t$-variation of B_z which clearly indicates the presence of coherent waves with frequency close to the proton cyclotron frequency Ω_p (Sauer and Dubinin, 2001).

is well-known, the addition of a second ion population into proton-electron plasma strongly modifies the dispersion of low-frequency waves by producing a new cutoff frequency and the appearance of new modes. Modes that 'intersect' in the $U - k$ space ($U = \omega/k$, phase velocity) may be split which is usually accompanied by mode conversion and the appearance of two adjacent wavenumbers for which phase and group velocity coincide. The splitting may lead to 'gaps' in the dispersion diagram which are understood as phase velocity intervals which are not accessible for any of the small amplitude wave types for any real wave number. It is known from the Hall-MHD model of single-ion plasma that such 'gaps' exist even in absence of a second ion species and that they represent the domain of existence for MHD solitons (Baumgärtel, 1999). The 'gaps' are occupied by evanescent stationary waves ($\sim \exp(\pm ik(x - Ut)$, $k^2 < 0$) which, including nonlinear effects, give rise to periodic nonlinear wave structures, especially genuine solitons, as a result of the balance between nonlinear steepening and dispersive effects.

4.2.4. *Two-Ion Solitons ('Oscillitons')*
A mechanism that particularly accounts for the generation of coherent wave packets is provided by the capability of a two-ion plasma to allow for particular solitary wave structures. Whereas solitons or the elements of periodic nonlinear wave structures in single-ion plasma exhibit sech-type space profile, the presence of a second ion species may create a new type of solitary waves in which a shorter wavelength spatial oscillation is superimposed the conventional sech-type shape ('oscillitons') (Sauer *et al.*, 2001). To examine such solitary wave structures, one has to start with linear dispersion analysis to identify parameter regions (propagation velocity, angle of propagation, relative drift velocity, temperatures of the ion populations), which satisfy the 'gap' condition. Subsequently the system of ordinary differential equations need be solved, to which the basic system is reduced after introducing the stationary wave ansatz, i.e., the requirement that all variables depend on space and time only in the combination $x - Ut$. This procedure, however, does not address the question of stability of the solitary solution which has to be tested separately by numerical solutions of the fully nonlinear time- and space-dependent basic system. In two-ion plasma, 'gaps' may be found with a non-zero real part of the wave number, which is responsible for the periodic substructure of a solitary wave. In numerical studies on the space-time evolution of perturbations in proton plasma superimposed by a thin proton beam, 'oscillitons' appeared as a rather robust structure (Sauer *et al.*, 2001, 2002a, b). Figure 28 shows two examples of a stationary wave consisting of a train of solitary wave packets in two-ion plasma. An observer in the plasma frame sees a time variation due to the relative motion of the structure as a whole. The waves are intrinsically left-hand polarized, are found for oblique propagation relative to the ambient field, and the small-scale variation is determined by the real part of the complex wave number within the 'gap'. The coherent oscillations near Ω_p seen in the mass loading simulations are consistent with the presence of such structures. It should be noted here, that 'oscillitons' can exist also in unstable two-proton plasmas (different proton temperatures, relative drift, etc.). The philosophy of interpretation of Ω_p-oscillations near Mars in terms of 'oscillitons' is based on the combination of two circumstances: the presence of a potentially unstable two-ion plasma with relative drift or temperature anisotropy as free energy source and the occurrence of mode splitting producing a wavenumber for which phase and group velocity coincide. It is suggested, that in this peculiar reference frame the plasma tends to develop a stable stationary structure (like an 'oscilliton') as a result of a frequency selection process from a possibly weak broadband source. When the spatially periodic structure sweeps over an observer, a signal periodic in time is recorded. The numerical findings so far predict that for conditions realized near Mars only oblique propagation guarantees the presence of the required singular point $v_{ph} = v_{gr}$ in the dispersion diagram, further, that the frequency seen by the observer at rest is always near to Ω_p and that the waves are left-hand polarized. These properties qualify this mechanism as one of the favored candidates for the excitation of PCWs at Mars.

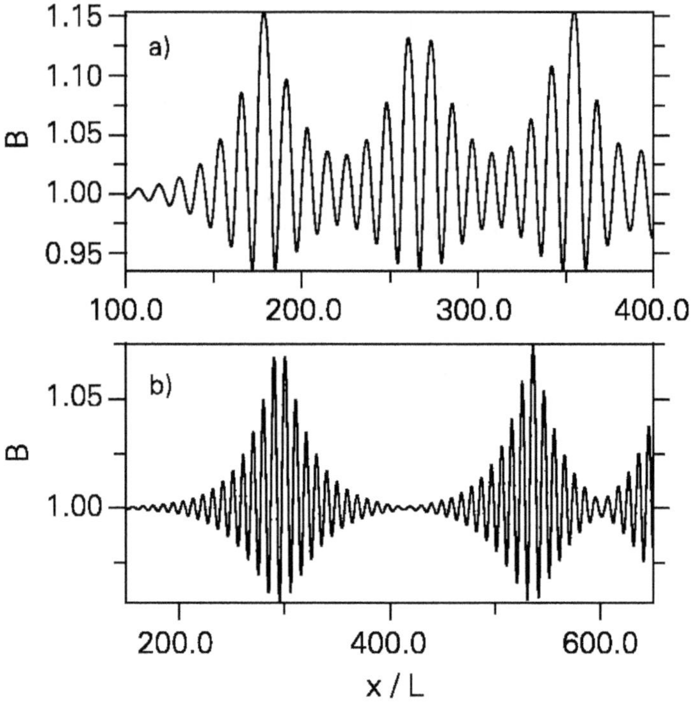

Figure 28. Two examples of bi-ion solitons with oscillating substructure ('oscillitons'). Top: cold proton plasma with an admixture of a cold heavier ($m_h = 3m_p$) minor ion population ($n_h = 0.25n_p$); Propagation velocity $U = 0.95v_A$, propagation angle θ relative to the magnetic field 30°. Bottom: Warm proton plasma with an admixture of a second minor proton population with different temperature; $U = 1.66$, $\theta = 25°$, $n_{p2} = 0.1n_{p1}$, $\beta_e = 0.5$, $\beta_{p1} = 1.0$, $\beta_{p2} = 5.0$ (Sauer *et al.*, 2001).

4.2.5. *Comparison Of The Theoretical Models*

The observed wave features of PCW's near Mars which must be addressed by any theoretical model for their interpretation include (1) frequency near to the local proton frequency, (2) left-hand polarization, (3) propagation at small to moderate angles to the ambient magnetic field and (4) spatial variation of wave intensity. There is general consensus that the PCW's can be attributed to the pick-up process of protons from the hydrogen corona of Mars. This is supported by the different amplitudes observed by Phobos and MGS which may be related to the different number densities of the neutral hydrogen atmosphere as a consequence of different cycles of solar activity. Phobos made measurements during solar maximum while MGS during solar minimum when the hydrogen density is almost two orders of magnitude higher. The conventional pick-up instabilities with peak growth at the gyrofrequency of the pick-up ions can likely be ruled out since they do not support the required generation of obliquely propagating waves. Both the high frequency whistler mechanism and the oscilliton model satisfy the requirements with respect to polarization and propagation direction, but they differ in the wavenumber of the waves thought to be responsible for the Ω_p-signal. Whereas the high-frequency pick-up instability involves short wavelength ($kv_{ph}/\Omega_p > 1$) whistler waves, the oscilliton scheme predicts much longer wavelengths $kv_{ph}/\Omega_p < 1$). Since no wave number estimate is provided by the observations, one cannot argue in terms of wavelength to discriminate between the two candidates. Also the question whether or not the PCW's are observed at their source cannot clearly be answered by the observations. The whistler instability predicts nearly group-standing waves in the observer frame which supports the idea of an observation in the source region; however, it is based on the assumption of unmagnetized beam ions which is not a good approximation for pick-up protons. The oscilliton model, on the other hand, does not predict group-standing waves in the observer frame but assumes the Ω_p-signal to be produced by a spatial periodic wave structure, stationary in a reference frame with $v_{ph} = v_{gr}$, which is convected with (in general) super-Alfvénic velocity over the observer. The problem as to what extent details of the observed spatial variation of the PCW features (altitude variation of the wave intensity, existence of two regions of high wave concentration near the subsolar point and at the flanks, with slightly different wave properties) may help to judge the theoretical models, needs further investigations.

5. Localized Disturbances in the Upstream Solar Wind

As a preliminary remark it should be mentioned that a controversially discussed issue is addressed here which is still not well understood. Actually, there is no commonly accepted physical mechanism for interpretation which covers all the events. In several of the observations, some evidence is suggested that the Martian moons Phobos or Deimos act as source for the perturbations, either the bodies itself through their miniatmoshere, produced by weak outgassing or dust ejection, or via a gas/dust ring along their orbits. The first interpretation of disturbances as the interaction of Deimos with the solar wind was made by Bogdanov (1981). On the other hand, during the events the IMF was always in an orientation such as to place the spacecraft in the ion foreshock, i.e., in a region in which one would expect reflected ions, indicating a close relation to the Martian bow shock. The latter is further supported by similarities in the structure of certain events with the properties of hot flow anomalies (HFAs, hot diamagnetic cavities) observed upstream of Earth's bow shock, which suggest that comparable mechanisms may be at work upstream of Mars to produce the events as counterpart of the terrestrial HFAs without involving the presence of the Martian moons.

5.1. PHOBOS/DEIMOS EVENTS AS FORESHOCK FEATURES

It was discussed in the previous section that the reflection of pickup ions at the Martian bow shock could significantly enhance the wave activity in the foreshock region. A simple 'test particle approach' based on a reasonable model of the Martian hydrogen corona and the photoion motion in the prescribed electric and magnetic fields with electrostatic barrier at the bow shock provided the distribution of the proton number density along the spacecraft trajectories at which the Phobos–Deimos events were observed. Significant enhancements of the plasma number density were found when the spacecraft occurred in the bunch of reflected from the bow shock exospheric protons. The position of this bunch is controlled by the interplanetary magnetic field and approximately coincides with the region occupied by the foreshock. Dubinin *et al.* (1995) suggested that 'overreflection' of the protons at the Martian bow shock due to a contribution of pickup ions may lead to a strong interaction between the solar wind and reflected protons and be responsible for the observed events. Figure 12 compares, for example, the model calculations and observations made on February 8, 1989 during the crossing of the hypothesized gas/dust Phobos torus. A reasonable agreement indicates that the appearance of the event may be related to such interaction. However it is important to note that the model is not self-consistent. The magnetic field configurations used for the calculation of ion trajectories were taken from the observations as the prescribed ones. On the other hand, the disturbed magnetic fields (the perturbations reach large amplitudes in the Phobos–Deimos events) may be caused by other sources, e.g., by the interplanetary disturbances, by ion currents of reflected particles and, at last, but not at least, by the solar wind – Phobos/Deimos/tori interaction.

5.2. Hot Diamagnetic Cavities (HDC)

It is known, for example, that when an interplanetary current sheet intersects the Earth's bow shock a structure termed a hot flow anomaly (HFA), or hot diamagnetic cavity, can form at the intersection point (Schwartz *et al.*, 1985; Thomsen *et al.*, 1986). Similar events with almost identical characteristics have been observed upstream of the Martian bow shock by MGS (Øieroset *et al.*, 2001). An example of HFAs observed upstream of the Martian bow shock is shown in Figure 29(a), which displays magnetometer (MAG) and electron reflectometer (ER) observations from 13:13–14:42 UT on July 22, 1998. MGS crossed the bow shock from the magnetosheath side and entered the undisturbed solar wind at 14:06 UT. Ten minutes later (at 14:16 UT) the MAG/ER instrument recorded a $\sim$2-min disturbance in both the energy flux (panel a) and the magnetic field magnitude and direction (panels d, e, and f). Fourteen minutes after the first disturbance ended a second interval (at 14:32 UT) with similar characteristics was recorded. Figure 29(b) shows the first disturbance in detail. The event is characterized by a turbulent magnetic field (panel d) and enhanced electron temperature (panel c), while the density (panel b) remains at the same level as the undisturbed solar wind. The interval of turbulent field is flanked by large magnetic field and electron density enhancements. The inner and outer boundaries of this edge region are marked by the red and the black dashed lines, respectively. Øieroset *et al.* (2001) concluded that hot flow anomalies

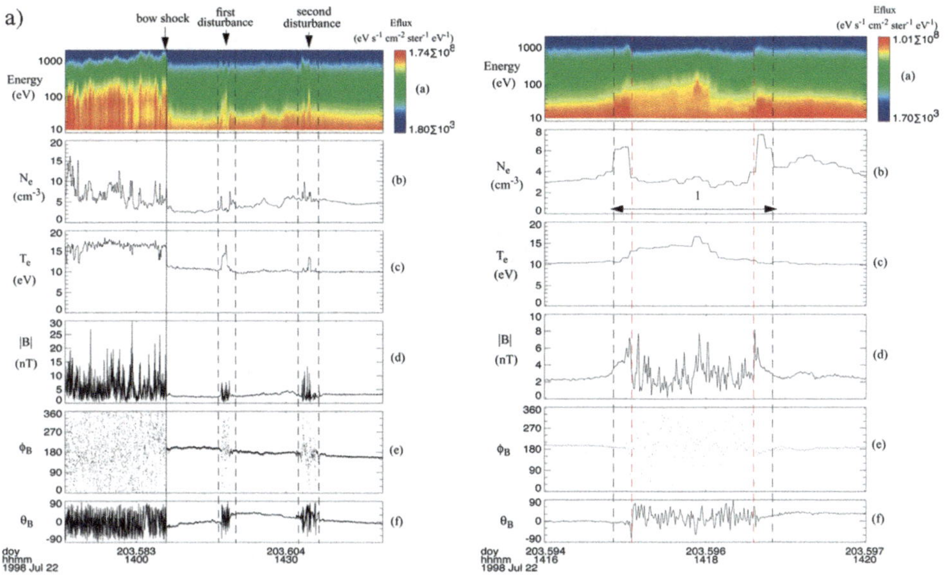

Figure 29. (a) MGS MAG/ER data from orbit (SPO) 443 on July 22, 1998. The electron energy spectrogram (energy flux) is shown in panel (a), while panel (b) and (c) show the electron density and temperature, respectively, both obtained by Maxwellian fitting of the electron spectra. The magnetic field observations are given in panels (d)–(f) as the magnetic field magnitude, the phi angle, and the theta angle. (b) Zoom-in of the first disturbance in (a). The hot central region is marked by the red dashed lines while the black dashed lines mark the outer boundaries of the edge regions. From Øieroset *et al.* (2001).

(HFAs) are a likely explanation for the events in Figure 29. They reached this conclusion based on the close similarity between the MGS events and HFAs observed upstream of the Earth's bow shock. For the Earth's case HFAs are believed to be caused by interplanetary current sheets interacting with the bow shock (Schwartz *et al.*, 1985, 2000; Thomas *et al.*, 1991). Øieroset *et al.* (2001) suggested that the same process could generate HFAs upstream of Mars. The presence of flow deflection, typical of terrestrial HFAs, cannot be verified for the MGS events because of the 2D nature of the MGS electron measurements.

A cartoon showing the generation of an HFA by an interplanetary current sheet interacting with the Martian bow shock is shown in Figure 30. A hot plasma region form outside the bow shock when ions initially reflected at the bow shock are channeled back along the current sheet by the motional electric field, which is directed toward the current sheet on both sides (Thomas *et al.*, 1991; Lin, 1997). The high temperature region then expands, which leads to a decrease in the density and magnetic field magnitude inside this region, while the incoming solar wind is deflected with weak shock waves.

Before reaching their conclusion Øieroset *et al.* (2001) considered several candidate causes for the disturbances described above: (1) multiple bow shock cross-

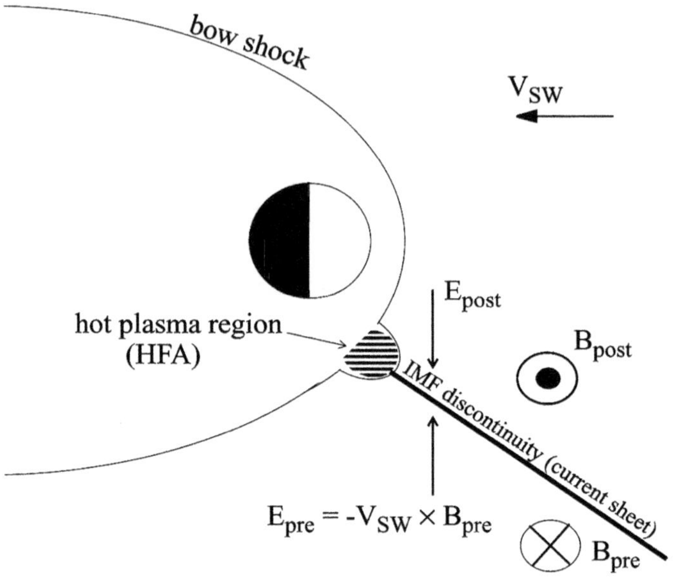

Figure 30. Cartoon showing the interaction of an interplanetary current sheet with the Martian bow shock. A hot flow anomaly (HFA) can form at the intersection point between the current sheet and the bow shock (figure adapted from Schwartz *et al.*, 2000).

ings, (2) foreshock turbulence, and (3) outgassing and/or dust escape from the Martian moons. Figure 31 shows the location of the disturbances displayed in Figure 29 along the MGS orbit, as well as the Phobos and Deimos orbits and the location of the moons at the time of the events. The MGS spacecraft was located downstream of Deimos at the time when the electron and magnetic field disturbances were recorded. During the first disturbance MGS was located 9 degrees off the Deimos-Sun line, and for the second disturbance the spacecraft was 13 degrees off. Solar wind disturbances observed by the Phobos-2 spacecraft have been suggested to be signatures of a Deimos Mach cone (Sauer *et al.*, 1995b). Although the geometry for the MGS events suggests a causal relationship with the Deimos location, Øieroset *et al.* (2001) argued that other features of the observations are not consistent with such an interpretation. In particular, in a Mach cone crossing one would not expect the temperature enhancement to be flanked by a layer of magnetic field magnitude and density enhancements. On the contrary, for a Mach cone crossing the density, temperature, and magnetic field magnitude are expected to increase simultaneously (e.g., Sauer *et al.*, 1995b). Øieroset *et al.* (2001) concluded that HFAs were a more likely explanation for the MGS events.

5.3. Models of Moon-Related Event Interpretation (MRE)

The discussion whether or not small celestial bodies such as magnetized asteroids, weakly outgassing planetary moons or clouds/trails of charged dust can interact

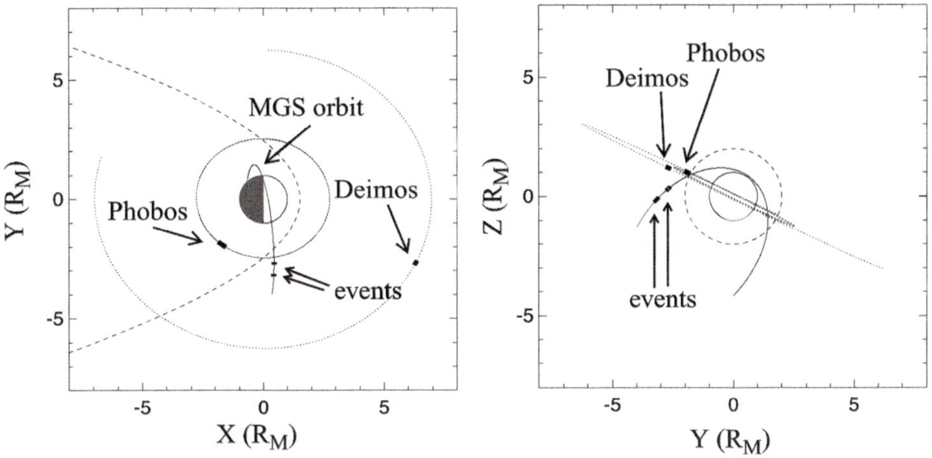

Figure 31. The MGS orbit in $x - y$ and $y - z$ sun synchronous ecliptic (SSE) coordinates (solid line). The two intervals of disturbances shown in Figure 1 are marked along the MGS orbit track. The orbits of Phobos and Deimos are shown with the dotted lines, and the positions of the moons at the time of the disturbances are also marked. In the SSE coordinate system X points from the center of the planet toward the Sun, the Z direction is normal to the ecliptic plane, pointing north, and Y is directed opposite to the planet's orbital velocity (Øieroset *et al.*, 2001).

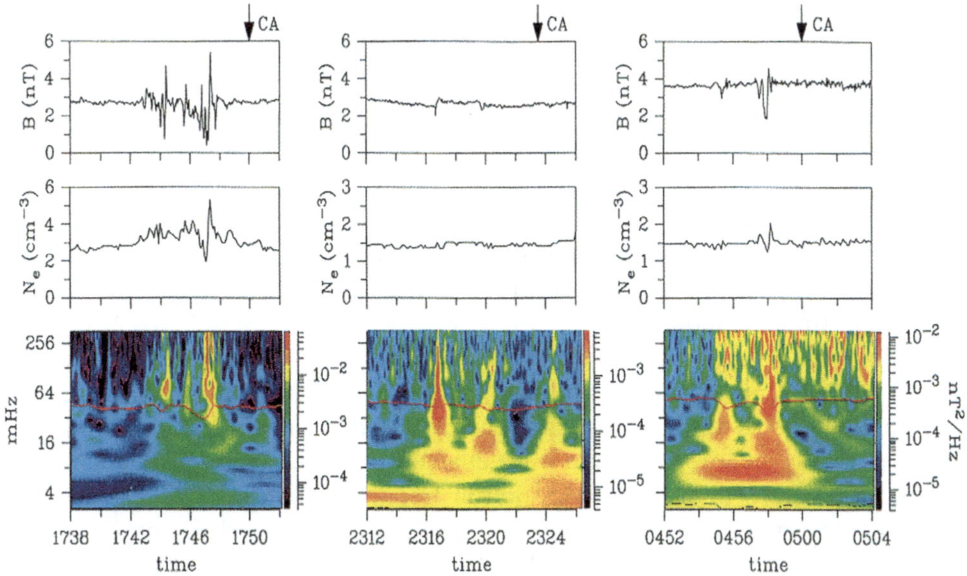

Figure 32. Phobos events observed by the Phobos-2 spacecraft during the first (a), the second (b) and the third (c) elliptical orbit (February 1, 4, 8, 1989) on just before crossing the Phobos orbit. The arrow indicates the time of closest approach. The bottom panel shows the low-frequency magnetic field power as a result of a wavelet analysis of the magnetic field turbulence. The proton cyclotron frequency is traced by a red line. (Baumgärtel *et al.*, 1998).

with the solar wind to produce observable effects, was initiated with a theoretical investigation of Greenstadt (1971) who predicted that a sufficiently magnetized asteroid may create interplanetary field perturbations that spacecraft magnetometers might detect. Observations by the Galileo satellite at its encounters with asteroids Gaspra and Ida and subsequent Hall-MHD simulations (Kivelson *et al.*, 1993; Baumgärtel *et al.*, 1994, 1997; Wang and Kivelson, 1996) supported an interpretation in terms of solar wind interaction with a magnetic dipole. The interest in this subject was considerably reinforced by remarkable Phobos-2 observations in 1989. During its first three elliptical orbits around Mars, the spacecraft observed significant, localized plasma and magnetic field perturbations upstream of the Martian bow shock. Several events were detected near to the crossing points of the spacecraft with the Phobos orbit whereas one of them occurred between the orbits of Phobos and Deimos ('mysterious' event). The events are seen as isolated perturbations in the magnitude of the interplanetary magnetic field, most pronounced at the edges of the perturbed region, accompanied by variations in the electron number density (Figure 32) in an otherwise relatively quiet solar wind. The events near the crossings of the Phobos orbit have been suggested as giving indirect evidence for the presence of either a gas ring or a dust ring along the Phobos orbit, and several theories have been worked out to explain these events in terms of solar wind interaction with a dust/gas torus (Dubinin *et al.*, 1990, 1991, 1993; Krymski *et al.*, 1992; Baumgärtel *et al.*, 1998). This interpretation fails for the 'mysterious' event, for which, however, a particular position of Deimos relative to the spacecraft was realized which favours an interpretation in terms of a Mach cone crossing. Numerical simulations of the solar wind interaction with a weakly outgassing small body (Sauer *et al.*, 1995b) confirmed the suggestion that the spacecraft crossed a fast wave Mach cone, with Deimos on top, about 15 000 km downstream of it, as the event was recorded. We can therefore distinguish between two types of moon-related events: those which have been observed near to the orbits of Phobos or Deimos and those which are not related to the orbit of any of the moons but might be attributed to the presence of a Mach cone-like stationary wave pattern originating from the moons.

Figure 32 depicts magnetic field and electron density disturbances observed by the Phobos-2 spacecraft during the first three elliptical orbits on just before crossing the Phobos orbit. Shown are records of the magnetic field amplitude, the electron density and the result of a wavelet analysis of the low-frequency magnetic field turbulence, with the proton cyclotron frequency indicated. A dust torus as well as a gas ring along the Phobos orbit has been proposed as source for the events. The existence of a Phobos dust torus was already postulated in 1971 by Sloter, but it has to date not been observed directly. The crucial parameter for any theory which ascribes the Phobos events to a dust torus is the magnitude of the total charge density established by the presence of charged dust grains of different size, because the solar wind responds to the dust only if its total charge density is not very small compared to the particle density of the solar wind protons. On the basis of a numerical simulation study on the dust ejecta from Phobos (Krivov and Hamilton, 1997), the total dust charge density was estimated to be about six orders of magnitude smaller than the solar wind proton density (Baumgärtel *et al.*, 1998) which suggests that the Phobos dust torus can very likely be ruled out as source of the events. On the other hand, the possible presence of a gas ring with neutral number density exceeding that of the hot atomic oxygen background was pointed out by Ip and Banaskiewicz (1990) if a certain level of outgassing is maintained by the Phobos moon. The presence of a heavy ion population, generated by ionization from the neutral gas halo, with enhanced density near to the Phobos orbit, may give rise to electromagnetic ion beam-whistler instabilities which have been discussed in the context of proton cyclotron wave generation outside of the Martian bow shock earlier in this review. Unlike the scenario of solar wind mass loading from the hydrogen atmosphere of Mars, we are here confronted with a rather localized interaction region, with a size of only a fraction of the gyroradius of an exospheric ion, which justifies much better the approximation of unmagnetized beam ions. On the other hand, the higher mass of the beam ions ($m_b = 16m_p$) requires a correspondingly higher beam density to reach growth rates comparable to those of a proton beam. According to a scenario for the neutral gas halo near Mars (Dubinin *et al.*, 1991; Krymski *et al.*, 1992) the oxygen ion density in the putative Phobos gas ring is of the order of 1% of the background proton density, as a result of a balance between ionization from the neutral gas torus and convective losses, on the basis of an outgassing rate of 10^{23} s^{-1} for Phobos. Warm plasma dispersion theory (Baumgärtel *et al.*, 1998; Sauer *et al.*, 1998a, 1999) predicts instability of the whistler mode supported by the solar wind plasma, for phase velocities near to the beam velocity projected on the whistler propagation direction. Peak growth is reached for oblique propagation ($\theta \approx 10°$) with real frequencies well above the proton cyclotron frequency and growth rates of the order of $0.1\Omega_p$. The wavenumber near peak growth is $kc/\Omega_p \approx 5$, corresponding to a wavelength $\lambda \approx 200$ km. In the beam ($\approx$ observer) frame the frequency of the most unstable wave is downshifted to a value near to the proton gyrofrequency, consistent with the observed enhanced wave activity near to Ω_p, seen in the bottom panel of Figure 32. The group velocity of the fastest growing wave in the beam frame is almost zero which stretches the convective growth time, allowing for the unstable mode to acquire measurable amplitudes even in a limited area of interaction.

There are several mechanisms that could explain how a small body could perturb the solar wind (Herbert, 1993) and give rise to the generation of a wake in the surrounding plasma: cometary outgassing, inductive interaction of a conducting body, remnant magnetization, or a charged dust sphere around the body. Among them the comet-like interaction appears to be the most likely mechanism for Phobos and Deimos to act as obstacle because both moons are thought to outgas water, and thus, via ionization, may be capable to maintain a thin heavy ion cloud around them. According to a hypothesis of Fanale and Salvail (1998) the Martian moons contained a significant amount of water in their interior and have not lost

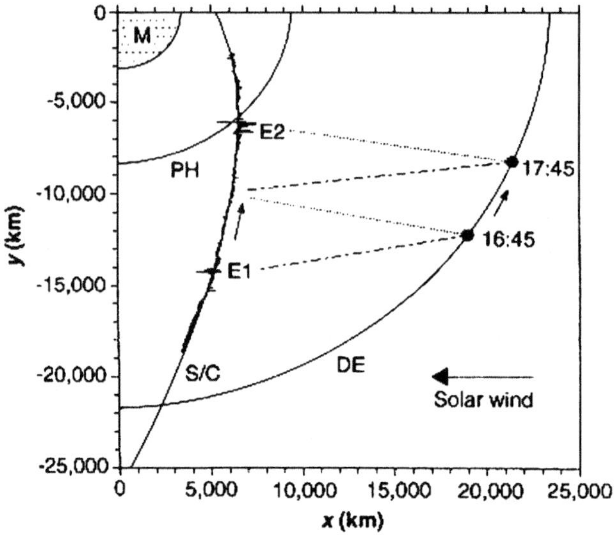

Figure 33. Location of the magnetic events E1 and E2 (February 1, 1989) in relation to the motion of Deimos. The spacecraft (S/C) trajectory and the orbits of the Martian moons Phobos (PH) and Deimos (DE) are plotted as solid lines. The curve along the spacecraft trajectory represents the relative deviation of the magnetic field amplitude from its mean value over three hours (resolution 10 s). The dotted and dashed-dotted lines illustrate that E1 and E2 may be related to spacecraft crossings of the left and right borders, respectively, of a Mach cone formed by the solar wind-Deimos interaction (Sauer *et al.*, 1995).

it over geological times. The present water flux may reach values of $\sim 10^{23}$ s^{-1}. It results in a gaseous envelope around the moons from which a plasma cloud is generated by ionization. It can be estimated that the above production rate is sufficient to maintain a heavy ion cloud with a peak density comparable with the number density of solar wind protons and a characteristic dimension of about twice the proton inertial length ($\approx$ 320 km) as balance between newly produced ions and those carried away by the solar wind. Unlike the solar wind interaction with 'large' bodies such as magnetized or nonmagnetized planets, where a recognizable cavity against the solar wind ram pressure is maintained, the dominant feature in the solar wind response to a 'small' obstacle is wave generation. Therefore it seems appropriate to employ a linear response model which describes the interaction in terms of the stationary pattern of perturbations originating from the small body in the solar wind stream. The simplest and well-known wave pattern is the Mach cone formed behind a small body moving through air with supersonic velocity. The dispersionless waves continuously generated by the moving point-like source constructively interfere with each other to form a conical wave crest with the body on top, whereas elsewhere they destructively interfere to almost zero disturbance. This well-understood process may appear different where the ambient medium is anisotropic, such as the solar wind, so that waves propagate at different speeds in different directions. In the conventional MHD description, there are three different modes, each of which propagates with different speed for different angles with respect to the magnetic field, and each produces distinct bow/wake waves. The wave crest associated with the fast magnetosonic wave still resembles the standard Mach cone in that it appears only if the body's velocity exceeds the fast sound speed, although its geometry is no longer axisymmetric with respect to the flow direction. Wake waves supported by the Alfvén wave, however, are generated by any source speed and have little similarity to a conventional Mach cone; the Alfvén wave crest degenerates to a one-dimensional entity, called 'Alfvén wing' or 'Alfvén line'. These features are relaxed somewhat if the small-amplitude waves exhibit not only anisotropy but also dispersion, i.e., if they propagate at different speeds for different wavelengths. This arises in the solar wind if the wavelengths become comparable to or less than the proton inertial length c/Ω_{pi}. In this case, the waves generated by the moving point source interfere to produce finite amplitude at more than a single wave crest surface. Thus the resulting stationary wave pattern may exhibit a rippled structure and may be thought of as a dispersed generalized Mach cone. The Mach cone philosophy has first been applied to the interpretation of events observed by Phobos-2 by Sauer *et al.* (1995b). Figure 33 illustrates the location of the 'mysterious' event (E1) and another event E2 near to the Phobos orbit (event (a) in Figure 32) in relation to the motion of Deimos. The spacecraft position relative to Deimos suggests that the events E1 and E2 might be the result of crossings of a fast wave Mach cone emanating from Deimos. The magnetosonic Mach number $v_{sw}/(v_{\mathrm{A}}^2 + v_S^2)^{1/2}$ (v_{A} is the Alfvén velocity, v_s is the acoustic velocity) was about 9, corresponding to an opening angle for the Mach cone of about 13°, with little deviation from axisymmetry with respect to the flow direction. Because the satellite moved faster than Deimos by about 1 km s^{-1}, it crossed the cone twice, inbound and outbound. Thereby the most pronounced event E2, which exhibits a rather complicated structure, appears to come about as a superposition of perturbations resulting from both the Phobos torus and the Deimos Mach cone. The diameter of the cone along the spacecraft orbit is about 4000 km and the spatial extent of the events is of the order of 300–500 km when considered as stationary entities. To account for nonlinear effects in the Mach cone evolution, simulations of the interaction have been carried out on the basis of the fully nonlinear MHD equations (Sauer *et al.*, 1995b). Figure 34 depicts the result of a run which describes the stationary solar wind wake perturbations behind an excess charge cloud represented by a cloud of immobile ions with a size of $\approx$ 300 km for a more realistic Mach number. The formation of Mach cones is clearly seen. In addition to the observations in the solar wind Dubinin *et al.* (1991) have found the distinct variations in the magnetic field and plasma when the spacecraft, being in the magnetosheath, occurred in the Phobos-moon tail. The draping features were well correlated to the spacecraft position in the moon frame. Although similar perturbations may be also caused by a dense detached 'cloud' of planetary plasma, the bi-ion MHD simulations confirm that the disturbances generated by the Mach cone behind the moon can propagate downstream the bow shock (Figure 35).

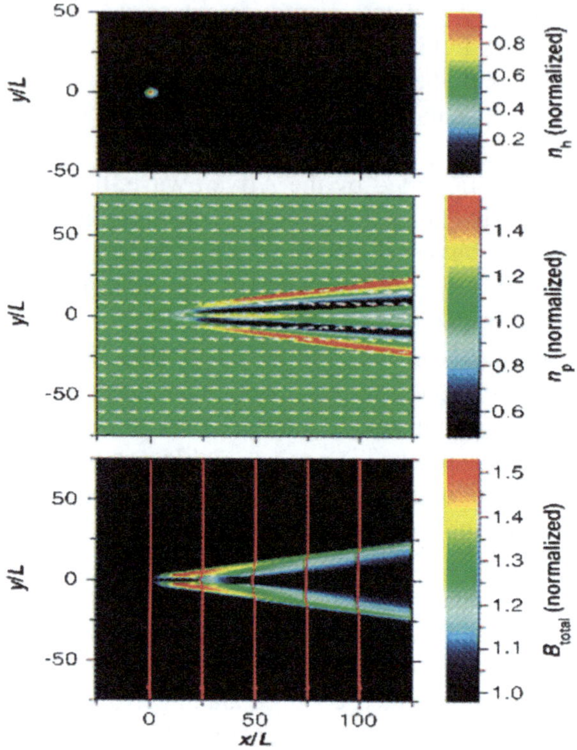

Figure 34. Solar wind interaction ($M_f = 9$, IMF perpendicular to the flow direction) with an immobile ion cloud of diameter ≈ 300 km calculated from the fully nonlinear MHD equations in 2-D approximation. Density and magnetic perturbations are seen to be concentrated along Mach cones. Arrows (middle panel) indicate the flow direction, red lines (bottom panel) correspond to IMF field lines (Sauer and Dubinin, 2001).

However, the direct relevance of the Mach cone pictures to the observed events is questionable, at least when only the moons itself and their plasma envelope is taken into consideration. Since the size of the Martian moons is of the order of 10 km, the expected effective obstacle cross section does hardly exceed the proton inertial length (in the solar wind of the same order as the thermal proton gyroradius). In this case, the wake produced by the obstacle is dominated by the generation of short wavelength whistler waves and results in a dispersed Mach cone without long-ranging wings and separated magnetic and density perturbations.

On the final stage of the Phobos-2 mission the spacecraft was settled on the quasi-synchronous with the Phobos–Moon orbit. The spacecraft periodically approached the moon to the distance of 200 km and went away to the distance of 400 km. The close encounters in the solar wind repeated every ∼ 8 hours. The magnetic field measurements with time resolution of 45 s were made on 11 encounters. Figure 36 shows data for 8 closest approaches (CAs) (two orbits were not included because of gaps in the data). The distinct feature of the most CAs is

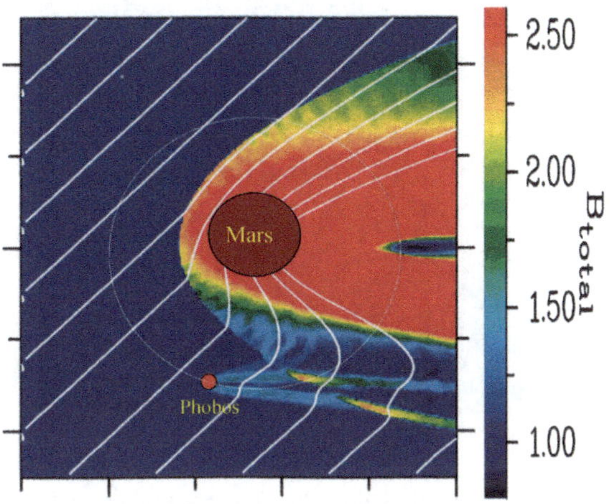

Figure 35. Solar wind interaction with the Phobos-moon and Mars calculated from the fully nonlinear MHD equations in 2-D approximation. Distinct magnetic field perturbations in the wake of the moon propagate downstream the Martian bow shock (Sauer and Dubinin, 2001).

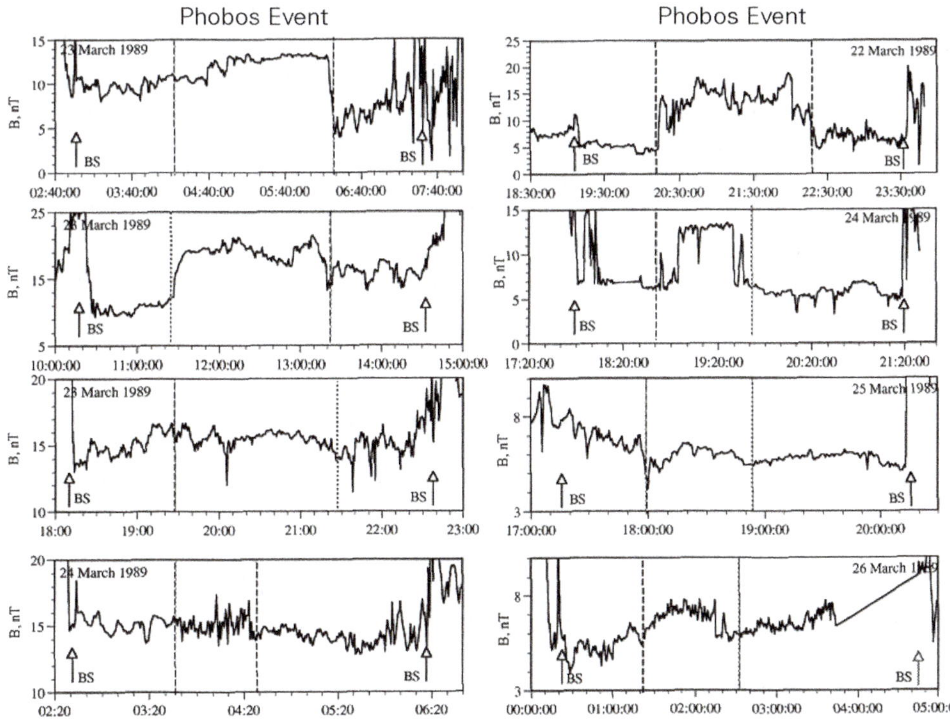

Figure 36. The set of the magnetic field measurements during the encounters with the Phobos moon on the final state of the Phobos-2 mission. Dotted curves show the intervals of the closest approaches when the distinct 'Phobos events' were observed (Sauer and Dubinin, 2001).

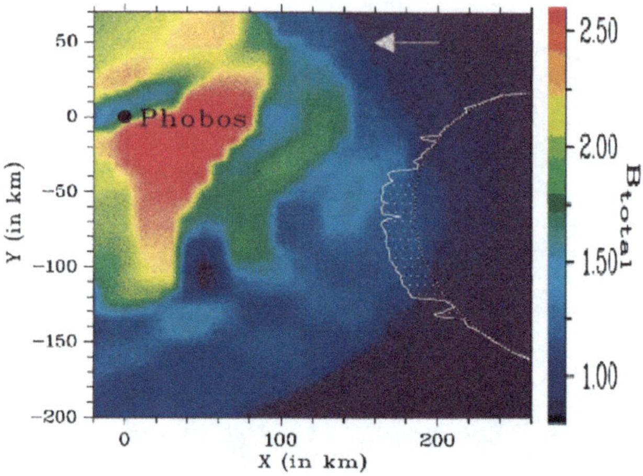

Figure 37. Composition of the magnetic field value in the neighborhood of the Phobos-moon from the bi-ion MHD simulations and the magnetic field strength along the spacecraft trajectory on the final state of the Phobos-2 mission. (Sauer and Dubinin, 2001).

a significant distortion of the magnetic field value. The perturbations range from shock-like enhancements to distinct strongly increased magnetic field fluctuations. Although some of the events were probably caused by the interplanetary shocks (e.g., at ~21 UT on 22 March or at ~12 UT on 23 March) or by outward motion of the bow shock (at ~19 UT on 24 March) the coincidence of all events with the closest approach is puzzling and therefore the interpretation in the terms of the moon events remains 'living'. Figure 37 shows the simulation results of the magnetic field disturbances in front of the outgassing moon. The white curve gives the magnetic field value along the spacecraft trajectory on March 24 in the moon reference frame. It is observed that the spacecraft could enter Phobos's 'minimagnetosphere'.

6. Conclusions and Open Questions

A few main points should be emphasised from the preceding sections. First, the bow shock at Mars is physically smaller than at all other solar system planets. This has implications for the plasma processes active at the shock and the amount of space available for the deceleration of the solar wind and for its collisionless heating via the wave-particle interaction processes in the downstream magnetosheath. At Mars, the thermalization of the ion populations is not expected to be complete in the dayside magnetosheath since the subsolar shock altitude is comparable to a typical solar wind proton gyroradius.

The Martian shock standoff distance is also larger relative to the size of the planet than at Venus. This was one argument often used in the past to infer the existence of a global dynamo magnetic field. We now know that Mars lacks such an internal magnetic field like Venus and present the same type of 'atmospheric' interaction with the solar wind as for active comets. This larger relative size of the shock comes mostly from the much more extended exosphere at Mars relative to the planet size, due to its weak gravity. This extension of the exosphere appears as the dimensioning parameter for the global plasma environment in such an interaction with the solar wind as it is for comets. This aspect is more widely discussed in a companion paper (Nagy *et al.*, this issue).

The Martian shock is also much more variable on a day-to-day basis than at Venus, as has been clearly revealed by the MGS observations. However its mean location and shape seems very steady and is surprisingly at least only very weakly dependent on the solar cycle variations contrary to what is observed for Venus. The influence of the pickup ion dynamics creates a reliable but small asymmetry for the bow shock location along the direction of the convection electric field. The location of the shock is also locally influenced by the presence of the remnant crustal sources but only at a very small level.

The major cause for the day-to-day variability likely lays in the physical parameters of the shock itself. The Alfvén Mach number M_{A} is above 10 for typical upstream solar wind parameters at Mars and the magnetosonic Mach number (more suitable for the properties of such a fast-mode shock) is $M_{Ms} \sim 6$, typically. With such values of M_{A} and M_{Ms}, the bow shock is extremely electromagnetically active. This is revealed both in the Phobos 2 and MGS observations. Moreover this is also a higher beta shock than at Venus. Another aspect comes from the possible effect of the curvature radius of the shock surface compared to the ion gyroradius. The major consequence is that the separation between quasi-parallel and quasi-perpendicular shocks is not so relevant than for the Earth or Venus, which may have strong implications on the local processes at the bow shock such as particle reflection and acceleration.

A great variety of upstream waves have been detected at Mars, providing valuable clues about the physics at this planet. Mars differ from Earth and from Venus in that upstream waves are caused not only by particles reflected at the bow shock or leaked upstream through the shock, but waves are also generated by interactions between the solar wind and the exosphere. The most striking observation to that respect is the existence of highly coherent waves with very large amplitude at the cyclotron proton frequency.

Several outstanding questions should be addressed in future analyses and observations. What processes account for the differences between the shocks at Venus and Mars? If mass-loading makes the Martian shock standoff distance slightly larger than for Venus, why is there not a significant variation in this altitude with the solar cycle as there is at Venus? Why is the observed Martian bow shock not as asymmetric as predicted from kinetic simulations? What more can we learn about the different particle populations (incident solar wind, reflected particles, shocked solar wind particles leaking from magnetosheath, exosphere) from upstream waves? About the observed strong local disturbances, how can we discriminate between the interpretations of interplanetary current sheet interaction with the shock and moon effects? Answer to these questions may come from continued analysis of existing data sets and theoretical investigations. Answers may also come from future observations made at Mars, such as those that will be made by Mars Express.

Acknowledgements

The authors thank the International Space Science Institute (ISSI), Bern, Switzerland, for providing support to the International Team 'Mars Magnetism and its Interaction with the Solar Wind'. This review was realised from the work of this team.

References

Acuña M. H., Connerney, J. E. P., Wasilewski, P., Lin, R. P., Anderson, K. A., Carlson, W., McFadden, C. J., Curtis, D. W., Rème, H., Cros, A., Medale, J. L., Sauvaud, J. A., d'Uston, C., Bauer, S. J., Cloutier, P., Mayhew, M. and Ness, N. F.: 1992, 'The Mars Observer Magnetic Fields Investigations', *J. Geophys. Res.* **97**, 7799–7814.

Acuña, M. H., Connerney, J. E. P., Wasilewski, P., Lin, R. P., Anderson, K. A., Carlson, C. W., McFadden, J., Curtis, D. W., Mitchell, D., Rème, H., Mazelle, C., Sauvaud, J. A., d'Uston, C., Cros, A., Médale, J. L., Bauer, S. J., Cloutier, P., Mayhew, M., Winterhalter, D. and Ness, N. F.: 1998, 'Magnetic Field and Plasma Observations at Mars: Initial Results of the Mars Global Surveyor Mission', *Science* **279**, 1676–1680.

Acuña, M. H., Connerney, J. E. P., Ness, N. F., Lin, R. P., Mitchell, D., Carlson, C. W., McFadden, J., Anderson, K. A., Rème, H., Mazelle, C., Vignes, D., Wasilewski, P. and Cloutier, P.: 1999, 'Global Distribution of Crustal Magnetization Discovered by the Mars Global Surveyor MAG/ER Experiment', *Science* **284**, 790–793.

Acuña, M. H., Connerney, J. E. P., Wasilewski, P., Lin, R. P., Mitchell, D., Anderson, K. A., Carlson, C. W., McFadden, J., Rème, H., Mazelle, C., Vignes, D., Bauer, S. J., Cloutier, P. and Ness, N. F.: 2001, 'Magnetic field of Mars: Summary of Results from the Aerobraking and Mapping Orbits', *J. Geophys. Res.* **106**(10), 23403–23417.

Akimoto, K. and Gary, S. P.: 1987, 'Electron/ion Whistler Instabilities and Magnetic Noise Bursts', *J. Geophys. Res.* **92**, 11209–11214.

Alexander, C. J. and Russell, C. T.: 1985, 'Solar Cycle Dependence of the Location of the Venus Bow Shock', *Geophys. Res. Lett.* **12**, 369–371.

Alexander, C. J., Luhmann J. G. and Russell C. T.: 1986, 'Interplanetary Field Control of the Location of the Venus Bow Shock: Evidence for Comet-like Ion Pickup', *Geophys. Res. Lett.* **13**, 917–920.

Axford, W. I.: 1991, 'A Commentary on Our Present Understanding of the Martian Magnetosphere', *Planetary Space Sci.* **39**, 167.

Barabash S. and Lundin, R.: 1993, 'Reflected Ions Near Mars, Phobos-2 Observations', *Geophys. Res. Lett.* **20**, 787.

Barabash, S., Dubinin, E., Pissarenko, N., Lundin, R. and Russell, C. T.: 1991, 'Picked-up Protons near Mars: Phobos Observations', *Geophys. Res. Lett.* **18**, 1805–1808.

Barabash, S., Lundin, R. and Norberg, O.: 1993, '*Upstream Regions at Mars', Plasma Environment of Non-Magnetic Planets, COSPAR colloq. Ser.* **4**, in T. I. Combosi, (ed.), 285, Pergamon, New York.

Baumgärtel, K.: 1999, 'Soliton Approach to Magnetic Holes', *J. Geophys. Res.* **104**, 28295–28308.

Baumgärtel, K., Sauer, K. and Bogdanov, A.: 1994, 'A Magnetohydrodynamic Model Solar Wind Interaction with Asteroid Gaspra', *Science*, **263**, 653–655.

Baumgärtel, K., Sauer, K., Story, T. R. and McKenzie, J. F.: 1997, 'Solar Wind Response to a Magnetized Asteroid: Linear Theory', *Icarus*, **129**, 94–105.

Baumgärtel, K., Sauer, K., Dubinin, E., Tarasov, V. and Dougherty, M.: 1998, "Phobos events" – Signatures of Solar Wind Interaction with a Gas Torus?', *Earth, Planets and Space* **50**, 453–462.

Bertucci, C., Mazelle, C., Crider, D. H., Vignes, D., Acuña, M. H., Mitchell, D. L., Lin, R. P., Connerney, J. E. P., Rème, H., Cloutier, P. A., Ness, N. F. and Winterhalter, D.: 2003a, 'Magnetic Field Draping Enhancement at the Martian Magnetic Pileup Boundary from Mars Global Surveyor Observations', *Geophys. Res. Lett.* **30**(2), 1099.

Bertucci, C., Mazelle, C., Slavin, J. A., Russell, C. T. and Acuña, M. H.: 2003b, 'Magnetic Field Draping Enhancement at Venus: Evidence for a Magnetic Pileup Boundary', *Geophys. Res. Lett.* **30**(17), 1876.

Bogdanov, A. V.: 1981, 'Mars Satellite Deimos Interaction with the Solar Wind and its Influence on Flow Around Mars', *J. Geophys. Res*, **86**, 6926.

Bogdanov, A. V., Sauer, K., Baumgärtel, K. and Shrivastava, K.: 1996, 'Plasma Structures at Weakly Outgassing Comets – Results from Bi-ion fluid Analysis', *Planetary Space Sci.* **44**, 519–528.

Bogdanov, T. and Motschmann, U.: 2002, 'From a weak to a strong comet-3D global hybrid simulation results', *Earth, Moon and Planets*, **90**, 305–321, 2002.

Bonifazi C., Moreno, G., Lazarus, A. J. and Sullivan, J. D.: 1980, 'Deceleration of the Solar Wind in the Earth's Foreshock Region: ISSE-2 and IMP-8 Observations', *J. Geophys. Res.* **85**, 6031.

Brain, D. A., Bagenal, F., Acuña, M. H., Connery, J. E. P., Crider, D. H., Mazelle, C., Mitchell, D. L. and Ness, N. F.: 2002, 'Observations of Low Frequency Electromagnetic Plasma Waves Upstream from the Martian shock', *J. Geophys. Res.* **107**(A6), pp. 148–227 (10.1029/2000JA000416).

Brain, D. A., Bagenal, F., Acuña, M. H. and Connery, J. E. P.: 2003, 'Martian Magnetic Morphology: Contribution of Solar Wind and Crust', *J. Geophys. Res.* (in press).

Brecht, S. H.: 1990, 'Magnetic Asymmetries of Unmagnetized Planets', *Geophys. Res. Lett.* **17**, 1243 –1246.

Brecht S. H.: 1997, 'Hybrid Simulations of the Magnetic Topology of Mars', *J. Geophys. Res.* **102**, 4743–4750.

Brecht, S. H.: 2002, 'Numerical Techniques Associated with Simulations of Solar Wind Interactions with Non-Magnetized Bodies', *Comparative Aeronomy in the Solar System*, AGU Press, New York (in press).

Brecht, S. H. and Ferrante, J. R.: 1991, 'Global Hybrid Simulation of Unmagnetized Planets: Comparison of Venus and Mars', *J. Geophys. Res.* **96**, 11209–11220.

Brecht, S. H., Ferrante, J. R. and Luhmann, J. G.: 1993, 'Three-Dimensional Simulations of the Solar Wind Interaction with Mars', *J. Geophys. Res.* **98**, 1345–1357.

Brinca, A.L.: 1991, 'Cometary Linear Instabilities: from Profusion to Prospective', *Cometary Plasma Processes*, in A. D. Johnstone (ed.), pp. 211–222, *Geophysical Monograph* **61**, 211–222.

Cloutier, P. A., Chen, Y., Crider, D. H., Law, C. C., Walker, P. W., Acuña, M. H., Connerney, J. E. P., Lin, R. P., Anderson, K. A., Mitchell, D. L., Reme, H., Mazelle, C., Sauvaud, J. A., d'Uston, C., Vignes, D., Bauer, S. J. and Ness, N. F.: 1999, Venus-like Interaction of the Solar Wind with Mars', *Geophys. Res. Lett.* **26**, 2685–2688.

Connerney, J. E. P., Acuña, M. H., Wasilewski, P. J., Ness, N. F., Rème, H., Mazelle, C., Vignes, D., Lin, R. P., Mitchell, D. L., Cloutier, P. A.,: 1999, 'Magnetic Lineation in the Ancien Crust of Mars', *Science* **284**, 794–798.

Cravens, T. E., Shinagawa, H. and Luhmann, J. G.: 1997, in Stephen W. Bougher, D. M. Hunten and R. J. Philips (eds.), *'Magnetohydrodynamic Processes: Magnetic Fields in the Ionosphere of*

Venus, Venus II : Geology, Geophysics, Atmosphere, and Solar Wind Environment. University of Arizona Press, Tucson, AZ, p. 61.

Crider, D. H.: 1999, 'Evidence of Electron Impact Ionization in the Magnetic Pileup Boundary of Mars – Observations and modeling results, Ph.D. Thesis', Rice University, 109 pp.

Delva, M. and Dubinin, E.: 1998, 'Upstream ULF Fluctuations Near Mars', *J. Geophys. Res.* **103**, 317.

Dubinin, E. M.: 1993, 'The Phobos and Deimos Effects', *Adv. Space Res.* **13**, 271–130.

Dubinin, E. and Sauer, K.: 1999, 'The Martian Magnetosphere – Alaboratory for Bi-ion Plasma Investigations', *Astrophys. Space Sci.* **264**, 273.

Dubinin E., Lundin, R., Koskinen, H. and Norberg, O.: 1993, 'Cold Ions at the Martian Bow Shock Phobos Observations', *J. Geophys. Res.* **98**, 5617.

Dubinin, E. M., Lundin, R., Pissarenko, N. F., Barabash, S. V., Zakharov, A. V., Koskinen, H., Schwingenschuh, K. and Yeroshenko, Y. G.: 1990, 'Indirect Evidence for a Gas/Dust Torus Along the Phobos Orbit', *Geophys. Res. Lett.* **17**, 861–864.

Dubinin, E. M., Pissarenko, N. F., Barabash, S. V., Zakharov, A. V., Lundin, R., Koskinen, H., Schwingenschuh, K. and Yeroshenko, Y. G.: 1991, 'Tails of Phobos and Deimos in the Solar Wind and in the Martian Magnetosphere', *Planetary Space Sci.* **39**, 123–130.

Dubinin E. M., Obod, D., Pedersen, A. and Grard, R.: 1994, 'Mass-Loading Asymmetry in Upstream Region near Mars', *Geophys. Res. Lett.* **21**, 2769.

Dubinin E. M., Obod, D., Lundin, R., Schwingenschuh, K. and Grard, R.: 1995, 'Some Features of the Martian Bow Shock', *Adv. Space Res.* **15**(8/9), 423.

Dubinin, E., Sauer, K., Baumgärtel, K. and Srivastava, K.: 1998, 'Multiple Shocks at Mars', *Earth Planets Space* **50**, 279–287.

Dubinin E. M., Sauer, K., Delva, M. *et al.*: 2000a, 'Multi-instrument Study of the Upstream Region near Mars: The Phobos-2 Observations', *J. Geophys. Res.* **105**, 7557.

Dubinin E., Sauer, K. *et al.*: 2000b, 'Deceleration of the Solar Wind Upstream of the Martian Bow Shock. Mass-loading or Foreshock Features?', *Adv. Space Res.* **26**(10), 1627.

Fanale, F. P. and Salvail, J. R.: 1989, 'Loss of Water from Phobos', *Geophys. Res. Lett.* **16**, 287–290.

Fairfield, D. H.: 1974, 'Whistler Waves Observed Upstream from Collisionless Shocks', *J. Geophys. Res.* **79**(10), 1368–1378.

Gleaves, D. G., Southwood, D. J., Dunlop, M. and Mier-Jedrzejowicz, W.: 1987, 'Low-Frequency Magnetic Wave Spectra Associated with the AMPTE Solar Wind Barium Release of 27 December 1984', *Planetary Space Sci.* **35**, 493–500.

Goldstein, M. L. and Wong, H. K.: 1987, 'A Theory for Low-Frequency Waves Observed at Comet Giacobini-Zinner', *J. Geophys. Res.* **92**, 4695–4700.

Grard, R., Klinge, D., Klimov, S., Savin, S. and Trotignon, J. G.: 1989, 'The Plasma Wave System on Phobos', *J. Phys. E Sci. Instrum.* **22**, 888–894.

Grard, R., Nairn, C., Pedersen, A., Klimov, S., Savin, S., Skalsky, A. and Trotignon, J. G.: 1991, 'Plasma and Waves around Mars', *Planetary Space Sci.* **39**, 89–98.

Grard, R., Skalsky, S. and Trotignon, J. G.: 1993, in T. I. Combosi (ed.), 'Selected Wave and Plasma Features of the Martian Environment', in Plasma Environment of non-Magnetic Planets, *COSPAR Colloq. Ser.* **4**, 321–327.

Greenstadt, E. W.: 1971, 'Conditions for Magnetic Interaction of Asteroids with the Solar Wind, *Icarus* **14**, 374–381.

Greenstadt, E. W., Le, G. and Strangeway, R. J.: 1995, 'ULF Waves in the Foreshock', *Adv. Space Res.* **15**(8–9), 71–84.

Gringauz, K. I., Kurt, V. G., Moroz, V. I. and Shklovskii, I. S.: 1981, 'Results of Observations of Charged Particles Observed Out to $R = 100\,000$ km, with the Aid of Charged-Particle Traps on Soviet Space Rockets', *Soviet Astron.* **4**, 680.

Herbert, F.: 1993, 'Solar Wind Interaction with Asteroids', *Adv. Space Res.* **13**(10), 249–258.

Hopcroft, M. W. and Chapman, S. C., '2D Hybrid Simulations of the Solar Wind Interaction with a Small Scale Comet in High Mach Number Flows', *Geophys. Res. Lett.* **28**(6), 1115.

Ip, W.-H.: 1988, 'On Hot Oxygen Corona at Mars', *Icarus* **76**, 135.

Ip, W.-H.: 1990, 'The Fast Atomic Oxygen Corona Extent of Mars', *Geophys. Res. Lett.* **17**, 2289–2292.

Ip, W.-H. and Banaszkiwicz, M.: 1990, 'On the Dust/Gas Tori of Phobos and Deimos', *Geophys. Res. Lett.* **17**, 857.

Ip, W.-H.: 1992, 'Neutral Particle Environment of Mars: The Exosphere-Plasma Interaction Effects', *Adv. Space Res.* **12**(9), 205.

Kim, J., Nagy, A. F., Fox, J. L. and Cravens, T. E.: 1998, 'Solar Cycle Variability of Hot Oxygen Corona at Mars', *J. Geophys. Res.* **103**, 29339.

Kivelson, M. G., Bargatze, L. F., Khurana, K. K., Southwood, D. J., Walker, R. J., Coleman P. J. jr.: 1993, 'Magnetic Field Signatures Near Galileo's Closest Approach to Gaspra', *Science* **261**, 331 –334.

Kotova, G. *et al.*: 1997, 'Study of the Solar Wind Deceleration Upstream of the Martian Terminator Bow Shock', *J. Geophys. Res.* **102**, 2165.

Krivov, V. A. and Hamilton, D. B.: 1997, 'Martian Dust Belts: Waiting for Discovery', *Icarus* **128**, 335–353.

Krymski, A. M., Breus, T. K., Dougherty, M. K., Southwood, D. J. and Axford, W. I.: 1992, 'The Electromagnetic Effects of the Solar Wind Interaction With the Phobos Neutral Gas Halo and Dust Torus', *Planetary Space Sci.* **40**, 1033–1041.

Le, G., Blanco-Cano, X., Russell, C. T., Zhou, X.-W., Mozer, F., Trattner, K. J., Fuselier, S. A. and Anderson, B. J.: 2001, 'Electromagnetic Ion Cyclotron Waves in the High-Altitude Cusp: Polar Observations', *J. Geophys. Res.* **106**, 19067–19079.

Lin, Y.: 1997, 'Generation of Anomalous Flows Near the Bow Shock by its Interaction with Interplanetary Discontinuities', *J. Geophys. Res.* **102**, 24265.

Lipatov, A., Sauer, K. and Baumgärtel, K.: 1997, '2.5D Hybrid Code Simulation of the Solar Wind Interaction with Weak Comets and Related Objects', *Planetary Space Sci.* **20**, 279–282.

Lipatov, A. S., Motschmann, U. and Bagdonat, T.: 2002, '3D Hybrid Simulations of the Interaction of the Solar Wind with a Weak Comet', *Planetary Space Sci* **50**(4), 403–411.

Ma Y., Nagy, A. F., Hansen, K. C., DeZeeuw D. L. and Gombosi, T. I.: 2002, 'Three-Dimensional Multispecies MHD Studies of the Solar Wind Interaction with Mars in the Presence of Crustal Fields', *J. Geophys. Res.* **107**(A10), 1282.

Mazelle, C. and Neubauer, F. M.: 1993, 'Discrete Wave Packets at the Proton Cyclotron Frequency at Comet P/Halley', *Geophys. Res. Lett.* **20**, 153–156.

Mazelle, C., Rème, H., Vignes, D., Acuña, M. H., Brain, D., Connerney, J. E. P., Crider, D. H., Lin, R. P., Mitchell, D. L., Cloutier, P. A., Winterhalter, D. and Ness, N. F.: 2000, 'Nonlinear Low Frequency Waves Observations by Mars Global Surveyor', *Eos Trans. AGU* **81**(48), Fall Meeting Suppl.

Mazelle, C., Bertucci, C., Sauer, K., Brain, D. A., Mitchell, D. L., Rème, H. and Acuña, M. H.: 2001, 'Properties of Upstream Waves at the Proton Cyclotron Frequency at Mars from MGS Observations', *Eos Trans. AGU* **82**(47), Fall Meet. Suppl.

Mitchell, D. L., Lin, R. P., Mazelle, C., Rème, H., Cloutier, P. A., Connerney, J. E. P., Acuña, M. H. and Ness, N. F.: 2001, 'Probing Mars'Crustal Magnetic Field and Ionosphere with the MGS Electron Reflectometer', *J. Geophys. Res.* **106**(E10), 23419–23428.

Mohlmann, D., Riedler, W., Rustenbach, J., Schwingenschuh, K., Kurths, J., Motschmann, U., Roatsch, T., Sauer, K. and Lichtenegger, H. T. M.: 1991, 'The Question of an Internal Martian Magnetic Field, *Planetary Space Sci.* **39**, 83–88.

Moore, K. R., Thomas, V. A. and McComas, D. J.: 1991, 'Global Hybrid Simulation of the Solar Wind Interaction with Venus', *J. Geophys. Res.* **96**, 7779–7791.

Moses S. L., Coroniti, F. V. and Scarf, F. L.: 1988, 'Expectations for the Microphysics of the Mars-Solar Wind Interaction', *Geophys. Res. Lett.* **15**, 429–432.

Øieroset, M., Mitchell, D. L., Phan, T. D., Lin, R. P. and Acuña, M. H.: 2001, 'Hot Diamagnetic Cavities Upstream of the Martian Bow Shock', *Geophys. Res. Lett.* **28**, 887.

Orlowski, D. S., Crawford, G. K. and Russell, C. T.: 1990, 'Upstream Waves at Mercury, Venus and Earth: Comparison of the Properties of One Hertz Waves', *Geophys. Res. Lett.* **17**(12), 2293–2296.

Orlowski, D. S., Russell, C. T., and Lepping, R. P.: 1992, 'Wave Phenomena in the Upstream Region of Saturn', *J. Geophys. Res.* **97(A12)**, 19187–19199.

Orlowski, D. S. and Russell, C. T.: 1995, 'Comparison of Properties of Upstream Whistlers at Different Planets', *Adv. Space Res.* **16**(4), 137–141.

Pedersen, A., Nairn, C., Grard, R. and Schwingenschuh, K.: 1991, 'Deviation of Electron Densities from Differential Potential Measurements Upstream and Downstream of the Bow Shock and in the Magnetosphere of Mars', *J. Geophys. Res.* **96**, 11243.

Price, C. P., Gaffey, J. D. and Dong, J. Q.: 1988, 'Excitations of Low-Frequency Hydromagnetic Waves by Freshly Created Ions in the Solar Wind', *J. Geophys. Res.* **93**, 837–844.

Riedler W., *et al.*: 1989, 'Magnetic Fields Near Mars: First Results', *Nature* **341**, 604.

Russell, C. T., Childers, D. D. and Coleman, P. J.: 1971, 'Ogo5 Observations of Upstream Waves in the Interplanetary Medium: Discrete Wave Packets', *J. Geophys. Res.* **76**, 845–861.

Russell, C. T.: 1985, in B. Tsurutani and G. Stone (eds.), 'Planetary Bow shocks', Collisionless Shocks in the Heliosphere: Reviews of Current Research, *Geophys. Monograph* **35**, 109, AGU.

Russell, C. T., Chou, E., Luhmann, J. G., Gazis, P., Brace, L. H. and Hoegy, W. R.: 1988, 'Solar and Interplanetary Control of the Location of the Venus Bow Shock', *J. Geophys. Res* **93**, 5461–5469.

Russell, C. T., Luhmann, J. G., Schwingenschuh, K., Riedler, W. and Yeroshenko, Y.: 1990, 'Upstream Waves at Mars: Phobos Observations', *Geophys. Res. Lett.* **17**, 897–900.

Sagdeev, R. Z., Shapiro, V. D., Shevchenko, V. I., Zacharov, A., Kiraly, P., Szego, K., Nagy, A. F. and Grard, R.: 1990, 'Wave Activity in the Neighborhood of the Bow Shock of Mars', *Geophys. Res. Lett.* **17**, 893–896.

Sauer, K. and Dubinin, E.: 2000, 'The Nature of the Martian 'Obstacle Boundary', *Adv. Space Res.* **26**, 1633–1637.

Sauer, K., Bogdanov, A. and Baumgärtel, K.: 1994, 'Evidence of an Ion Composition Boundary (Protonopause) in Bi-ion Fluid Simulations of Solar Wind Mass Loading', *Geophys. Res. Lett.* **21**, 2255–2258.

Sauer, K., Bogdanov, A. and Baumgärtel, K.: 1995a, 'The Protonopause – an Ion Composition Coundary in the Magnetosheath of Comets, Venus and Mars', *Adv. Space Res.* **16**, 153–156.

Sauer, K., Dubinin, E., Baumgärtel, K., Bogdanov, A.: 1995b, 'Deimos: An obstacle to the Solar Wind', *Science* **269**, 1075.

Sauer, K., Dubinin, E. and Baumgärtel, K.: 1997, 'Bi-ion Structuring in the Magnetosheath of Mars. Theoretical Modeling'. *Adv. Space Res.* **20**, 137–140.

Sauer K., Baumgärtel, K., Dubinin, E. and Tarasov, V.: 1998a, 'Low-Frequency Electromagnetic Waves and Instabilities within the Martian Bi-ion Plasma', *Earth, Planets Space* **50**, 269–278.

Sauer, K., Dubinin, E. and Baumgärtel, K.: 1998b, 'Nonlinear MHD Waves and Discontinuities in the Martian Magnetosheath. Observations and 2D Bi-ion MHD Simulations', *Earth Planets Space*, **50**, 793–801.

Sauer, K., Dubinin, E., Dunlop, M., Baumgärtel, K. and Tarasov, V.: 1999, 'Low-Frequency Electromagnetic Waves Near and Below the Proton Cyclotron Frequency at the AMPTE Ba Release: Relevance to Comets and Mars', *J. Geophys. Res.* **104**, 6763–6771.

Sauer, K., and Dubinin, E.: 2003, 'The Magnetic Pile-Up Boundary at Mars and Related Ion Dynamics: Observations and Simulations', *Geophys. Res. Lett.* (in press).

Sauer, K., Dubinin, E. and McKenzie, J. F.: 2001, 'New Type of Soliton in Bi-ion Plasmas and Possible Implications', *Geophys. Res. Lett.* **28**, 3589–3592.

Sauer, K., Dubinin, E. and McKenzie, J. F.: 2002a, 'Coherent Waves in Multi-Ion Plasmas', *Physica Scripta* **T98**, 52.

Sauer, K., Dubinin, E. and McKenzie, J. F.: 2002b, 'Wave Emission by Whistler Oscillitons: Application to Coherent Lion Roars', *Geophys. Res. Lett.* **29**, 24, 2226, doi: 10.1029/2002 GLO15771, 2002.

Sauer, K., Dubinin, E., Phobos/Deimos events: observations and simulations, MPAE (Max–Planck-Institut für Aeronomie) – Report: 3 – 2001.

Schwartz, S. J. *et al.*: 1985, 'An Active Current Sheet in the Solar Wind', *Nature* **318**, 269.

Schwartz, S. J. *et al.*: 2000, 'Conditions for the Formation of Hot Flow Anomalies at Earth's Bow Shock', *J. Geophys. Res.* **105**, 12639.

Schwingenschuh, K., Riedler, W., Lichtennegger, H., Yeroshenko, Y., Sauer, K, Luhmann, J. G., Ong, M. and Russell, C. T.: 1990, 'Martian Bow Shock: Phobos Observations', *Geophys. Res. Lett.* **17**, 889.

Shimazu, H.: 1999, 'Three-Dimensional Hybrid Simulations of Magnetized Plasma Flow Around an Obstacle', *Earth Planets Space* **51**, 383–393.

Shimazu, H.: 2001, 'Three-Dimensional Hybrid Simulation of Solar Wind Interaction With Unmagnetized Planets', *J. Geophys. Res.* **106**, 8333–8342.

Skalsky, A., Grard, R., Klimov, S., Nairn, S. M. C., Trotignon, J. G., and Schwingenschuh, K.: 1992, 'The Martian Bow Shock: Wave Observations in the Upstream Region', *J. Geophys. Res.* **97**, 2927–2933.

Skalsky, A., Grard, R., Kiraly, P., Klimov, S., Kopanyi, V., Schwingenschuh, K. and Trotignon, J. G.: 1993, Simultaneous Plasma Wave and Electron Flux Observations Upstream of the Martian Bow Shock', *Planetary Space Sci.* **41**, 183–188.

Skalsky A., Dubinin, E., Petrukovich, A., Sauer, K., Kiraly P. and Schwingenschuh, K.: 1998, 'Ion and Electron Heating at the Martian Bow Shock, Common for Bow Shocks or Not?', *Earth Planets Space* **50**, 289.

Slavin, J. A., Elphic, R. C., Russell, C. T., Wolfe, J. H. and Intriligator, D. S.: 1979, 'Position and Shape of the Venus Bow Shock – Pioneer Venus Orbiter Observations', *Geophys. Res. Lett.* **6**, 901–904.

Slavin J. A. and Holzer, R. E.: 1981, 'Solar Wind Flow About the Terrestrial Planets: 1. Modeling Bow Shock Position and shape', *J. Geophys. Res.* **86**, 11401–11418.

Slavin J. A. and Holzer, R. E., 1982, 'The Solar Wind Interaction with Mars Revisited, *J. Geophys. Res.*87(B12), 10285–10296.

Slavin, J. A., Holzer, R. E. Spreiter, J. R. Stahara, S. S. and Chaussee, D. S.: 1983, 'Solar Wind Flow About the Terrestrial Planets. II – Comparison With Gas Dynamic Theory and Implications for Solar-Planetary Interactions', *J. Geophys. Res.* **88**, 19–35.

Slavin, J. A., Schwingenschuh, K., Riedler, W. and Eroshenko, E.: 1991, 'The Solar Wind Interaction with Mars – Mariner 4, Mars 2, Mars 3, Mars 5 and PHOBOS 2 Observations of Bow Shock Position and Shape', *J. Geophys. Res.*96, 11235–11241.

Sloter, S.: 1971, *'The Dust Belts of Mars*, Cornell Center for Radiophys. and Space Phys., Rep. No. 462.

Smith, E. J., Davis, L. Jr., Coleman, P. J. Jr. and Jones, D. E.: 1965, 'Magnetic Field Measurements Near Mars', *Science* **149**, 1241–1242.

Spangler, S., Fey, A., Anderson, G. and Fuselier, S.: 1988, 'An Observational Study of MHD Wave-induced Density Fluctuations Upstream of the Earth's Bow Shock', *J. Geophys. Res.* **93**, 845–857.

Spreiter, J. R. and Stahara, S S.: 1980, 'A New Predictive Model for Determining Solar Wind-Terrestrial Planet Interactions', *J. Geophys. Res.* **85**, 6769–6777.

Strangeway, R. J. and Crawford, G. K.: 1995, 'Comparison of Upstream Phenomena at Venus and Earth', *Adv. Space Res.* **16**, (4), 125–136.

Szegö, K., Glassmeier, K.-H., Bingham, R., Bogdanov, A., Fischer, C., Haerendel, G., Brinca, A., Cravens, T., Dubinin, E., Sauer, K., Fisk, L., Gombosi, T., Schwadron, N., Isenberg, P., Lee, M., Mazelle, Ch., Möbius, E., Motschmann, U., Shapiro, V. D., Tsurutani, B. and Zang, G.: 2000, 'Physics of Massloaded Plasmas', *Space Sci. Rev.* **94**(3/4).

Tatrallyay, M., Gevai, G. and Apathy, I. *et al.*: 1997, 'Magnetic Field Overshoots in the Martian Bow Shock', *J. Geophys. Res.* **102**, 2157.

Thomas, V. A. and Brecht, S. H.: 1987, 'Angular Dependence of High Mach Number Plasma Interactions', *J. Geophys. Res.* **92**, 3175.

Thomas, V. A. *et al.*: 1991, 'Hybrid Simulation of the Formation of a Hot Flow Anomaly', *J. Geophys. Res.* **96**, 11625.

Thomsen, M. F. *et al.*: 1986, 'Hot, Diamagnetic Cavities Upstream from the Earth's Bow Shock', *J. Geophys. Res.* **91**, 2961.

Trotignon, J. G., Hamelin, M., Grard, R., Pedersen, A., Klimov, S., Savin, S. and Skalsky, A.: 1991a, 'A Comparison of the Earth's and Mars' Bow Shocks Detected by the Phobos Plasma-Wave System', *Planetary Space Sci.* **39**, 99–112.

Trotignon, J. G., Grard, R. and Klimov, S.: 1991b, 'Location of the Martian Bow Shock Measurements by the Plasma Wave System on Phobos 2', *Geophys. Res. Lett.* **18**, 365–368.

Trotignon, J. G., Grard, R. and Savin, S.: 1991c, Plasma Wave System Measurements of the Martian Bow Shock from the Phobos 2 Spacecraft', *J. Geophys. Res.* **96**, 11253–11264.

Trotignon, J. G., Skalsky, A., Grard, R., Nairn, C. and Klimov, S.: 1992, 'Electron Density in the Martian Foreshock as a By-product of the Electron Plasma Oscillation Observations', *J. Geophys. Res.* **97**, 10831–10840.

Trotignon, J. G., Grard, R. and Skalsky, A.: 1993, 'Position and Shape of the Martian Bow Shock: the Phobos 2 Plasma Wave System Observations', *Planetary Space Sci.* **41**, 189–198.

Trotignon, J. G., Trotignon, A., Dubinin, E., Skalsky, A., Grard, R. and Schwingenschuh, K.: 2000, 'Electron Plasma Waves Beyond the Mars' Bow Shock: the PWS/Phobos 2 Observations', *Adv. Space Res.* **26**(10), 1619–1622.

Tsurutani, B. T.: 1991, in A.D. Johnstone (ed.), Comets: A Laboratory for Plasma Waves and Instabilities', *Cometary Plasma Processes*, pp. 189–209, *Geophysical Monograph* **61**, 189–209, American Geophysical Union.

Tsurutani, B. T., Brinca, A. L., Buti, B., Smith, E. J., Thorne, R. M. and Matsumoto, H.: 1989, 'Magnetic Pulses with Duration Near the Local Proton Cyclotron Period: Comet Giacobini-Zinner', *J. Geophys. Res.* **94**, 29–35.

Vaisberg, O. L. and Zelenyi, L. M.: 1984, 'Formation of the Plasma Mantle in the Venusian Magnetosphere', *ICARUS* **58**, 412–430.

Vaisberg, O. L.: 1992a, 'The Solar Wind Interaction with Mars: a Review of Results from Previous Soviet Missions to Mars', *Adv. Space Res.* **12**(9), 137–161.

Vaisberg, O. L.: 1992b, 'The Solar Wind Interaction with Mars: a Review of Results from Previous Soviet Missions to Mars, in: Venus and Mars: Atmospheres, Ionospheres, and Solar Wind Interactions, Proceedings of 1990 Chapman conference, *Geophysical Monograph* **66**, pp. 311–326, American Geophysical Union.

Verigin, M. I., Gringauz, K. I. *et al.*: 1991, 'On the Problem of the Martian Atmosphere Dissipation: Phobos-2 TAUS Spectrometer Results', *J. Geophys. Res.* **96**, 19315.

Verigin, M. I. *et al.*: 1993, 'The Dependence of the Martian Magnetopause and Bow Shock on Solar Wind Ram Pressure According to Phobos 2 TAUS Ion Spectrometer Seasurements', *J. Geophys. Res.* **98**, 1303.

Verigin, M., Kotova, G., Shutte, N., Remizov, A., Szegö, K., Tátrallyay, M., Apáthy, I., Rosenbauer, H., Livi, S., Richter, A. K., Schwingenschuh, K., Zhang, T.-L. and Slavin, J.: 1997, 'Quantitative

Model of the Martian Magnetopause Shape and its Variation with the Solar Wind Ram Pressure Based on Phobos 2 Observations', *J. Geophys. Res.* **102**(A2), 2147–2156.

Verigin, M. I., Kotova, G. A., Remizov, A. P., Styazhkin, V. A., Schutte, N. M., Zhang, T.-L., Riedler, W., Rosenbauer, H., Szegö, K., Tatrallyay, M. and Schwingenschuh, K., 'Shape and Location of Planetary Bow Shocks', 1999 *Cosmic Research* **37**(1), 34.

Vignes, D., Mazelle, C., Rème, H., Acuña, M. H., Connerney, J. E. P., Lin, R. P., Mitchell, D. L., Cloutier, P., Crider, D. H. and Ness, N.VF.: 2000, The Solar Wind Interaction with Mars: Locations and shapes of the Bow Shock and the Magnetic Pile-up Boundary from the Observations of the MAG/ER Experiment Onboard Mars Global Surveyor', *Geophys. Res. Lett.* **27**(1), 49–52.

Vignes, D., Acuña, M. H., Connerney, J. E. P., Crider, D. H., Mazelle, C. and Rème, H.: 2002 'Factors Controlling the Location of the Bow Shock at Mars', *Geophys. Res. Lett.* **29**(9), 42-1–42-4.

Wang, Z. and Kivelson, M. G.: 1996, 'Asteroid Interaction with Solar Wind', *J. Geophys. Res.* **101**, 24479–24493.

Wong, H. K. and Goldstein, M. L.: 1988, 'Proton Beam Generation of Oblique Whistler Waves', *J. Geophys. Res.* **93**, 4110–4114.

Wu, C. S. and Davidson, R. C.: 1972, 'Electromagnetic Instabilities Produced by Neutral-particle Ionization in Interplanetary Space', *J. Geophys. Res.* **77**, 5399.

Zakharov, A. V.: 1992, 'The Plasma Environment of Mars: Phobos Mission Results. A 1990 status', *Adv. Space Res.* **12**(9), 169–189.

Zelenyi, L. M. and Vaisberg, O. L.: 1985, in Buti (ed.), 'Venusian Interaction with the Solar Wind Plasma Flow as a Limiting Case of the Cometary Type Interaction', *Advances of Space Plasma Physics* World Scientific, Singapore, pp. 59–76.

Zhang, T.-L., Luhmann, J. G. and Russell, C. T.: 1990, 'The Solar Cycle Dependence of the Location and Shape of the Venus Bow Shock', *J. Geophys. Res.* **95**, 14961–14967.

Zhang, T. L., Schwingenschuh, K., Lichtenegger, H., Riedler, W. and Russell, C. T.: 1991, 'Interplanetary Magnetic Field Control of the Mars Bow Shock – Evidence for Venuslike Interaction', *J. Geophys. Res.* **96**, 11265–11269.

Zhang T. L., Schwingenschuh, K. and Russell, C. T.: 1995, 'A Study of the Solar Wind Deceleration in the Earth's Foreshock Region', *Adv. Space Res.* **15**(8/9), 137.

Zhang T. L. *et al.*: 1997, 'Solar Sind Deceleration at Mars and Earth: A Comparison', *Adv. Space Res.* **15**(2), 133.

Part II: Contributed Papers

THE MAGNETIC FIELD PILE-UP AND DENSITY DEPLETION IN THE MARTIAN MAGNETOSHEATH: A COMPARISON WITH THE PLASMA DEPLETION LAYER UPSTREAM OF THE EARTH'S MAGNETOPAUSE

MARIT ØIEROSET[1], DAVID L. MITCHELL[1], TAI D. PHAN[1], ROBERT P. LIN[1], DANA H. CRIDER[2] and MARIO H. ACUÑA[1]

[1]*Space Sciences Laboratory, University of California, Berkeley, CA 94720, U.S.A.*
[2]*NASA Goddard Space Flight Center, Code 695, Greenbelt, MD 20771, U.S.A.*

(Accepted in final form 14 March 2003)

Abstract. Using magnetometer and electron observations from the Mars Global Surveyor (MGS) and theWind spacecraft we show that the region of magnetic field pile-up and density decrease located between the Martian ionosphere and bow shock exhibit strong similarities with the plasma depletion layer (PDL) observed upstream of the Earth's magnetopause in the absence of magnetic reconnection when the magnetopause is a solid obstacle in the solar wind. A PDL is formed upstream of the terrestrial magnetopause when the magnetic field piles up against the obstacle and particles in the pile-up region are squeezed away from the high magnetic pressure region along the field lines as the flux tubes convect toward the magnetopause. We here discuss the possibility that at least part of the region of magnetic field pile-up and density depletion upstream of Mars may be formed by the same physical processes which generate the PDL upstream of the Earth's magnetopause. More complete ion, electron, and neutral measurements are needed to conclusively determine the relative importance of the plasma depletion process versus exospheric processes.

Abbreviations: MGS – Mars global surveyor; MPB – magnetic pileup boundary; PDL – plasma depletion layer; PEB – photo electron boundary.

1. Introduction

The nature of the Martian obstacle boundary has been a major topic of debate during the last three decades (e.g., Gringauz *et al.*, 1973; Slavin and Holzer, 1982; Breus *et al.*, 1991; Dubinin *et al.*, 1997; Acuña *et al.*, 1998; Sauer and Dubinin, 2000; Crider *et al.*, 2000). Using observations from the Phobos-2 spacecraft Lundin *et al.* (1991) and Kallio *et al.* (1994) reported that the solar wind plasma flow becomes stagnant several hundred km above the ionosphere, implying the existence of a solar wind obstacle located between the bow shock and the ionosphere, and the possibility that the obstacle was created by an intrinsic magnetic field. However, measurements from MGS revealed that the Martian intrinsic magnetic field consists only of scattered crustal sources and would not be an effective magnetospheric obstacle (Acuña *et al.*, 1998).

The boundary where the proton flow stops was found to be located near, and possibly coincide with the ion composition boundary (ICB) (Breus *et al.*, 1991;

Space Science Reviews **111:** 185–202, 2004.

Sauer *et al.*, 1995). The ICB was defined as a boundary separating solar wind (H^+) dominated plasma from planetary (O^+) dominated plasma (Breus *et al.*, 1991). The ICB was also found to coincide with a boundary termed the planetopause, marked by a disappearance of strong electric field fluctuations (Grard *et al.*, 1989; Trotignon *et al.*, 1996).

Measurements from the MGS spacecraft demonstrated that a permanent feature characterized by strong magnetic field pile-up accompanied by a rather sharp drop in electron fluxes exists upstream of the ionosphere in the vicinity of the Phobos-2 ICB/planetopause (Acuña *et al.*, 1998; Vignes *et al.*, 2000). This feature was termed the magnetic pileup boundary (MPB), after a similar feature observed at comets (Neubauer *et al.*, 1986; Mazelle *et al.*, 1989). The altitude of the MPB has been found to be located closer to the planet when the solar wind pressure is high (Crider et al., 2003). Also, Crider *et al.* (2002) have shown that the altitude where the magnetic field is piling up is higher in the southern hemisphere where the crustal magnetic fields are stronger (Acuña *et al.*, 1999).

Crider *et al.* (2000) interpreted the sharp drop in electron phase space density in this region to be a consequence of electron impact ionization of atmospheric neutrals since the steepest decrease was observed in the 40–150 eV energy range, which is where the cross sections for neutral hydrogen and oxygen are high (Shah *et al.*, 1987; Thompson *et al.*, 1995). In the proposed electron impact ionization process incident electrons lose kinetic energy when they impact an exospheric particle and a secondary electron with lower energy is produced.

The magnetic field pile-up in the region of electron depletion has recently been attributed to charge exchange between solar wind protons and atmospheric neutrals (Chen *et al.*, 2001). In the Chen *et al.* model a magnetic field increase in good agreement with that observed by MGS is produced by assuming that ion pressure lost by charge exchange is compensated for by an increase in magnetic pressure.

At Earth a similar increase in magnetic field magnitude and an associated decrease in electron fluxes have been observed immediately upstream of the Earth's magnetopause, where it has been linked to the existence of a plasma depletion layer (PDL) (e.g., Paschmann *et al.*, 1978; Crooker *et al.*, 1979; Fuselier *et al.*, 1991; Song *et al.*, 1993; Anderson and Fuselier, 1993; Anderson *et al.*, 1994; Phan *et al.*, 1994). A PDL is formed by the pile-up of magnetosheath field lines against the magnetopause in the absence of magnetopause reconnection, when the magnetopause becomes a solid obstacle in the solar wind, and is favored for northward interplanetary magnetic field (when the magnetosheath and the day-side magnetospheric fields are parallel) and high solar wind pressure (Anderson *et al.*, 1997). Particles in the field pile-up region are squeezed away from the high magnetic pressure region along the field lines as the flux tubes convect toward the magnetopause, leading to a decrease first in the higher energy fluxes, later in the lower energies (Zwan and Wolf, 1976; Paschmann *et al.*, 1993). Increasing ion temperature anisotropy ($T_\perp \gg T_\parallel$) and enhanced magnetosheath flow perpendicular to the magnetic field are observed in the PDL (Phan *et al.*, 1994). The enhancement of the perpendicular flow results from acceleration by magnetic tension associated with field lines being draped around the solid obstacle (Phan *et al.*, 1994). Mirror mode waves, common in the high beta magnetosheath, tend to be damped in the low beta PDL where the mirror condition $T_{\perp}/T_{\parallel} - (1 + /\beta_{\perp}) \geq 0$ is no longer satisfied (Anderson and Fuselier, 1993; Phan *et al.*, 1994). Zwan and Wolf (1976) postulated that a PDL should form in front of any planetary boundary, for example upstream of Venus and Mars, if the obstacle is effectively impenetrable. The PDL process is not related to impact ionization.

In this paper we compare the magnetic field pile-up and density depletion observed upstream of Mars by MGS with observations of the terrestrial PDL made by the Wind spacecraft. We show that the two regions have quite similar particle and field characteristics and raise the possibility that at least part of the region of magnetic field pile-up observed upstream of Mars may be a PDL. More complete ion and electron measurements are needed to conclusively determine the relative importance of plasma depletion versus exospheric processes.

2. Instrumentation

Observations from the Martian plasma environment are obtained by the Mars Global Surveyor (MGS) spacecraft. The magnetometer/electron reflectometer (MAG/ER) experiment onboard the MGS spacecraft is described in Acuña *et al.* (1998). In the present study a magnetic field resolution of 0.3 s has been used. The electron distribution function is measured in the range from $\sim$ 10 eV to 20 keV every 2 to 12 s, depending on the energy channel. The ER does not provide full 3D measurements, but the electron density above $\sim$ 10 eV can be reliably estimated. The electron temperature above the ionosphere have been estimated by fitting the measurements to a Maxwellian distribution, assuming isotropy. The electron bulk flow velocity cannot be reliably estimated from the ER measurements.

The Wind spacecraft provides full 3D plasma and magnetic field measurements from the Earth's magnetosheath. Averaged ion distribution functions are obtained every 51 s from the three-dimensional 'top-hat' electrostatic analyzer, operating in a mode to detect particles with energies from 80 eV to 27 keV. Electron distribution functions are measured every 51 s, covering energies from 10 eV to 27 keV. The magnetic field instrument onboard Wind measures the magnetic field at a rate of 10.9 samples s^{-1}, but for our analysis the magnetic field data are averaged over the 3-s spin period. Details of the Wind plasma and magnetic field instruments are described by Lin *et al.* (1995) and Lepping *et al.* (1995), respectively.

To analyze the region of magnetic field pile-up upstream of the obstacles we use the LMN boundary normal coordinate system (see Figure 1). The N axis points outward away from the planet along the boundary normal and the (L,M) plane is tangential to the boundary with L oriented in the direction of maximum variance of the magnetic field and M along the intermediate variance direction. λ is the angle

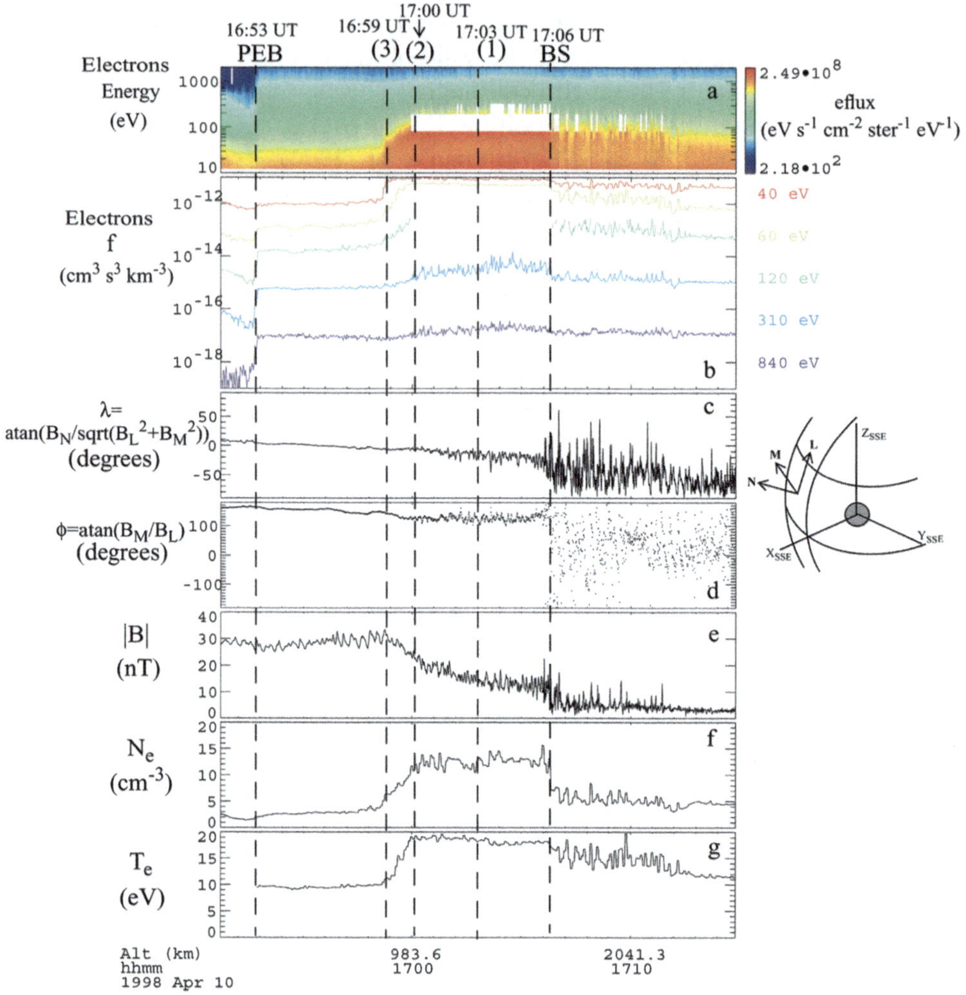

Figure 1. MGS observations from SPO 231 on April 10, 1998. The LMN coordinate system defined in the text is shown to the right in the figure. (a) Electron energy spectrogram. Some data points are missing due to instrument saturation; (b) electron phase space density for the 40–840 eV energy bins; (c) the λ magnetic field angle (angle between **B** and the (L,M) plane); (d) the ϕ magnetic field angle (angle of **B** in the (L,M) plane); (e) the magnitude of the magnetic field; (f) the electron density; (g) the electron temperature. The bow shock and the photoelectron boundary are marked with the vertical dashed lines 'BS' and 'PEB'. The vertical lines (1), (2), and (3) are explained in the text.

between **B** and the (L,M) plane while ϕ is the angle of **B** in the (L,M) plane, with $0°$ along L. The boundary normal direction was found using minimum variance analysis of the magnetic field (Sonnerup and Cahill, 1967).

3. MGS Observations

We here present two typical examples of magnetic field pile-up and density depletion in the prenoon sector upstream of Mars. Both examples are from the elliptical science phasing orbits (SPO) where the Martian magnetosheath was sampled at high latitudes. The first example has been discussed previously by Crider *et al.* (2000).

3.1. First MGS Example

Figure 1 shows data from SPO 231 on April 10, 1998, when MGS was in a polar elliptical 10 LT orbit with perigee near the north pole. Going back in time from right to left in the plot, MGS encountered the bow shock at 17:06 UT (marked by the dashed line 'BS') and entered the turbulent (Figure 1(e)), high density (Figure 1(f)) magnetosheath. Until ~ 17:03 UT (marked by the dashed line '(1)' in Figure 1) the magnetosheath magnetic field stayed rather constant at a ~ 13 nT magnitude with ~ 5 nT amplitude fluctuations (Figure 1(e)). These magnetic field fluctuations are anti-correlated with the density fluctuations and hence likely to be mirror mode waves, common in high beta plasmas (Tsurutani *et al.*, 1982). The amplitude of the waves diminishes when the magnetic field piles up (Figure 1(e)). From 17:03 UT to 16:59 UT (marked by '(3)' in Figure 1) a steady increase in magnetic field magnitude from 13 nT to 30 nT was observed. After 16:59 UT the magnetic field magnitude stayed rather constant. The spacecraft encountered the photoelectron boundary (PEB) in the vicinity of the ionopause (Mitchell *et al.*, 2000) at 16:53 UT, marked by the sharp drop in high energy (120–840 eV) electron phase space density (Figure 1(b)). The angle λ (Figure 1(c)) was close to zero during the interval of magnetic field pile-up (between (1) and PEB), implying that the normal component of the magnetic field stayed rather constant and small. The gradual rotation in λ from ~ 10° at (1) to ~ 0° at the PEB indicates that the magnetic field became increasingly more draped. A slow gradual rotation in ϕ (Figure 1(d)) was also seen.

The electron density (Figure 1(f)) dropped from ~ 13 cm^{-3} at (1) to ~ 8 cm^{-3} at (3) while the electron temperature (Figure 1(g)) decreased from ~ 19 eV to ~ 11 eV. The density and temperature stayed rather constant between (3) and the PEB. The variations in flux levels depend strongly on the energy. The high-energy levels (310 eV and 840 eV) started to decrease smoothly already near the bow shock and reached a stable level after 16:59 UT (3). The 60 eV channel, on the other hand, showed only little variation until 17:00 UT when the flux levels dropped rather sharply (marked by '(2)' in Figure 1) and continued to drop until the spacecraft reached the PEB. There is a data gap in the 120 eV channel between 17:06 UT and 17:00 UT due to saturation, but this channel also dropped sharply from (2) to (3). The 40 eV channel started to decrease even later, at ~ 16:59 UT. The region between (2) and (3) was termed the magnetic pile-up boundary (MPB)

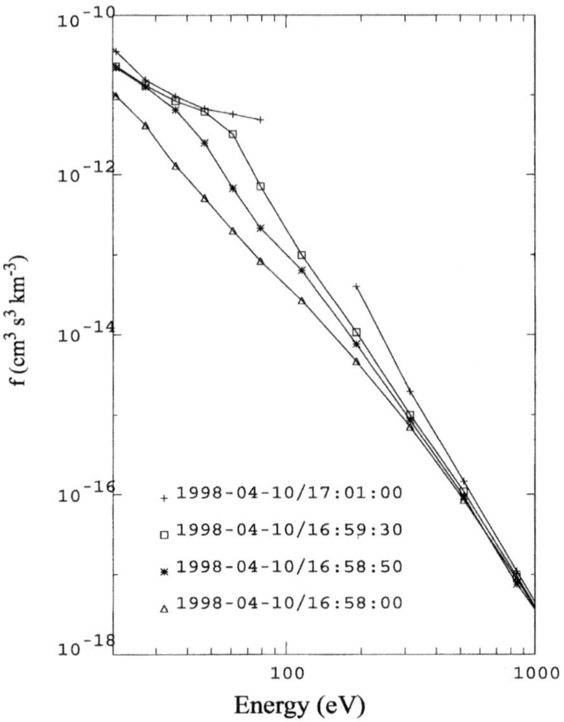

Figure 2. Development of electron spectra in the region where the magnetic field piles up observed by MGS on SPO 231 on April 10, 1998. After Crider *et al.* (2000).

and the steep drop in the 40–120 eV phase space density between (2) and (3) was interpreted to be a signature of electron impact ionization by Crider *et al.* (2000). Mazelle *et al.* (2002) also suggested that the term 'MPB' be used for the relatively steep increase in magnetic field magnitude seen between (2) and (3), analogous to what is observed at comets (Mazelle *et al.*, 1989), while they called the region planetward of (3) the magnetic pile-up region.

The development of the electron distribution function through the region where the magnetic field is piling up is shown in Figure 2. This figure is similar to Figure 2 in Crider *et al.* (2000). At 17:01:00 UT, right before the density depletion started, the spectrum was a typical post-shock magnetosheath spectrum. At 16:58:00 UT, when the magnetic field had stopped increasing, electrons with energies between ~20 eV and ~200 eV had been significantly depleted (almost two orders of magnitude at ~80 eV).

3.2. Second MGS Example

Figure 3 shows data from SPO 233 on April 11, 1998, when MGS was still in its polar elliptical 10 LT orbit with perigee near the north pole. Again going backwards in time, MGS encountered the bow shock at 16:29 UT (marked by the dashed line

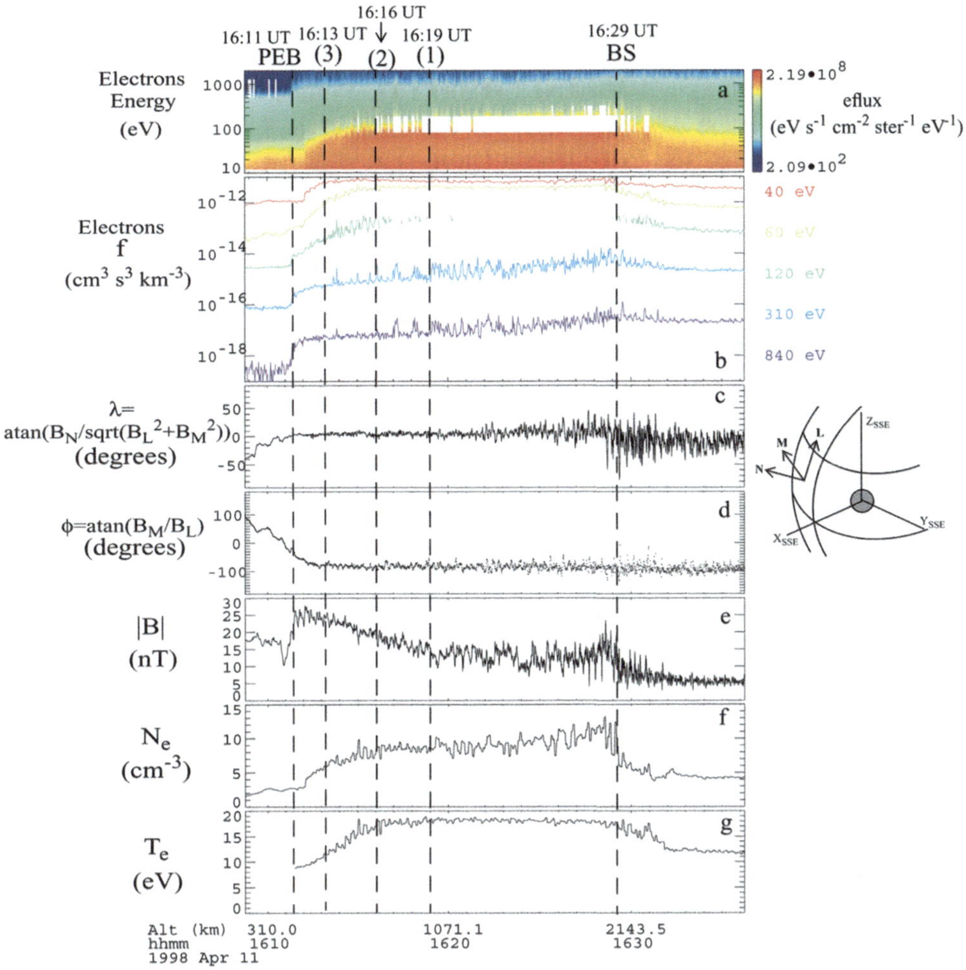

Figure 3. MGS observations from SPO 233 on April 11, 1998. The panels are explained in the caption of Figure 1. The LMN coordinate system defined in the text is shown to the right in the figure.

'BS') and entered the magnetosheath. Until ~16:19 UT (1) the magnetosheath magnetic field magnitude (Figure 2e) stayed rather constant at ~12 nT. Magnetic field fluctuations anti-correlated with density fluctuations were again observed and the wave amplitude diminished as the magnetic field piled up (Figure 2(e)). As mentioned before these waves are likely to be mirror mode waves. From 16:19 UT (1) to 16:13 UT (3) a steady increase in magnetic field magnitude from 12 nT to 24 nT was observed. After 16:13 UT the magnetic field magnitude stayed rather constant until the spacecraft encountered the PEB in the vicinity of the ionosphere at 16:11 UT, marked by the sharp drop in the 120–840 eV electron phase space density (Figure 2(b)). The angle λ (Figure 2(c)) was small and approaching zero during the interval of magnetic field pile-up (between (3) and PEB), implying that the normal component of the magnetic field stayed rather constant and small and that the draping of the magnetic field became increasingly stronger (see also Bertucci *et al.*, 2002). In contrast to the first example (Figure 1) a gradual rotation of $\sim 40°$ was seen in the ϕ angle (Figure 2(d)) from 16:13 UT to 16:11 UT when the spacecraft entered the PEB. The rotation started at (3), at the time when the magnetic field stopped increasing.

The electron density (Figure 2(f)) dropped from ~ 9 cm^{-3} at (1) to ~ 7 cm^{-3} at (3) while the electron temperature (Figure 2(g)) decreased from ~ 18 eV to ~ 11 eV. The density and temperature continued to decrease until the PEB. As in the previous example (Figure 1) the variations in flux levels depend strongly on the energy (Figure 2b). The high energy levels (310 eV and 840 eV) started to decrease smoothly near the bow shock and continued to decrease until the PEB. The 120 eV and 60 eV channels, on the other hand, showed only little variation until 16:16 UT (marked by '(2)' in Figure 1) when the flux levels dropped rather sharply and continued to drop until the spacecraft reached the PEB. The 40 eV channel started to decrease even later, at 16:14 UT.

3.3. Key Features of the Region of Magnetic Field Pile-up Upstream of Mars

As shown in the two examples presented above the following features are typically present in the region between the bow shock and the ionosphere at Mars:

(1) The magnetic field piles up and when the piling up has reached a certain point the electron density and temperature decreases.

(2) The electron phase space density drops earlier at higher energies than at lower energies.

(3) The magnetic field becomes increasingly more draped around the Martian obstacle as the magnetic field piles up.

(4) The magnetic field lines are nearly perfectly draped in the region planetward of where the maximum magnetic field magnitude is observed.

(5) In the second example (Figure 3) a rotation in the ϕ angle coincides with where the magnetic field stops increasing. This could imply a current sheet at this location.

4. Wind Observations

For comparison with the MGS observations we here review two examples of plasma depletion layer (PDL) observed upstream of the Earth's magnetopause. The first example is from aWind flank crossing, ideal for comparison with the MGS flank crossings presented above. The second example is from a Wind subsolar crossing.

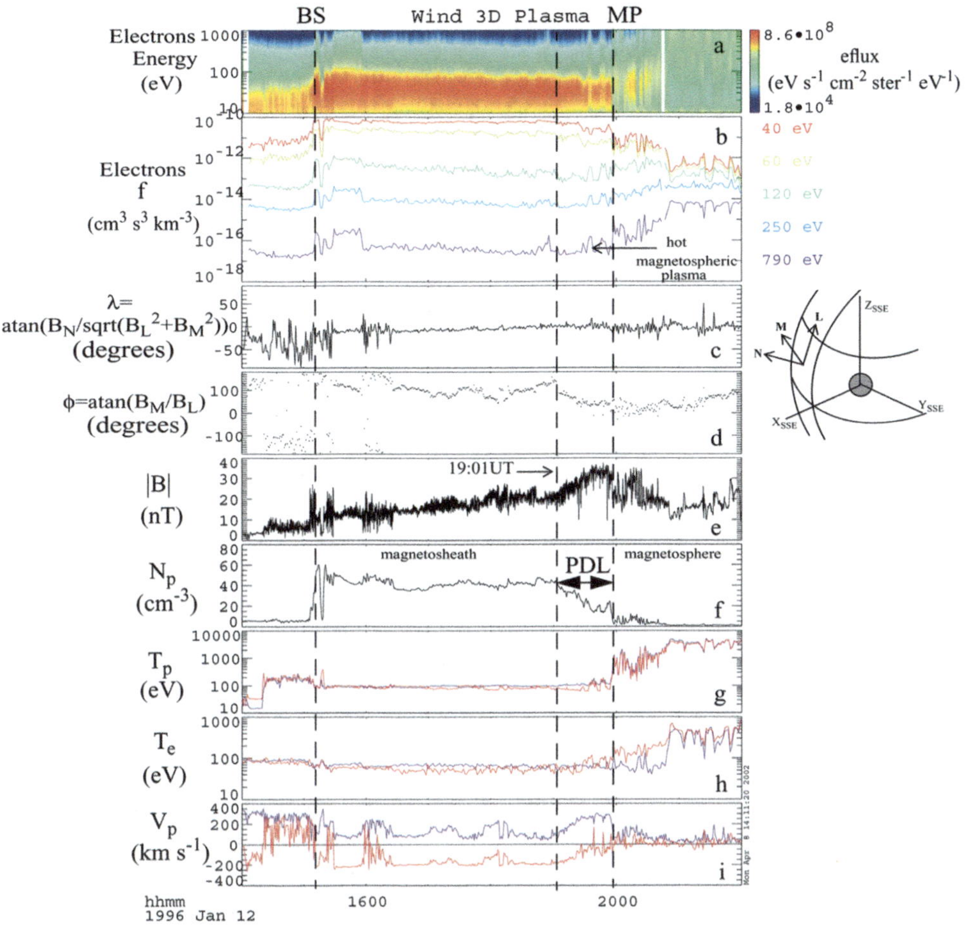

Figure 4. Wind observations from January 12, 1996. (a) Electron energy spectrogram; (b) Electron phase space density for the 40–790 eV bins; (c) the λ magnetic field angle; (d) the ϕ magnetic field angle; (e) the magnitude of the magnetic field; (f) the proton density; (g) the parallel (red) and perpendicular (blue) proton temperature; (h) the parallel (red) and perpendicular (blue) electron temperature; (i) proton velocity shown as the component perpendicular (blue) and parallel (red) to the magnetic field. The bow shock (BS), magnetopause (MP), and the onset of the PDL are marked with vertical dashed lines. Multiple crossings of the bow shock took place on this day and the bow shock (BS) is marked at the time of the first bow shock crossing. The LMN coordinate system defined in the text is shown to the right in the figure.

4.1. First Wind example: flank PDL

On January 12, 1996, Wind crossed the low shear dusk flank magnetopause near 18 LT. The event was analyzed in detail by Phan *et al.* (1997) and in Figure 4 we present an overview of the measurements relevant for the MGS comparison. The spacecraft encountered the bowshock at 15:08 UT and spent almost 5 hours in the magnetosheath before it crossed the magnetopause at 19:55 UT. The PDL

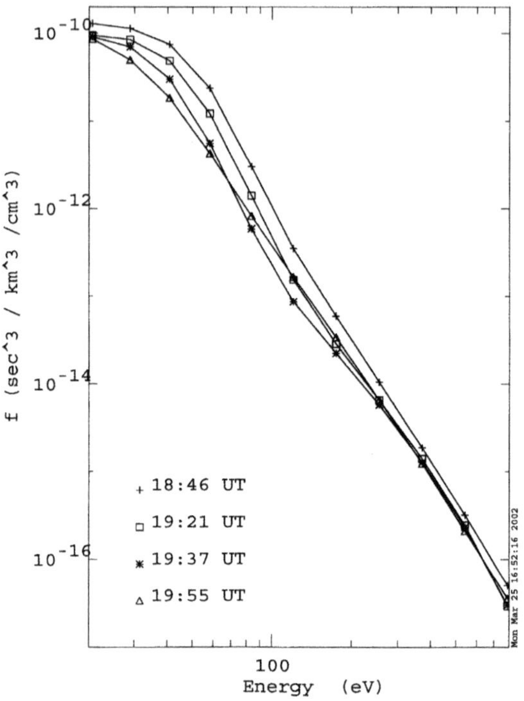

Figure 5. Development of the electron spectra observed by Wind in the PDL on January 12, 1996.

was observed from ~19:01 UT when the proton density (Figure 4(f)) started to drop and the magnetic field magnitude (Figure 4(e)) started to increase strongly (Phan *et al.*, 1997). The density had dropped more than 60%, from ~44 cm^{-3} in the pre-pile-up magnetosheath to ~15 cm^{-3} when the spacecraft encountered the magnetopause, and the simultaneous magnetic field increase was ~30% (from ~20 nT to ~33 nT). The λ angle (Figure 4(c)) gradually moved towards zero as the spacecraft approached the magnetopause, implying that the magnetic field lines became more and more draped around the magnetopause. The ϕ angle (Figure 4(d)) exhibited a rather abrupt rotation at the start of the PDL associated with a rotation of the interplanetary magnetic field while no significant rotation was seen across the magnetopause, indicating a low magnetic shear magnetopause. In the PDL the parallel proton temperature (red line) dropped while the perpendicular temperature (blue line) remained constant (Figure 4(g)). No significant anisotropy was observed in the electron temperature (Figure 4(h)). The perpendicular proton velocity increased in the PDL while the parallel velocity decreased to ~zero (Figure 4(i)). Mirror mode waves were observed throughout the magnetosheath and the PDL (Figure 4(e)). Mirror mode waves were identified based on the fact that the mirror condition ($T_\perp/T_\parallel - (1 + 1/\beta_\perp) \geq 0$) was marginally satisfied (Phan *et al.*, 1997).

The decrease in flux levels was different for the low and high-energy channels (Figure 4(b)). The 250 eV and 790 eV flux levels started to drop right after the

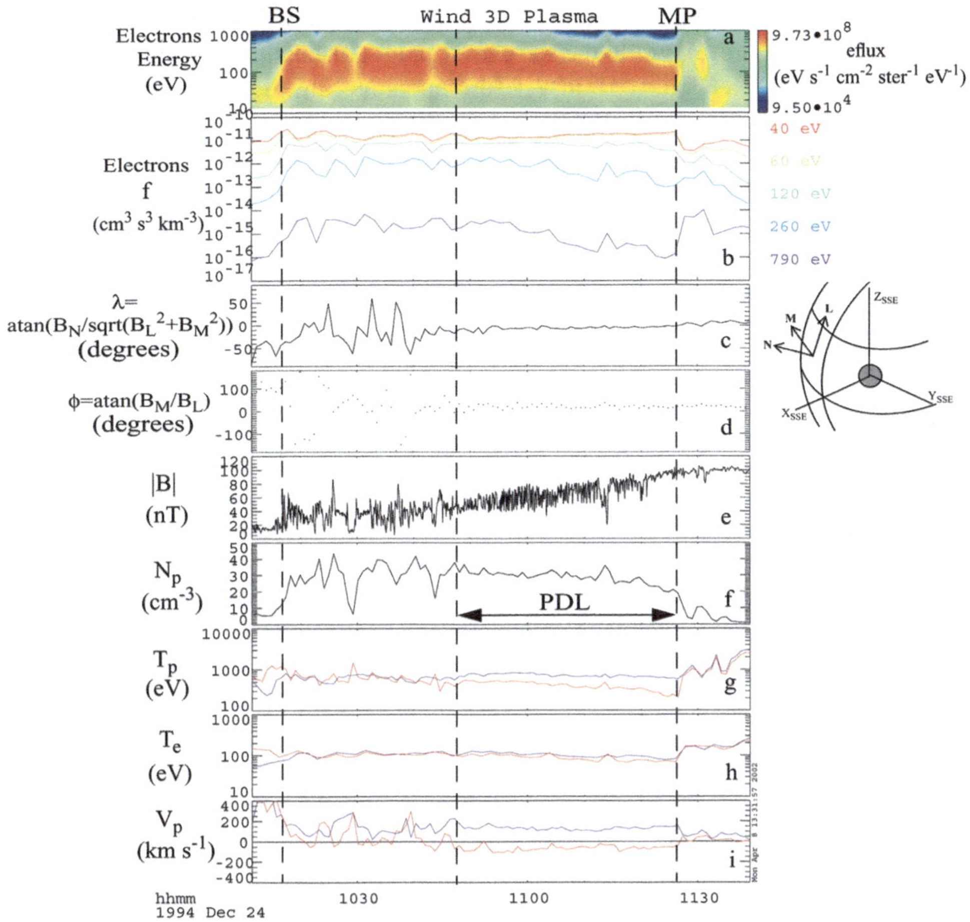

Figure 6. Wind observations from December 24, 1994. The panels are explained in the caption of Figure 4.

bow shock and the flux levels had reached a rather constant level inside the PDL. The 60 eV and 120 eV channels also decreased smoothly from the bow shock but exhibited a steeper drop inside the PDL. The 40 eV flux level stayed rather constant throughout the magnetosheath, started to decrease inside the PDL, and finally dropped sharply at 19:26 UT, approximately midway through the PDL. Hot plasma leaking from the magnetosphere (marked in Figure 4) was seen in the PDL as the spacecraft moved closer to the magnetopause.

The development of the electron distribution function as Wind moved through the PDL is shown in Figure 5. At 18:46 UT, before the density depletion started, the spectrum was a typical terrestrial post-shock magnetosheath spectrum. The shape of the post-shock spectrum at Mars (Figure 2) and Earth are somewhat different, with an enhancement in the 50–200 eV phase space density at Mars compared to the Earth. At Mars the drop in the 50–200 eV electron phase space density across the region of strong magnetic field pile-up was between one and two orders of magnitude (Figure 2), while the terrestrial magnetosheath spectrum drops more evenly. At 19:55 UT (Figure 5), just before encountering the magnetopause, electrons with energies between $\sim$20 eV and $\sim$200 eV had been significantly depleted (almost an order of magnitude between 30 eV and 100 eV). The distribution function declined first at higher energies, then later at lower energies, similar to what is observed in this region upstream of Mars (Figure 2). For the last spectrum, obtained at 19:55 UT near the magnetopause, the effect of leakage of hot magnetospheric plasma is evident in the higher energies.

4.2. Second Wind Example: Subsolar PDL

The data shown in Figure 6 are from December 24, 1994, whenWind crossed the subsolar terrestrial magnetosheath in the ecliptic plane. This event was presented in detail by Phan *et al.* (1996) and we here summarize some aspects of the event.

The Wind spacecraft encountered the bow shock at 10:16 UT and spent 70 min in the magnetosheath before it crossed the Earth's magnetopause at 11:26 UT. From approximately 10:49 UT a gradual decrease in the magnetosheath plasma density was seen (Figure 6(f)), marking the beginning of the PDL. This decrease in density continued until the spacecraft encountered the magnetopause where the density had dropped by 30% from $\sim$30 cm^{-3} to $\sim$20 cm^{-3}. During the interval of density decrease strong piling up of the magnetic field was observed (Figure 6(e)). The magnetic field magnitude increased from 45 nT at 10:49 UT to 97 nT at the magnetopause 37 min later. The λ angle (Figure 6(c)) gradually moved towards zero as the spacecraft approached the magnetopause, implying that the magnetic field lines became more and more draped around the magnetopause. The ϕ angle (Figure 6(d)) stayed close to zero throughout the PDL. No significant rotation was seen across the magnetopause, consistent with a low magnetic shear (no reconnection) magnetopause. The ion and electron temperature (Figure 6(g) and 6h) exhibited strong anisotropies in the interval of density decrease/magnetic field pile-up. The parallel proton velocity decreased inside the PDL while the perpendicular velocity stayed rather constant (Figure 6(i)).

Mirror mode waves were seen throughout the magnetosheath and the fluctuation amplitude decreased as Wind entered deeper into the PDL (Phan *et al.*, 1996). The damping of the mirror mode waves took place when the mirror condition ($T_\perp/T_\parallel - (1+1/\beta_\perp) \geq 0$) was no longer satisfied in this low beta region (Phan *et al.*, 1996).

The decrease in flux levels was again different for the low and high energy channels (Figure 6(b)). The 250 eV and 790 eV flux levels started to decrease earlier than the 120 eV channel. In this case the 40 eV and 60 eV phase space density stayed approximately constant throughout the PDL.

4.3. Key features of the PDL

The flank example and the subsolar example presented above are qualitatively similar. Based on the two examples we find the following features to be characteristic for the PDL upstream of the Earth's magnetopause when magnetic reconnection is not operating:

(1) Magnetic field increase

(2) Density decrease

(3) Enhanced ion temperature perpendicular to the magnetic field

(4) Enhanced flow perpendicular to the magnetic field

(5) Electron phase space density drops earlier at higher energies than at lower energies

(6) Magnetic field becomes increasingly more draped around the magnetopause as the magnetic field piles up

(7) Mirror mode waves, observed throughout the magnetosheath, are damped in the PDL when the mirror condition is no longer satisfied.

5. Discussion

5.1. Comparison between the magnetic field pile-up and density decrease observed upstream of Mars and Earth

The examples above show that the magnetic field and electron properties in the region where the magnetic field is piling up upstream of Mars strongly resemble those observed in the PDL upstream of the Earth's magnetopause. At Mars the electron density dropped 40% (Figure 1) and 20% (Figure 3) from the post-shock magnetosheath to the end of the magnetic field pile-up, while the corresponding density decrease upstream of the Earth was 60% for the flank example (Figure 4) and 30% for the subsolar PDL example (Figure 6). The magnetic field increase upstream of both Mars and Earth was a factor of ~ 2. Mirror waves were observed in both the terrestrial and the Martian magnetosheath. These waves tend to be damped in the subsolar terrestrial PDL (Anderson and Fuselier, 1993; Phan *et al.*, 1994). However, mirror waves may persist in the flank PDL where the magnetic field magnitude is small and the mirror condition still satisfied (e.g., Figure 4). As shown in Figures 1 and 3 the mirror mode waves are damped but still present in the region of strong magnetic field pile-up upstream of Mars, consistent with observations from the flank PDL at Earth (Figure 4). For both Mars and Earth the magnetic field becomes increasingly more draped in the region where the magnetic field is piling up.

The behavior of the electron phase space density levels in front of the two obstacles (Mars and Earth) is remarkably similar, with a gradual decrease in the higher energy channels starting already near the bow shock with the lower energy channels dropping only later, inside the region where the magnetic field increases

(Figures 1–3). In other words, upstream of both planets the distribution function first declines at the highest energy (~1 keV) and the decrease in phase space density becomes stronger as the energy gets lower (Figures 2 and 5). At the lowest energy (~20 eV), however, the decrease is not as strong. At Mars the largest drop in phase space density is seen in the 50–150 eV energy bins with a decrease of almost two orders of magnitude for the 80 eV bin. In the Earth's PDL one order of magnitude decrease in electron phase space density is seen for the 40–80 eV bins.

The slow decrease in the higher energy phase space density from the bow shock throughout the magnetosheath can be explained in terms of a gas-dynamic effect (Spreiter *et al.*, 1966), while the sharper drop in phase space density for the lower (< 120 eV) energy bins is not predicted by gas-dynamics.

Crider *et al.* (2000) interpreted the sharp drop in the 40–150 eV electron phase space density observed in the region where the magnetic field is piling up between the Martian ionosphere and bow shock to be a consequence of electron impact ionization of atmospheric neutrals since the 40–150 eV energy range is where the cross sections for neutral hydrogen and oxygen are high (Shah *et al.*, 1987; Thompson *et al.*, 1995). The increase in low energies that should result from the secondary electrons was not observed, although it is possible that most of the secondary electrons may fall below the detection limit of the ER instrument onboard MGS. The fact that the low-energy secondary electrons were not observed could be interpreted as evidence for the plasma depletion process being important at Mars since plasma depletion could account for some of the observed density decrease.

Although electron impact ionization may be important in the solar wind interaction with Mars (Crider *et al.*, 2000) and comets (Cravens *et al.*, 1987) this process cannot explain the rather sharp decrease in the lower energy bins (40–80 eV) observed in the PDL outside the Earth's magnetopause, where the effects of exospheric processes are absent. In the Earth's PDL the density depletion is generated when particles are squeezed away along the field lines from the strong magnetic field region with the higher energy particles disappearing first, followed later by the lower energies (e.g., Song *et al.*, 1993; Paschmann *et al.*, 1993). This depletion does not depend on the process in which the strong magnetic field region was created, i.e., whether the increase is due to a simple compression of the magnetic field against the solar wind obstacle boundary (Zwan and Wolf, 1976), or whether the strong magnetic field has been produced by charge exchange, as suggested by Chen *et al.* (2001).

5.2. Predictions of PDL Ion Signatures for Future Mars Missions

We can exclude that exospheric phenomena produce the PDL observed upstream of the Earth's magnetosphere since the Earth lacks an extended exosphere with significant density. However, the question of whether the magnetic field pile-up and density depletion upstream of Mars is an 'Earth-like' PDL or generated by exospheric processes cannot be fully answered using the existing MGS measurements.

Full three-dimensional ion and electron measurements, as well as observations of the neutrals are needed.With threedimensional ion measurements one will be able to determine whether the ion temperature and flow perpendicular to the magnetic field is enhanced in the region of magnetic field pile-up. Such an observation would imply the existence of a PDL upstream of Mars.

A possible scenario would be a magnetic field pile-up and density decrease which is partly caused by plasma depletion and partly caused by exospheric processes. A one order of magnitude decrease in the 80 eV phase space density is observed upstream of the Earth's magnetopause (Figure 5) and hence the plasma depletion process could explain a similar decrease upstream of Mars. However, the corresponding decrease in the 80 eV electron phase space density upstream of Mars is two orders of magnitude, suggesting that additional processes, like e.g. electron impact ionization (Crider *et al.*, 2000), are operative in depleting the electrons (Figure 2). Future missions to Mars carrying full three-dimensional ion and electron instruments will be able to determine the relative contribution from the plasma depletion process and from exospheric phenomena. ESA's Mars Express is equipped to study the possible existence of a PDL upstream of Mars.

5.3. Consequences of a Martian PDL on the Location of the Martian Obstacle Boundary

If a PDL exists upstream of Mars it implies that the Martian solar wind obstacle boundary is located at the inner edge of this PDL, i.e., where the magnetic field reaches its maximum value (Zwan and Wolf, 1976). The observed rotation in the magnetic field at this location (see Figure 3) could indicate the presence of a boundary. Phobos-2 observations suggest that the ion composition boundary (ICB), a boundary separating solar wind (H^+) dominated plasma from planetary (O^+) dominated plasma, is located in the region where the proton flow becomes stagnant (Breus *et al.*, 1991; Kallio *et al.*, 1994; Sauer *et al.*, 1995). Hence the flow of plasma is slowed down by mass loading of atmospheric neutrals and the magnetic field carried by this plasma will become nearly stagnant, giving rise to an (induced) solar wind obstacle with an obstacle boundary located at some altitude above the PEB. The average location of strong magnetic field pile-up upstream of Mars is indeed found to be located near the ICB, several hundred km above the PEB (Vignes *et al.*, 2000).

6. Summary

The piling up of magnetic field and the associated decrease in density upstream of Mars exhibit strong similarities with the plasma depletion layer (PDL) observed upstream of the Earth's magnetopause when reconnection is absent. The plasma depletion process can account for at least one order of magnitude decrease in the electron phase space density (Figure 5). It is therefore possible that the observed density decrease upstream of Mars is in part caused by this process. Electron impact ionization may account for the observed additional decrease in phase space density. The exact nature of the Martian solar wind obstacle remains unknown.

Acknowledgements

We thank Ron Lepping, the principal investigator of the Wind magnetic field experiment, for making the Wind magnetic field data available. We would like to thank Christian Mazelle, David Sibeck, Laura Peticolas,Yue Chen, Esa Kallio, and Konrad Sauer for helpful discussions. This work benefited greatly from the workshop on Mars magnetism and its interaction with the solar wind, held at the International Space Science Institute (ISSI) in Bern, Switzerland on October 22 –26, 2001, sponsored by ISSI and NASA/JPL/MGS. This research was funded in part by NASA grants NAG5-959, NAG5-6928, and NAG5-10428 at U.C. Berkeley.

References

Acuña, M. H. *et al.*: 1998, 'Field and Plasma Observations at Mars: Initial Results of the Mars Global Surveyor Mission', *Science* **279**, 1676.

Acuña, M. H. *et al.*: 1999, 'Global Distribution of Crustal Magnetization Discovered by the Mars Global Surveyor MAG/ER Experiment', *Science* **184**, 790.

Anderson, B. J. and Fuselier, S. A.: 1993, 'Magnetic Pulsations from 0.1 to 4.0 Hz and Associated Plasma Properties in the Earth's subsolar Magnetosheath and Plasma Depletion Layer', *J. Geophys. Res.* **98**, 1461.

Anderson, B. J., Fuselier, S. A., Gary, S. P. and Denton, R. E.: 1994, 'Magnetic Spectral Signatures in the Earth's Magnetosheath and Plasma Depletion Layer', *J. Geophys. Res.* **98**, 1461.

Anderson, B. J., Phan, T. D. and Fuselier, S. A.: 1997, 'Relationships Between Plasma Depletion and Subsolar Reconnection', *J. Geophys. Res.* **102**, 9531.

Bertucci, C. *et al.*: 2002, 'Magnetic Field Draping Enhancement at the Magnetic Pileup Boundary of Mars from Mars Global Surveyor Observations', submitted to *Geophys. Res. Lett.*

Breus, T. K., Krymskii, A. M., Lundin, R., Dubinin, E. M., Luhmann, J. G., Yeroshenko, Y. G., Barabash, S. V., Mitnitskii, V. Y., Pissarenko, N. F. and Styashkin, V. A.: 1991, 'The Solar Wind Interaction with Mars: Consideration of Phobos 2 Mission Observations of an Ion Composition Boundary on the Dayside', *J. Geophys. Res.* **96**, 11 165.

Chen, Y., Cloutier, P. A., Crider, D. H., Mazelle, C. and Rème, H.: 2001, 'On the Role of Charge Exchange in the Formation of the Martian Magnetic Pile-up Boundary', *J. Geophys. Res.* **106**, 29 387.

Cravens, T. E., Kozyra, J. U., Nagy, A. F., Gombosi, T. I. and Kurtz, M.: 1987, 'Electron Impact Ionization in the Vicinity of Comets', *J. Geophys. Res.* **92**, 7341.

Crider, D. H. *et al.*: 2000, 'Evidence of Electron Impact Ionization in the Magnetic Pileup Boundary of Mars', *Geophys. Res. Lett.* **27**, 45.

Crider, D. H. *et al.*: 2002, 'Observations of the Latitude Dependence of the Location of the Martian Magnetic Pileup Boundary', *Geophys. Res. Lett.* (in press).

Crider, D. H., Vignes, D., Krymskii, A. M., Breus, T. K., Ness, N. F., Mitchell, D. L. et al.: 2003, 'A Proxy for Determining Solar Wind Dynamic Pressure at Mars using Mars Global Surveyor Data; *J. Geophys. Res.* **108** (A12): 1461, 10.1029/2003JA009875.

Crooker, N. U., Eastman, T. E. and Stiles, G. S.: 1979, 'Observations of Plasma Depletion in the Magnetosheath at the Dayside Magnetopause', *J. Geophys. Res.* **84**, 869.

Dubinin, E. M., Sauer, K., Baumgärtel, K. and Lundin, R.: 1997, 'The Martian Magnetosheath: Phobos-2 Observations', *Adv. Space Sci.* **20**, 149.

Fuselier, S. A., Klumpar, D. M., Shelley, E. G., Anderson, B. J. and Coates, A. J.: 1991, 'He(2^+) and H($^+$) Dynamics in the Subsolar Magnetosheath and Plasma Depletion Layer', *J. Geophys. Res.* **96**, 21 095.

Grard, R., Pedersen, A., Klimov, S., Savin, S. and Skalskii, A.: 1989, 'First Measurements of Plasma Waves near Mars', *Nature* **341**, 607.

Gringauz, K. I., Bezrukikh, V. V., Volkov, G. I., Breus, T. K., Musatov, I. S., Havkin, L. P. and Sloutchonkov, G. P.: 1973, 'Preliminary Results on Plasma Electrons from Mars-2 and Mars-3', *Icarus* **18**, 54.

Kallio, E., Koskinen, H., Barabash, S., Lundin, R., Norberg, O. and Luhmann, J. G.: 1994, 'Proton Flow in the Martian Magnetosheath', *J. Geophys. Res.* **99**, 23 547.

Lepping, R. P. *et al.*: 1995, 'The Wind Magnetic Field Investigation', *Space Sci. Rev.* **71**, 207.

Lin, R. P. *et al.*: 1995, 'A Three-Dimensional Plasma and Energetic Particle Investigation for the Wind Spacecraft', *Space Sci. Rev.* **71**, 125.

Lundin, R., Norberg, O., Dubinin, E. M., Pisarenko, N. and Koskinen, H.: 1991, 'On the Momentum Transfer of the Solar Wind to the Martian Topside Ionosphere', *Geophys. Res. Lett.* **18**, 1059.

Mazelle, C. *et al.*: 1989, 'Analysis of Suprathermal Electron Properties at the Magnetic Pile-up Boundary of Comet P/Halley', *Geophys. Res. Lett.* **16**, 1035.

Mazelle, C. *et al.*, 'The Magnetic Pileup Boundary at Mars: A Comet-like Feature in the Interaction of the Planet Atmosphere with the Solar Wind', this issue.

Mitchell, D. L., Lin, R. P., Rème, H., Crider, D. H., Cloutier, P. A., Connerney, J. E. P., Acuña, M. H. and Ness, N. F.: 2000, 'Oxygen Auger Electrons Observed in Mars' ionosphere', *Geophys. Res. Lett.* **27**, 1871.

Neubauer, F. M. *et al.*: 1986, 'First Results from the Giotto Magnetometer Experiment at Comet Halley', *Nature* **321**, 352.

Paschmann, G., Sckopke, N., Haerendel, G., Papamastorakis, J., Bame, S. J., Asbridge, J.R., Gosling, J. T., Hones, E. W. Jr and Tech, E. R.: 1978, 'ISEE Plasma Observations near the Subsolar Magnetopause', *Space Sci. Rev.* **22**, 717.

Paschmann, G., Baumjohann, W., Sckopke, N., Phan, T. D. and Lühr, H.: 1993, 'Structure of the Dayside Magnetopause for Low Magnetic Shear, *J. Geophys. Res.* **98**, 13 409.

Phan, T. D., Paschmann, G., Baumjohann, W., Sckopke, N. and Lühr, H.: 1994, 'The Magnetosheath Region Adjacent to the Dayside Magnetopause: AMPTE/IRM Observations', *J. Geophys. Res.* **99**, 121.

Phan, T. D. *et al.*: 1996, 'The Subsolar Magnetopause for High Solar Wind Ram Pressure: WIND Observations', *Geophys. Res. Lett.* **23**, 1279.

Phan, T. D. *et al.*: 1997, 'Low-latitude Dusk Flank Magnetosheath, Magnetopause and Boundary Layer for Low Magnetic Shear: Wind Observations', *J. Geophys. Res.* **102**, 19 883.

Sauer, K., Bogdanov, A. and Baumgartel, K.: 1995, 'The Protonopause-an Ion Composition Boundary in the Magnetosheath of Comets, Venus and Mars', *Adv. Space Res.* **16**(4), 153.

Sauer, K. and Dubinin, E.: 2000, 'The Nature of the Martian 'Obstacle Boundary'', *Adv. Space. Res.* **26**, 1633.

Shah, M. B., Elliot, D. S. and Gilbody, H. B.: 1987, 'Pulsed Cross-Beam Study of the Ionisation of Atomic Hydrogen by Electron Impact', *J. Phys. B.* **20**, 3501.

Slavin, J. A. and Holzer, R. E.: 1982, 'The Solar Wind Interaction with Mars Revisited', *J. Geophys. Res.* **87**, 10 285.

Song, P., Russell, C. T., Fitzenreiter, R. J., Gosling, J. T., Thomsen, M. F., Mitchell, D. G., Fuselier, S. A., Parks, G. K., Anderson, R. R., Hubert, D.: 1993, 'Structure and Properties of the Subsolar Magnetopause for Northward Interplanetary Magnetic Field - Multiple-instrument Particle Observations', *J. Geophys. res.* **98**, 11 319.

Sonnerup, B. U. Ö and Cahill Jr., L. J.: 1967, 'Magnetopause Structure and Attitude from Explorer 12 Observations', *J. Geophys. Res.* **72**, 171.

Spreiter, J. R., Summers, A. L. and Alksne, A. Y.: 1966, 'Hydromagnetic Flow around the Magnetosphere', *Planetary Space Sci.* **14**, 223.

Thompson, W. R., Shah, M. B. and Gilbody, H. B.: 1995, 'Single and Double Ionization of Atomic Oxygen by Electron Impact', *J. Phys.* **B28**, 1321.

Trotignon, J. G., Dubinin, E., Grard, R., Barabash, S. and Lundin, R.: 1996, 'Martian Planetopause as Seen by the Plasma Wave System Onboard Phobos 2', *J. Geophys. Res.* **101**, 24 965.

Tsurutani, B. T., Smith, E. J., Anderson, R. R., Ogilvie, K. W., Scudder, J. D., Baker, D. N. and Bame, S. J.: 1982, 'Lion Roars and Nonoscillatory Drift Mirror Waves in the Magnetosheath', *J. Geophys. Res.* **87**, 6060.

Vignes, D., Mazelle, C., Rème, H., Acuña, M. H., Connerney, J. .E. P., Lin, R. P., Mitchell, D. L., Cloutier, P., Crider, D. H. and Ness, N. F.: 2000, 'The Solar Wind Interaction with Mars: Locations and Shapes of the Bow Shock and the Magnetic Pile-up Boundary from the Observations of the MAG/ER Experiment Onboard Mars Global Surveyor', *Geophys. Res. Lett.* **27**, 49.

Zwan, B. J. and Wolf, R. A.: 1976, 'Depletion of Solar Wind Plasma Near a Planetary Boundary', *J. Geophys. Res.* **81**, 1636.

MARS GLOBAL SURVEYOR OBSERVATIONS OF SOLAR WIND MAGNETIC FIELD DRAPING AROUND MARS

DANA H. CRIDER[1], DAVID A.BRAIN[2], MARIO H. ACUÑA[3], DIDIER VIGNES[3], CHRISTIAN MAZELLE[4] and CESAR BERTUCCI[4]
[1]*Catholic University of America, Dept. of Physics., Washington, DC, U.S.A.*
[2]*University of Colorado, Boulder, CO, U.S.A.*
[3]*Code 695, NASA Goddard Space Flight Center, Greenbelt, MD, U.S.A.*
[4]*Centre d'Etude Spatiale des Rayonnements - CNRS/UPS, Toulouse, France*

Accepted in final form 14 March 2003

Abstract. We examine the magnetic field in the martian magnetosheath due to solar wind draping. Mars Global Surveyor provided 3-D vector magnetic field measurements at a large range of altitudes, local times, and solar zenith angles as the spacecraft orbit evolved. We choose orbits with very clean signatures of draping to establish the nominal morphology of the magnetic field lines at local times of near-subsolar and near-terminator. Next, using a compilation of data from Mars Global Surveyor, we determine the average magnetic field morphology in the martian magnetosheath due to the solar wind interaction. The topology of the field is as expected from previous observations and predictions. The magnetic field magnitude peaks at low altitude and noon magnetic local time and decreases away from that point. The magnetic field has an inclination from the local horizontal of 5.6° on average in the dayside magnetosheath and 12.5° on the nightside. The inclination angle is closest to zero at noon magnetic local time and low altitude. It increases both upward and to later local times. The magnetic field in the induced magnetotail flares out from the Mars–Sun direction by 21°. Finally, we compare the observations to gasdynamic model predictions and find that the shocked solar wind flow in the martian magnetosheath can be treated as a gasdynamic flow with the magnetic pileup boundary as the inner boundary to the flow.

1. Introduction

Although it was known from previous missions to Mars that Mars had at most a weak planetary magnetic field (Dolginov *et al.*, 1973; Dolginov, 1978), the arrival of Mars Global Surveyor (MGS) established Mars as primarily an ionospheric obstacle to the solar wind (SW) flow (Acuña *et al.*, 1998). Crustal magnetic fields contribute to the overall obstacle; however, their effects appear to be highly localized (Mitchell *et al.*, 2001; Crider *et al.*, 2002). Therefore, both on average and in regions away from the strong crustal features, the SW interaction with Mars has common elements with Venus (Cloutier *et al.*, 1999) and with comets (Mazelle *et al.*, 2002).

In such an interaction, the conducting ionosphere and mass loading limits act as the obstacle to the SW flow (Alfvén, 1957). The impinging interplanetary magnetic field (IMF) can not completely penetrate the conducting ionosphere. The

Space Science Reviews **111:** 203–221, 2004.

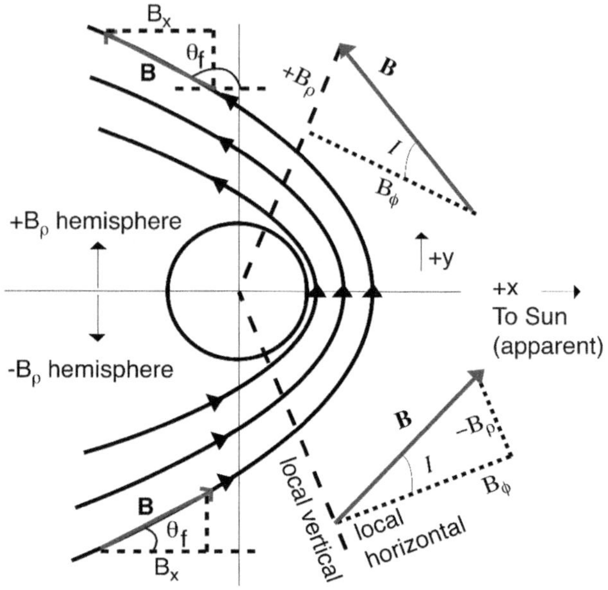

Figure 1. Cartoon depicting the geometry of magnetic field lines in the 'draping' regime. The coordinate system and angles are defined in Sections 2 and 4.

IMF 'hangs' in front of the planet. The result is that an induced magnetosphere is created that consists of magnetic field lines from the SW that are 'draped' around the obstacle (see Figure 1). The region immediately downstream of the bow shock is typically called the 'magnetosheath'. Behind the planet, the elongated, two-lobed magnetic feature is termed the 'magnetotail'. It is similar to the Earth's magnetotail except that this tail contains IMF flux tubes instead of planetary flux tubes.

The instantaneous, global morphology of the magnetic field in a draping configuration can not be measured with only one or even a few spacecraft. However, much work has been done to create models of the interaction that accurately describe the observed magnetic field topology (Spreiter and Stahara, 1980; Tanaka, 1993; Brecht, 1997; Kallio and Koskinen, 2000; Liu *et al.*, 2001). In this paper, two approaches are taken to present data that can be used to validate those models (Russell *et al.*, 1984; Slavin *et al.*, 1983). First, one can compare the model predictions to *selected data* that are representative of typical conditions to see if the model reproduces the observations. We present such comparisons in Section 3. Alternatively, one can compare the model predictions to a *compilation of data*. By compiling data from many orbital passes that occurred under different upstream conditions, one finds the average configuration of the magnetic field, and certain diagnostic parameters can be developed for comparison with models. For that purpose, we compile a global picture of the magnetic field in the SW interaction with Mars in this work using data from the MGS Magnetometer (MAG) instrument. We limit our study to the data from the martian magnetosheath and induced magnetotail, where crustal magnetic fields should only have minimal contributions. The results from this analysis are presented in Section 4.

2. Data Analysis

Mars Global Surveyor is not currently located in the magnetosheath; however, it did traverse that area earlier in its mission. Its first 1200 orbits were elliptical orbits that sampled altitudes from 100 km to well beyond the bow shock. Meanwhile the spacecraft trajectory was shaved down from the highly eccentric insertion orbit to the near-circular mapping orbit. MGS MAG data from the magnetosheath is available for the time period of September 13, 1997 to February 4, 1999, which we term the 'elliptical orbits'. Throughout the elliptical orbits, the spacecraft orbit was highly inclined (near-polar). Orbital parameters such as local time of orbit, latitude of periapsis, and altitude of apoapsis varied slowly. Throughout the entire elliptical orbit epoch, the spacecraft sampled a large span of altitudes and local times in the region of interest for the solar wind interaction with Mars. See Brain *et al.* (2003) and Albee *et al.* (2001) for more detailed accounts of MGS orbital coverage.

For the purpose of producing a picture of the magnetic field topology around Mars resulting from the draping of the IMF, one must ensure that the data compiled are self-compatible. Because the magnetic field in the magnetosheath is comprised of IMF lines and the direction of the upstream IMF changes with time, it is logical to align the data to a consistent upstream IMF direction. Therefore, we rotate the data on an orbit by orbit basis into a coordinate system defined by the upstream IMF direction as described below.

Data from MGS are converted to a coordinate system in which the $+x$ axis points from the center of Mars to the apparent origin of the SW. The SW appears to come from a direction of $\sim 3.5°$ off of the Mars–Sun line because of the orbital velocity of Mars around the Sun. After fixing the $\hat{x}$ direction, the $\hat{y}$ direction is determined based on the upstream IMF direction. We rotate the coordinate system around the x-axis such that the average upstream IMF lies in the $x - y$ plane and such that the y component of the magnetic field is positive. Mathematically:

$$\mathbf{B}_{SW} = \langle B_{\mathrm{IMF}x} \rangle \, \hat{\mathbf{x}} + \sqrt{\langle \mathbf{B}_{\mathrm{IMF}} \rangle^2 - \langle B_{\mathrm{IMF}x} \rangle^2} \hat{\mathbf{y}} + 0\hat{\mathbf{z}}, \quad (1)$$

where $\mathbf{B}_{SW}$ is the vector average upstream magnetic field in IMF-related coordinates. We refer to this coordinate system as 'IMF coordinates' or 'magnetic coordinates' interchangeably throughout this paper. This method does not account for differences in the Parker Spiral angle of the IMF.

MGS data are transformed orbit by orbit into magnetic coordinates to account for variations in IMF direction. However, the IMF also varies in magnitude. Larger IMF strengths lead to higher magnetic fields in the magnetosheath which artificially weights the data from strong IMF orbits higher in determining the average magnetic field configuration. Therefore we normalize all magnetic field values on an orbital pass by the magnitude of the average upstream IMF for that pass. In effect, we are assuming that the magnetic field magnitude throughout the interaction region scales with the upstream value as it would if the magnetic field were completely frozen to the plasma and convected along with the flow. This is not strictly accurate; but it is preferred to the uneven weighting scenario.

Naturally, there are limits to the accuracy of the transformation in Equation (1). There is no guarantee that the IMF is constant during a periapsis pass. In fact the IMF certainly changes during some orbits as a sector boundary is crossed, for example. However, we can only determine the upstream IMF direction when there is a spacecraft there to detect it. Therefore, we are limited to determining the IMF orientation at some point along the orbit and assuming that it remains constant throughout the orbit. We recognize that, on some orbits, this is inappropriate. By compiling data from many orbits, these errors will approach the value of the intrinsic variability of IMF direction (Ness *et al.*, 1971; Burlaga and King, 1979).

Note that there is another difficulty in determining the upstream IMF direction. The magnetometers on MGS measure the ambient magnetic field at their positions at the end of the solar array panels of the spacecraft. The magnetic field at that point is a composite of the IMF and magnetic fields originated by the spacecraft. The data we use have been calibrated to remove the contribution to the magnetic field from the spacecraft, however, some $\sim$ 1 nT contribution still exists. The IMF has a median magnitude of 4 nT during the elliptical orbit phase. This was determined by assigning an upstream IMF for each orbit then taking the median. Brain *et al.* (2003) have determined the median IMF at Mars by taking all of the vectors obtained well upstream of the bow shock and find an IMF of 2.5–3.0 nT. In any case, because the magnitude of the field we are trying to characterize is small, the vector components are also necessarily small. Then, an error of $\pm$ 1 nT in the magnetic field components can contribute to a large error in the derived direction of the IMF. For this reason, determining the direction of the IMF using the measurements upstream of the bow shock is problematic.

However, across the bow shock, the IMF strength is magnified. One can determine the direction of the upstream IMF by looking at the magnetic field in the $y - z$ plane just across the bow shock where the magnitude is much larger, thus lessening the relative contribution of the spacecraft field. According to the Rankine–Huginot shock jump relations, the direction of the magnetic field changes across the bow shock in that there is an amplification of the component of the magnetic field perpendicular to the shock normal and there is no change in the component of the magnetic field parallel to the shock normal. In the coordinate system with $B_z = 0$ upstream of the bow shock, $B_z = 0$ downstream of the bow shock also. However, the turbulence of the magnetic field is magnified inside the bow shock as well as the magnetic field magnitude. In order to rely on the directions determined by this method, we averaged the magnetic field vectors over 100 points starting 25 points downstream of the bow shock crossing. This corresponds to many wave cycles for most orbits and, therefore, mitigates the effects of turbulence. In this way, we use the magnetic field measured just inside of the bow shock to determine the upstream IMF direction for each orbit. In the following sections, the accuracy of this technique will be made evident.

In summary, to convert the MGS data from Mars-Solar-Ecliptic (MSE) (′) coordinates to IMF (unprimed) coordinates, we perform the following transformation:

$$\begin{bmatrix} B_x \\ B_y \\ B_z \end{bmatrix} = \frac{1}{|\mathbf{B}_{SW}|} \begin{bmatrix} 1 & 0 & 0 \\ 0 & \cos\delta & \sin\delta \\ 0 & -\sin\delta & \cos\delta \end{bmatrix} \begin{bmatrix} \cos\theta_a & \sin\theta_a & 0 \\ -\sin\theta_a & \cos\theta_a & 0 \\ 0 & 0 & 1 \end{bmatrix} \begin{bmatrix} B'_{x\mathrm{MSE}} \\ B'_{y\mathrm{MSE}} \\ B'_{z\mathrm{MSE}} \end{bmatrix}. \quad (2)$$

where $|\mathbf{B}_{SW}|$ is the upstream IMF magnitude for the orbit, the angle that rotates the $y' - z'$ component of the upstream IMF into the $+y$ direction is $\delta \equiv \arctan(B'_{zSW}/B'_{ySW})$, and θ_a is the 3.5° aberration of the apparent SW direction.

The MGS orbit is a near-polar orbit in MSE coordinates. With the slow evolution of the local time of the orbit coupled with the slow evolution of the line of apsides, MGS's sampling of low altitude space is limited in MSE coordinates to primarily one $y - z$ quadrant (see Figure 2). As a sector boundary passes, the sign of the IMF reverses. This results in an $\sim 180°$ rotation of IMF coordinates, which provides points in the opposite hemisphere. Further, the orbit departs from a polar geometry in the IMF coordinate system inasmuch as the IMF deviates from the ecliptic plane. For the 746 orbits for which we are able to determine the upstream IMF direction, 50% are within 32° of the ecliptic plane. The other half provide most of the spatial distribution of points in IMF coordinates. Without the changing IMF direction, there would be little sampling of alternate $y - z$ quadrants by MGS.

In the draping geometry, the symmetry outlined in Table I is expected. The behavior of the magnetic field x and y components is evident from inspection of Figure 1. Figure 2 displays the draping of magnetic field lines projected in the $y - z$ plane. The z component behavior comes from the fact that the draped magnetic field line that approaches the nose of the planet at the equator must go over the pole to get past the obstacle. The part of the field line that encounters the planet will be displaced in z more than the part of the field line far away from the planet that does not encounter the obstacle. Therefore, the B_z component describes the direction of the small deviation in z for the field line to slip over or under the obstacle.

3. Data from Representative MGS Orbits

We select 2 MGS orbits (P132 and P541) to demonstrate the nominal magnetic field configuration in the martian magnetosheath at various local times (LT). These orbits, from days February 15 and September 8 of 1998, are shown in Figures 3 and 4, respectively. The local times in MSE coordinates of these orbits are from

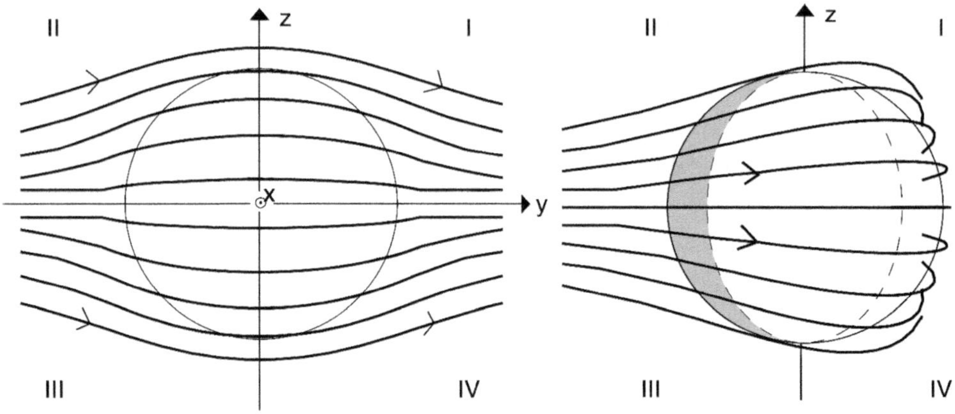

Figure 2. Cartoon depicting the geometry of magnetic field lines in the 'draping' regime. The field lines are shown in the $y - z$ plane on the left and rotated 45° around the z-axis on the right.

TABLE I

Expected draping magnetic field signs by $y - z$ quadrant

Quadrant	B_x	B_y	B_z
I	−	+	−
II	+	+	+
III	+	+	−
IV	−	+	+

noon and the terminator, respectively. Each figure displays the polar projection of MGS vector magnetic field measurements in IMF coordinates on the left-hand side. The right-hand panels in Figures 3 and 4 give the predicted values from the Spreiter and Stahara (1980, 1992) model for Mars under nominal conditions at the same points as given on the left. The Spreiter and Stahara model treats the SW flow as a gasdynamic (GD) flow with convected magnetic field. Although there are more sophisticated models of the interaction (e.g., Brecht, 1997; Liu *et al.*, 2001), the simplified GD representation of the interaction is very good on average, as is evident by comparing the right-hand panels to the left-hand panels.

MGS orbit P132 provides a rare glimpse of the magnetic field near the subsolar point. Figure 3 shows the projection of the magnetic field in the (x, y) plane from P132, revealing two important features of magnetic draping. First, at the nose of the planet, the magnetic field vectors have large magnitude and point in the $+y$ direction. The sign is as expected for IMF coordinates. The large magnitude stems from the pileup of field lines in front of the planet. Vectors are locally horizontal at

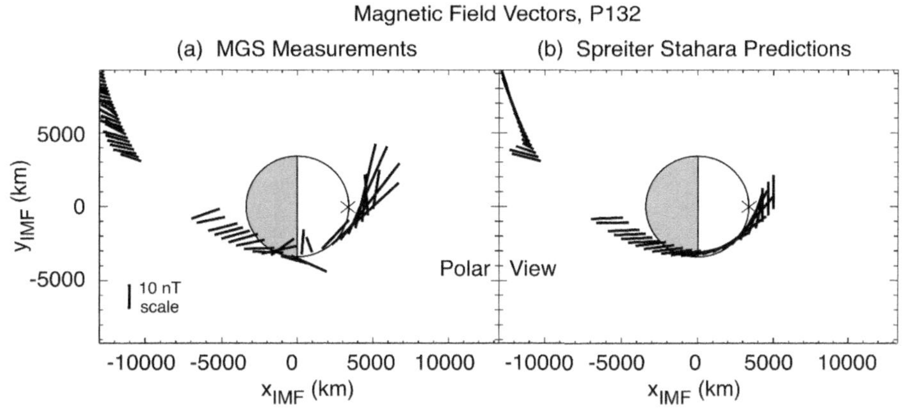

Figure 3. The $x - y$ plane projection of magnetic field vectors from MGS orbit p132 is shown. (a) is the polar view of the magnetic field vectors measured along the MGS orbital track. (b) is the predicted magnetic field by the GD model for the same points. The 'X' marks the subsolar point on the planet. The local time of the orbit was near noon, but transformation to IMF coordinates rotates the polar orbit to an orbit at 45° inclination. Periapsis occurred just sunward of the terminator.

the nose. As the trajectory moves away from the nose of the planet, an x component to the magnetic field enters as is necessary to maintain locally horizontal magnetic field (see Bertucci *et al.*, 2003). Secondly, in the tail, there are two obvious lobes to the tail. Although a data gap exists here, the two tail lobes are probably separated by the $y = 0$ plane. The vectors in the $-y$ hemisphere point sunward and the vectors in the $+y$ hemisphere point tailward.

Figure 4 shows the symmetry about the $y = 0$ plane of the magnetic field vectors. The draping of magnetic field is obvious in orbit P541 as B_x goes from tailward in the $+y$ hemisphere to a dawn-dusk orientation at $y = 0$ to sunward in the $-y$ hemisphere. Because periapsis occurred close to the equator in IMF coordinates, the inbound and outbound passes trace almost on top of one another in Figure 4. The symmetry between the $+z/-z$ hemispheres becomes clear in this view.

Direct comparison with GD model predictions suggest that the GD model does a good job of predicting the morphology of the magnetic field on representative orbits, especially when the data comes from an orbit with steady upstream conditions. For most orbits, the GD model does best when the inner boundary of the model is mapped to the Magnetic Pileup Boundary (MPB) in the data. When the inner boundary is placed elsewhere, e.g., at the ionopause, the GD model does not reproduce the observed increase of magnetic field at the MPB. Even so, the actual situation is far more complicated than indicated by the simple GD model. MGS consistently detects the IMF permeating to much lower altitudes (see Acuña *et al.*, 1998) indicating that the MPB is not the actual obstacle. Therefore, the interaction is probably better described by more sophisticated models. However, we continue

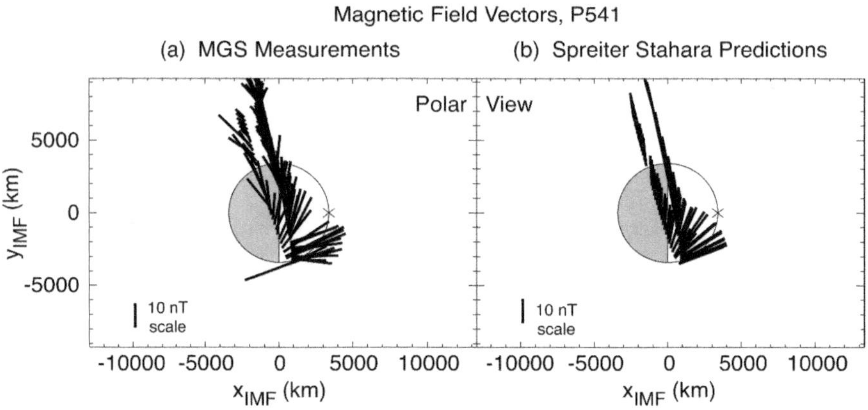

Figure 4. The projection of magnetic field vectors from MGS orbit p541 into the $x - y$ plane is shown in the same format as Figure 3. The local time of the orbit was near the terminator.

to compare the data to the GD model because of its simplicity and its success in the magnetosheath.

These selected MGS orbits depict the nominal form of magnetic field draping at Mars. The selected data were chosen because of their difference in local time that illuminates two different slices through the interaction region. Although crustal magnetic fields may contribute to the morphology at low altitudes, the magnetic field in the magnetosheath and magnetotail is best described by draping of IMF field lines around the planet. In the next section, we compile data to obtain an average view of the magnetic field morphology rather than this representative view.

4. Compilation of MGS Data

For a more complete view of the magnetic field morphology in the SW interaction with Mars, we compile a global picture from MGS data and investigate how the magnetic field varies as a function of position. First, we investigate the distribution of magnetic field values observed over the course of many spacecraft orbits. Then we average those to produce a single global picture.

We take all the MGS data from within the bow shock for the 746 orbits for which we are able to determine the upstream IMF direction. For each orbit, we remove the data obtained while the spacecraft was directly over a crustal source. This was accomplished by excluding data acquired at an areographic position for which the magnitude of the average magnetic field from the mapping orbits from Connerney *et al.* (2001) ($h \approx 400$ km) is greater than 10 nT. After excluding data that could be contaminated by crustal fields, we further decimate the data volume by computing an average magnetic field vector over each 100 km altitude bin for each orbit. From the elliptical orbits, that leaves 26,777 magnetic field vectors in magnetic coordinates. We then study the magnetic field morphology as a function of position

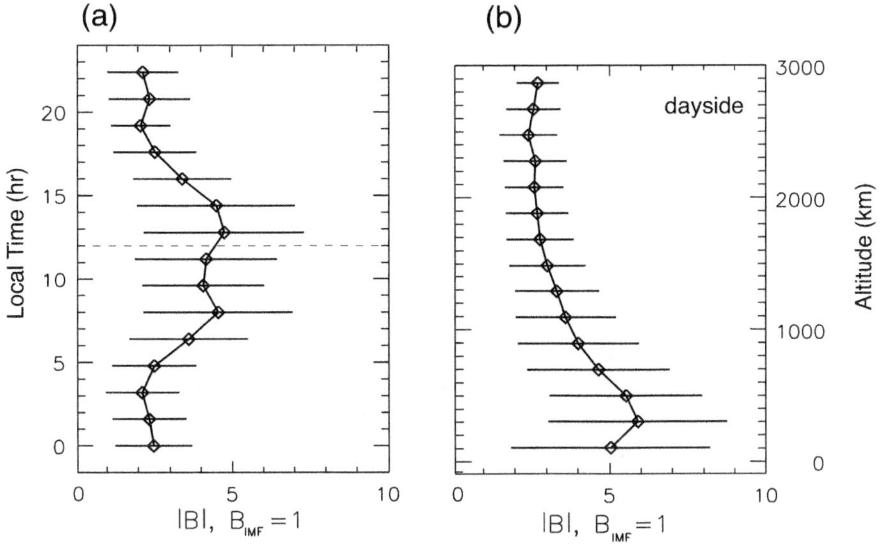

Figure 5. Average (diamond) and 1σ standard deviation (bar) of the magnetic field magnitude are plotted as a function of (a) local time and (b) altitude. The data are from 100 km altitude bins from each MGS elliptical orbit for which the upstream IMF could be determined. Only data from the dayside are included the averages for (b).

by plotting the evolution of the distribution of parameters of these vectors with position.

4.1. Magnetic Field Magnitude

First, we investigate the evolution of the magnetic field magnitude as a function of position. $|B|$ is expected to peak at the obstacle boundary at low solar zenith angles. Figure 5(a) shows the progression of magnetic field magnitude with local time. In the sector spanning 12:00 ±3, $\langle |B|/|B_{\rm IMF}| \rangle$ is highest at 4.0–4.5. Multiplying by the median normalization factor for the upstream IMF, $\langle |B(12{:}00)| \rangle = 16$–18 nT. These distributions average over all altitudes in the local time bin making the error bars large. The difference with altitude is discussed next. However, note that the magnetic field magnitude distributions are symmetric on the morning and evening sides. Also, note that the standard deviations are smaller on the nightside than on the dayside of Mars showing that the magnetic field magnitude does not vary much on the nightside. The larger error bars on the dayside indicate a wide range of values due to the dramatic pileup of magnetic field in front of the planet.

That pileup can be observed in Figure 5(b), which shows the evolution of $|B|$ with altitude. Figure 5(b) uses only points from the dayside of Mars. $\langle |B| \rangle$ decreases with increasing altitude above 300 km. The magnetic pileup boundary (MPB), defined as the position of steep increase of magnetic field (Vignes *et al.*, 2000), occurs at an altitude between 800–1200 km on the dayside. The magnetic field remains high in the magnetic pileup region (MPR) which spans the altitude range from the ionosphere to the MPB, generally from 400–1200 km. Figure 5(b) shows that the average magnetic field magnitude drops with increasing altitude even in this region whereas the definition would suggest it would be constant with altitude there. The reason for this is that the magnetic field magnitude at a given altitude drops with solar zenith angle and more low altitude data were obtained near the terminator than at low solar zenith angles.

Brain *et al.* (2003) have already found the MGS average magnetic field magnitude as a function of position in both cylindrical coordinates and in planetocentric coordinates. This investigation confirms their results that the magnetic field magnitude increases with decreasing altitude as the magnetic field piles up in front of the planet. Whereas the Brain *et al.* (2003) study focused on the relative contributions of planetary and SW magnetic fields, this study emphasizes only the SW contribution. Data taken near crustal magnetic fields has been eliminated and the data are presented in the IMF coordinate system.

4.2. Inclination Angle

One measure of draping of magnetic field around a planet is the average inclination of the magnetic field on the dayside. The inclination angle, I, is defined as the angle the magnetic field makes to the local horizontal, or $I \equiv \arcsin(B_\rho/|\mathbf{B}|)$, where B_ρ is the radial component of the magnetic field in spherical coordinates (see Figure 1). In a draping configuration, the magnetic field is mostly locally horizontal on the dayside, or $I = 0°$. However, a small radial component does exist simply because the field line is not at a constant altitude as it is wrapped around the planet. Further, the sign of the radial component changes from negative, indicating downward or towards the planet, in the $y < 0$ hemisphere to positive or outward in the $y > 0$ hemisphere.

Uncalibrated MGS data have already been used to show properties of the inclination angle on the dayside of Mars (Crider *et al.*, 2001). They found the average inclination angle was closer to 0° within the ionosphere than above it. However, the distribution was much broader within the ionosphere, allowing for the existence of large inclination angles in the martian ionosphere. Now with the calibrated data, we are able to quantify I. Figure 6 is a histogram showing the distribution of I on the dayside of Mars using the same data as in the previous section. The dayside values are divided into two groups depending on the sign of the y coordinate of the observation. The shaded histogram is from points in the western hemisphere where a positive angle is expected. The unshaded histogram is from the points where a negative angle is expected. Both histograms are peaked at an angle offset from zero by $\sim 5°$, indicating a small but finite inclination to magnetic field lines in the draping on the dayside. The distributions are both skewed towards 0° which indicates that the magnetic field is rarely nearly-radial and more often nearly-horizontal. Because the orbit by orbit transformation of data to IMF-coordinates is

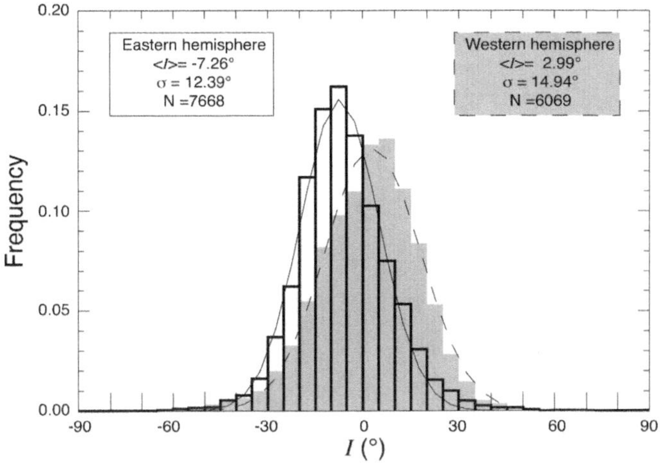

Figure 6. The distribution of observed inclination angles on the dayside of Mars is plotted for points from the eastern (white) and western (gray) hemispheres in magnetic coordinates. Parameters to the superposed Gaussian fit to the distributions are given.

not always accurate, there are some points that are placed in the wrong hemisphere here, especially points close to the $y = 0$ plane. For this reason, we present the evolution of the inclination angle as a function of local time in Figure 7(a).

The local time evolution of the inclination angle is demonstrative of the draping of magnetic field around Mars. Figure 7(a) depicts the constant increase in inclination angle with local time. I passes through zero at local noon and has odd function symmetry about noon. On the dayside, the standard deviations are small, showing that the draping geometry is well defined and the angle of inclination does not vary much with altitude. The distributions broaden on the nightside, however. The standard deviation is large approaching midnight from the a.m. and p.m. sides as would be expected for points in the magnetotail at multiple altitudes.

Although I does not change as much with altitude on the dayside as on the nightside, it has a systematic altitude dependence. We show the explicit evolution with altitude in Figure 7(b). This figure uses data from both the eastern and western hemispheres, but multiplies the inclination angles from the eastern hemisphere by -1 before finding the average and standard deviation for each altitude bin. $\langle I \rangle$ increases monotonically with altitude from 1° at 100 km to 13° at 1900 km. The standard deviation is minimal from 400–700 km. This is in agreement with the results of Bertucci *et al.* (2003) that find draping is most well-defined in the magnetic pileup region below the MPB. The higher standard deviation at high altitude reflects the lack of organized draping above the MPB. Note that turbulence exists at these altitudes, but averaging over large spatial bins removes the contribution from waves here. In contrast, below the MPR at ionospheric altitudes the high standard deviation is probably due to ionospheric effects on the magnetic field (Crider *et al.*, 2001). Flux ropes are known to form in the martian ionosphere with random

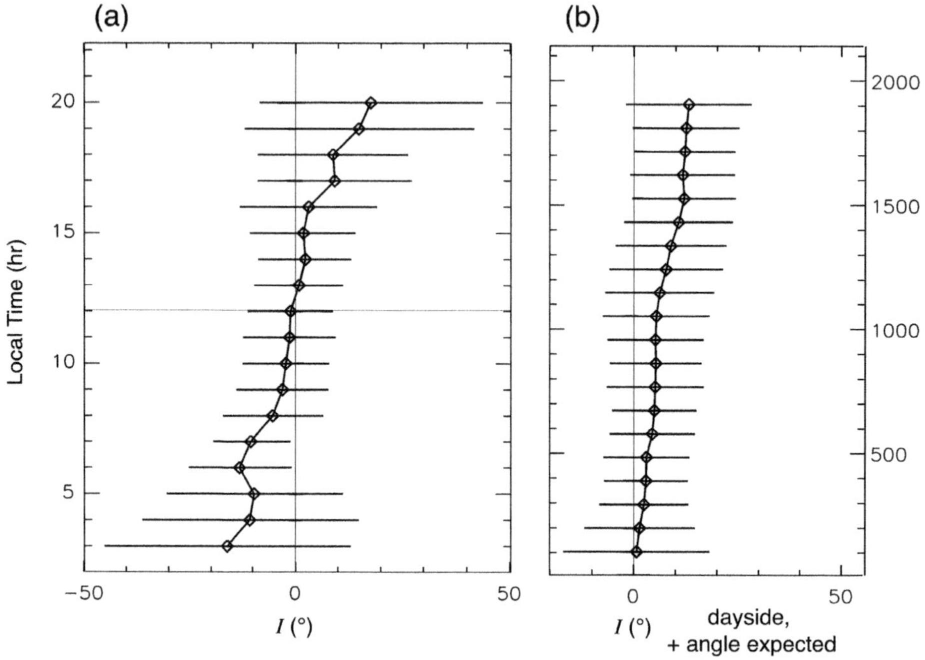

Figure 7. Average (diamond) and 1σ standard deviation (bar) of the angle of inclination of the magnetic field are plotted as a function of (a) local time and (b) altitude. The points used in the altitude plot are taken only from the dayside.

orientation to the local vertical (Vignes *et al.*, 2004). Also, although known crustal sources have been excluded from this data set, magnetic reconnection between IMF and crustal fields can lead to high $|I|$ at locations laterally displaced from the source itself (Brain *et al.*, 2001). Therefore we reaffirm that the ionospheric magnetic field at Mars has an average value that is very close to locally horizontal. However, highly inclined magnetic fields do exist in the ionosphere with high frequency. This fact must be included in ionospheric models of Mars because vertical transport is enabled where inclined magnetic field lines exist.

4.3. Flare of the Magnetotail

The flaring of magnetic field lines in the magnetotail is another diagnostic parameter of magnetic field draping. The flare angle is defined as the angle the magnetic field makes to the Mars–Sun line, or $\theta_f \equiv \arccos(B_x/|\mathbf{B}|)$ (see Figure 1). The flare angle is related to the width of the magnetotail and influences the shape of plasma boundaries.

As indicated in Figure 1, a hemispheric difference is expected for θ_f in the standard draping picture. The sunward magnetic field in the $-y$ hemisphere leads to a flare angle near 0° whereas the tailward field in the $+y$ hemisphere leads to a flare angle near 180°. Figure 8 shows the distribution of θ_f obtained for the points

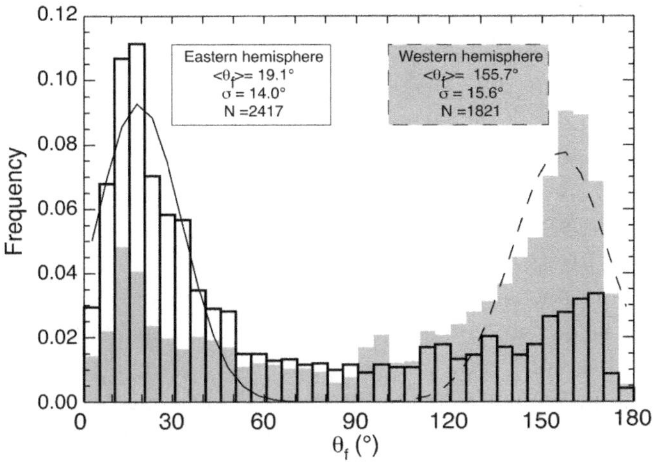

Figure 8. The distribution of observed flare angles on the dayside of Mars is plotted for points from the eastern (white) and western (gray) hemispheres in magnetic coordinates. Parameters to the superposed Gaussian fits to the distributions are given.

with $x < 0$ in the above data set. Two distributions are shown as the data have been separated into y hemispheres. The shaded histogram is the distribution of θ_f from the western hemisphere ($y > 0$ where it is expected that $B_x < 0$) while the unshaded histogram is from the eastern hemisphere. Also plotted are Gaussian fits to the distributions and their parameters. It is important to note that the spacecraft sampling in the tail is not uniform and that there are selection effects inherent in these large spatial groupings. Both distributions are peaked near 21.0° away from the Mars-Sun line and exhibit an excess of data in the same θ_f range as the other hemisphere. These, presumably, correspond to data points that are misplaced by the transformation into IMF coordinates or to places where the neutral sheet is offset from the $y = 0$ plane (Marubashi *et al.*, 1985).

The magnetotail of Mars has already been explored by the Phobos-2 spacecraft and found to have a flare angle of 23° (Luhmann *et al.*, 1991) and 27.2° on average (Zhang *et al.*, 1994; Rosenbauer *et al.*, 1994). The martian magnetotail is known to respond to changes in SW pressure (Verigin *et al.*, 1993; Dubinin *et al.*, 1996). The average angle observed by MGS is close to but slightly smaller than the angle determined from the Phobos-2 data probably due to the difference in spatial coverage of the two spacecraft. Most of the Phobos-2 data come from a distance of 9600 km from the planetary center and from low latitudes. The MGS data was obtained out of the equatorial plane and spans distances from 3700 km to past the Phobos-2 distance. MGS's orbital geometry and helps to fill in a large spatial extent of the tail compared to the region samples by Phobos-2. The spatial distribution of the flare angle is discussed in the following section.

4.4. Spatial Distribution

Finally, we present global maps of the magnetic field in the martian magnetosheath and magnetotail. To compile global maps of magnetic field morphology, we take cubic bins with sides of length 0.1 R_M in all three Cartesian directions. We average the magnetic field components from each orbital segment that falls within the bin. In doing so, the MGS data is found to cover 5.6% of these bins from the limits $-3\ R_M \leq x \leq 1.6\ R_M$; $|y| \leq -3\ R_M$; $|z| \leq -3\ R_M$. However, by mapping all the data to only one quadrant using the symmetry in Table I, coverage is improved. Still, there are not enough points to present a planar cut through the data set. Instead, we present a projection of points from several z values onto a single $x - y$ half plane in Figure 9. In that projection, the image represents the average morphology of the magnetic field in the martian magnetosheath from multiple z coordinates. The data at positions closer to the Mars–Sun line are from higher $|z|$ coordinates and data at positions far from the Mars–Sun line are from more equatorial locations. The left hand panels in Figure 9 show the average magnetic field morphology from the MGS data. The top panels illustrate the magnitude of the field; the middle panels give the inclination angle; and the bottom panels show the flare angle.

The right hand panels in Figure 9 give the predicted values from the Spreiter and Stahara (1980, 1992) model for the same data coverage as in the data panels. The Spreiter and Stahara model treats the SW as a gasdynamic flow with convected magnetic field. The model was run for $M = 8$. For ease of comparison, the size and color scales are the same for each pair of data and model plots.

The top panels show that the magnitude of the magnetic field is highest in front of the planet. The build up in magnetic field indicates the dominance of magnetic field pressure on the dayside of the planet. $|\mathbf{B}|$ decreases with increasing solar zenith angle. The bulk of the magnetic field pileup begins at the magnetic MPB and it remains high throughout the MPR. The MPR corresponds to the magnetic barrier at Venus (see Zhang *et al.*, 1991) although the magnetic field pileup occurs at higher altitudes at Mars. The GD model falls short of predicting the magnitude of pileup in front of the planet. However, it does a better job if the inner boundary of the model is elevated to the observed MPB position.

The middle panels in Figure 9 show a large green area close to the planet where the magnetic field is primarily horizontal. In contrast, the nightside magnetic field tends to be highly inclined to the local horizontal. This is the magnetic field geometry that is expected in an induced magnetotail (McComas *et al.*, 1986). Note, however, that the highest inclination angles are observed in the martian wake, whereas the gasdynamic model predicts their presence more in the flanks. Nearly radial magnetic fields, such as these, have strong implications for the existence of nightside ionospheric holes at Mars as were observed at Venus (Marubashi *et al.*, 1985). However, the possibility remains that the magnetic field in the martian wake

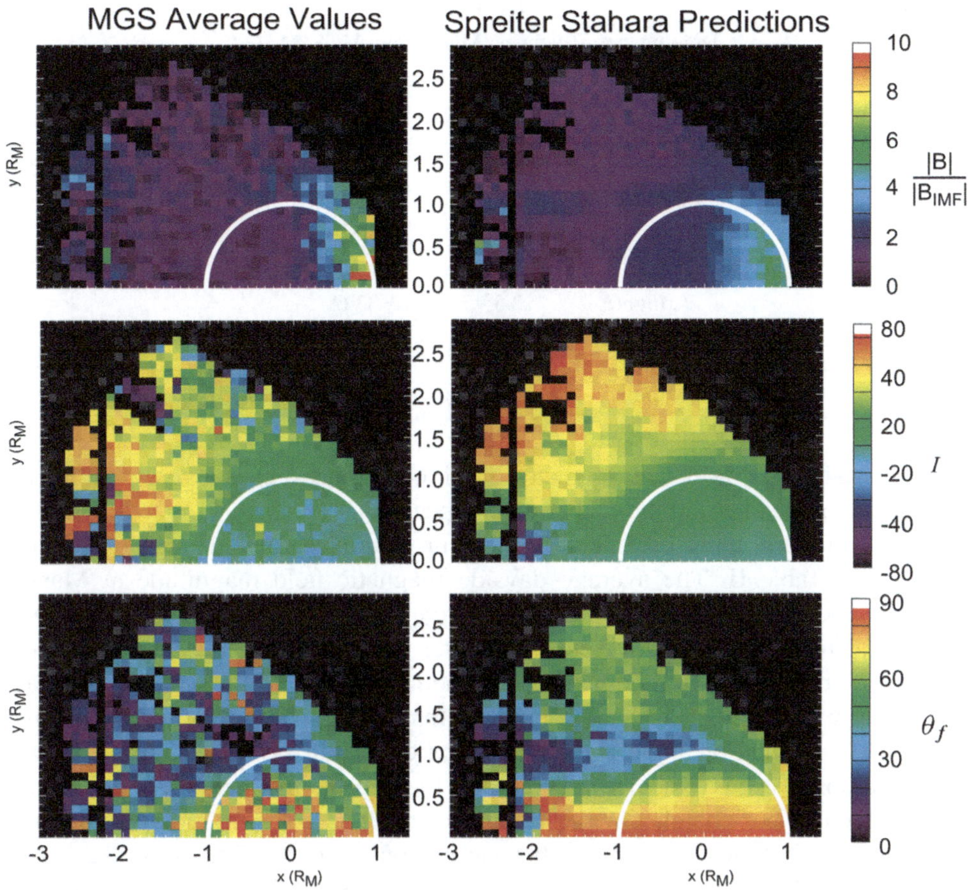

Figure 9. The left-hand panels are the MGS composite of the magnetic field around Mars. The right side shows the GD model predictions using the same coverage as the MGS data. The panels from top to bottom display the magnetic field magnitude, inclination angle, and flare angle.

contains a significant contribution from magnetic flux of planetary origin that leaks out into the tail (Krymskii *et al.*, 2002).

The bottom panels show how the magnetic field rotates from mostly cross-planet ($\theta_f \approx 90°$) close to the Mars–Sun line to a more sunward/tailward orientation ($\theta_f << 90°$) at larger $|y|$. Although only one half plane is shown here, referring to Figure 1 will show how this is the expected geometry as the sign of the B_x component changes at the Mars-Sun line. The gasdynamic model predicts a minimum in the flare angle tailward of the terminator at $y \sim 1\ R_M$. That minimum also exists in the data; however, it is not as clear. A major difference in the model and measured flare angle appears in the planetary wake, where the high flare angles predicted by the model are not observed. However, this is most likely due to errors in aligning the coordinate systems in the data set as discussed in Section 2. Another difference between the measured and predicted flare angle is that θ_f, especially at

TABLE II

Draping parameters at Mars from MGS MAG data

Parameter	Average	Standard dev.
$\|\mathbf{B}\|$ (day)	20.7 nT	$\pm$ 14.7 nT
$\|\mathbf{B}\|$ (night)	11.6 nT	$\pm$ 7.6 nT
I (day)	5.6°	$\pm$ 13.6°
I (night)	12.5°	$\pm$ 33.6°
θ_f (night)	21.0°	$\pm$ 14.8°

large y, is slightly lower than predicted at all local times. This suggests that the tail is more elongated behind the planet than expected.

The average draping parameters found for certain regions of interest are summarized in Table II. The average dayside magnetic field magnitude at Mars is 20.7 nT. The magnetic field values in this survey are all taken in positions far away from known crustal magnetic sources. Therefore, the magnitude is due primarily to the SW interaction. In the magnetotail, the magnetic field lines flare on average 20° to the x direction. This is slightly lower than predicted by Spreiter and Stahara (1992). The inclination angle of the magnetic field on the dayside is 5.6° on average (where a positive I is expected). Therefore, the dayside magnetic field is predominately locally horizontal. On the nightside, there is a larger radial component to the magnetic field which raises the inclination angle to 12.5° on average with an extremely large scatter about the mean. Again, errors in the mapping and tail asymmetries lead to the large standard deviation and the expectation of negative angles.

5. Conclusion

Compiling data in IMF coordinates from the MGS elliptical orbits shows that on average the magnetic field morphology of the martian magnetosheath is described well by a draping picture. Further, there are no surprises compared to previous work. On the dayside, the magnetic field lines are inclined on average $< 10°$ from the local horizontal. On the nightside, the magnetic field turns to a more radial orientation as the draped IMF produces an induced magnetotail.

We compare both selected data and the compiled data set to model magnetic field predictions from (Spreiter and Stahara, 1992). The GD model typically underestimates the magnitude of magnetic field pileup on the dayside of Mars. It also predicts a broader tail configuration than is observed. For the most part, however, the GD model is an adequate representation of the magnetic field in the SW interaction with Mars remote from magnetic crustal sources. It is especially good above the martian MPB.

However, regions of intense crustal magnetization do exist at Mars (Acuña *et al.*, 1998; Connerney *et al.*, 2001; Brain *et al.*, 2003). Now that we have developed a picture of the magnetic field morphology resulting from the SW alone, we will be better able to deconvolve the contributions of the solar wind and planetary magnetic fields in regions where crustal fields play a role. These crustal magnetic fields are the distinguishing feature of the Martian obstacle to the solar wind. Although the MAG/ER instrument has provided much insight into the complex interaction of the solar wind with the martian obstacle, more plasma data are needed to fully unravel the intricacies of the interaction.

Acknowledgement

DHC was supported by NASA grant NAG5-11225.

References

Acuña, M. H., Connerney, J. E. P., Wasilewski, P. *et al.*: 1998, 'Magnetic Field and Plasma Observations at Mars: Initial Results of the Mars Global Surveyor Mission', *Science* **279**(5357), 1676–1680.

Albee, A. L., Arvidson, R. E., Pallucioni, F. and Thorpe, T.: 2001, 'Overview of the Mars Global Surveyor Mission', *J. Geophys. Res.* **106**(E10), 23291–23316.

Alfvén, H.: 1957, 'On the Theory of Comet Tails', *Tellus* **9**, 92–96.

Bertucci, C., Mazelle, C., Crider, D.H. *et al.*: 2003, 'Magnetic Field Draping Enhancement at the Martian Magnetic Pileup Boundary from Mars Global Surveyor Observations', *Geophys. Res. Lett.* **30**(2), 1099, 10.1029/2002GL015713.

Brain, D. A., Bagenal, F., Crider, D. H., Acuña, M. H. and Connerney, J. E.: 2001, *Martian Magnetic Topology: Evaluation of Crustal Magnetization and Solar Wind Interaction Models Using MGS MAG Data*, American Geophysical Union Spring Meeting.

Brain, D. A., Bagenal, F., Acuña, M. H. and Connerney, J. E. P.: 2003, 'Martian Magnetic Morphology: Contributions from the Solar Wind and Crust', *J. Geophys. Res.* **108**(A12), 1424, 10.1029/2002JA009482.

Brecht, S. H.: 1997, 'Hybrid Simulations of the Magnetic Topology of Mars', *J. Geophys Res.* **102**(A3), 4743–4750.

Burlaga, L. F. and King, J. H.: 1979, 'Intense Inter-Planetary Magnetic-Fields Observed by Geocentric Spacecraft During 1963–1975', *J. Geophys. Res.* **84**(NA11), 6633–6640.

Cloutier, P. A., Law, C. C., Crider, D. H. *et al.*: 1999, 'Venus-Like Interaction of the Solar Wind with Mars', *Geophys. Res. Lett.* **26**(17), 2685–2688.

Connerney, J. E. P., Acuña, M. H., Wasilewski, P. J. *et al.*: 2001, 'The Global Magnetic Field of Mars and Implications for Crustal Evolution', *Geophys. Res. Lett.* **28**(21), 4015–4018.

Crider, D. H., Acuña, M. H., Connerney, J. E. P. *et al.*: 2001, 'Magnetic Field Draping around Mars: Mars Global Surveyor Results', *Adv. Space Res.* **27**(11), 1831–1836.

Crider, D. H., Acuña, M. H., Connerney, J. E. P. *et al.*: 2002, 'Observations of the Latitude Dependence of the Location of the Martian Magnetic Pileup Boundary', *Geophys. Res. Lett.* **29**(8), 10.1029/2001 GLO13860.

Dolginov, S. S., Yeroskenko, Y. G., Zhuzgov, L. N.: 1973, 'The Magnetic Field in the Very Close Neighborhood of Mars According to Data from the Mars-2 and 3 Spacecraft', *J. Geophys. Res.* **78**(22), 4779–4786.

Dolginov, S. S.: 1978, 'On the Magnetic-Field of Mars - Mars-5 Evidence', *Geophys. Res. Lett.* **5**(1), 93–95.

Dubinin, E., Sauer, K., Lundin, R. *et al.*: 1996, 'Plasma Characteristics of the Boundary Layer in the Martian Magnetosphere', *J. Geophys. Res.* **101**(A12), 27061–27075.

Kallio, E. J. and Koskinen, H. E. J.: 2000, 'A Semiempirical Magnetosheath Model to Analyze the Solar Wind-Magnetosphere Interaction', *J. Geophys. Res.* **105**(A12), 27469–27479.

Krymskii, A. M., Breus, T. K., Ness, N. F. *et al.*: 2002, 'Structure of the Magnetic Field Fluxes Connected with Crustal Magnetization and Top-Side Ionosphere at Mars', *J. Geophys. Res.* **107** (A9), 1245, doi: 10.1029/2001 JA000239.

Liu, Y., Nagy, A. F., Gombosi, T. I., DeZeeuw, D. L. and Powell, K. G.: 2001, 'The Solar Wind Interaction with Mars: Results of Three-Dimensional Three-Species MHD Studies', *Adv. Space Res.* **27**(11), 1837–1846.

Luhmann, J. G., Russell, C. T., Schwingenschuh, K. and Yeroshenko, Ye.: 1991, 'A comparison of Induced Manetotails of Planetary Bodies: Venus, Mars, and Titan', *J. Geophys. Res.* **96**(A7), 11199–11208.

Marubashi, K., Grebowsky, J. M., Taylor, H. A. *et al.*: 1985, 'Magnetic Field in the Wake of Venus and the Formation of Ionospheric Holes', *J. Geophys. Res.* **90**(NA2), 1385–1398.

Mazelle, C., Bertucci, C., Rème, H. *et al.*: 2002, 'The Magnetic Pileup Boundary at Mars: a Comet-Like Feature in the Interaction of the Planet Atmosphere with the Solar Wind' (in press).

McComas, D. J., Spence, H. E., Russell, C. T. *et al.*: 1986, 'The Average Magnetic-field Draping and Consistent Plasma Properties of the Venus Magnetotail', *J. Geophys. Res.* **91**(A7), 7939–7953.

Mitchell, D. L., Lin, R. P., Mazelle, C. *et al.*: 2001, 'Probing Mars' Crustal Magnetic Field and Ionosphere with the MGS Electron Reflectometer', *J. Geophys. Res.* **106**(E10), 23418–23427.

Ness, N. F., Hundhaus, A. J. and Bame, S. J.: 1971, 'Observations of Interplanetary Medium - VELA 3 and IMP 3, 1965–1967', *J. Geophys. Res.* **76**(28), 6643.

Rosenbauer, H., Verigin, M., Kotova, G., Livi, S. *et al.*: 1994, 'On the Correlation of the Magnetic Field in the Martian Magnetotail to the Solar Wind Parameters', *J. Geophys. Res.* **99**(A9), 17199 –17204.

Russell, C. T., Luhmann, J. G., Spreiter, J. R. and Stahara, S. S.: 1984, 'The Magnetic Field of Mars–Implications from Gas-Dynamic Modeling', *J. Geophys. Res.* **89**(NA5), 2997–3003.

Slavin, J. A., Holzer, R. E., Spreiter, J. R., Stahara, S. S. and Chausee, D. S.: 1983, 'Solar-Wind Flow About the Terrestrial Planets 2. Comparison with Gas-Dynamic Theory and Implications Flor Solar-Planetary Interactions', *J. Geophys. Res.* **88**(NA1), 19–35.

Spreiter, J. R. and Stahara, S. S.: 1980, 'A New Predictive Model for Determining Solar Wind-Terrestrial Planet Interactions', *J. Geophys. Res.* **85**(NA12), 6769–6777.

Spreiter, J.R. and Stahara, S. S.: 1992, in J. G. Luhmann, M. Tatrallyay and R. O. Pepin (eds), 'Computer Modeling of the Solar Wind Interaction with Venus and Mars", *Venus and Mars: Atmospheres, Ionospheres, and Solar Wind Interactions, Geophysical Monograph* **66**, 345.

Tanaka, T.: 1993, 'Configurations of the Soar-Wind Flow and Magnetic-Field Around the Planets with no Magnetic-Field - Calculation by a New MHD Simulation Scheme', *J. Geophys. Res.* **98**(A10), 17251–17262.

Verigin, M. I., Gringauz, K. I., Kotova, G. A. *et al.*: 1993, 'The Dependence of the Martian Magnetopause Shape and Bow Shock on Solar Wind Dynamic Pressure According to Phobos 2 TAUS Ion Spectrometer Measurements', *J. Geophys. Res.* **98**(A2), 1303–1309.

Vignes, D., Mazelle, C., Rème, H. *et al.*: 2000, 'The Solar Wind Interaction with Mars: Locations and Shapes of the Bow Shock and the Magnetic Pile-up Boundary from the Observations of the MAG/ER Experiment Onboard Mars Global Surveyor', *Geophys. Res. Lett.* **27**(1), 49–52.

Vignes, D., Acuña, M. H., Connerney, J. E. P., Crider, D. H., Rème, H. and Mazelle, C.: 2004, 'Magnetic Flux Ropes in the Martian Ionosphere: Global Characteristics', *Space Sci. Rev.* **111** (1–2).

Zhang, T. L., Luhmann, J. G. and Russell, C. T.: 1991, 'The Magnetic Barrier at Venus', *J. Geophys. Res.* **96**(A7), 11145–11153.

Zhang, T. L., Schwingenschuh, K., Russell, C. T. *et al.*: 1994, 'The Flaring of the Martian Magnetotail Observed by the Phobos-2 Spacecraft', *Geophys. Res. Lett.* **21**(12), 1121–1124.

MAGNETIC FLUX ROPES IN THE MARTIAN ATMOSPHERE: GLOBAL CHARACTERISTICS

D. VIGNES[1], M.H. ACUÑA[1], J.E.P. CONNERNEY[1], D.H. CRIDER[1], H. RÈME[2] and C. MAZELLE[2]

[1]*NASA Goddard Space Flight Center, Greenbelt, MD 20771, U.S.A.*
(e-mail: vignes@lepmom.gsfc.nasa.gov)
[2]*Centre d'Etude Spatiale des Rayonnements, Toulouse, France*

Accepted in final form 14 March 2003

Abstract. We report observations of magnetic fields amplitude, which consist of a series of individual spikes in the Martian atmosphere. A minimum variance analysis shows that these spikes form twisted cylindrical filaments. These small diameter magnetic filaments are commonly called magnetic flux ropes. We examine the global characteristics of magnetic flux ropes, which are observed on 5% of the elliptical orbits of Mars Global Surveyor. Flux ropes are more often observed in Venus' atmosphere (70% of the orbits). In this paper we report some of the global characteristics of the flux ropes identified in the Martian atmosphere. No flux ropes are observed in the southern hemisphere of Mars. Most of them occur at high solar zenith angles, close to the terminator plane, and at high latitude with altitudes below 400 km. The orientation of the flux ropes appears random while in the case of Venus the orientation is more horizontal near the terminator for altitudes greater than 200 km. We have identified fewer flux ropes for SZA between 40 to 60 deg and for SZA lower than 20 deg, like in the case of Venus (Elphic and Russell, 1983b). Statistically, Mars' ionosphere with SZA range between 40^{circ} to 60^{circ} is less magnetized than near the subsolar point. As the Martian ionosphere is quite often magnetized by the magnetic components of the crustal field, this crustal magnetic field seems to inhibit the flux ropes formation in the southern hemisphere. However, some orbits without crustal magnetic field, called magnetic cavities, were observed without flux ropes. So the flux ropes formation process seems to be suppressed by another factor, like the solar wind dynamic pressure for Venus (Krymskii and Breus, 1988).

1. Introduction

Before the Mars Global Surveyor (MGS) Mission, the nature of the solar wind interaction with Mars was not fully determined. The MAG/ER experiment on MGS, which includes two magnetometers and an electron reflectometer, has shown that the interaction of the solar wind with Mars is dominated by the planet's ionosphere and atmosphere. However, recent observations by the same experiment show intense crustal magnetic sources mainly confined to the southern highlands (Acuña *et al.*, 1999). The amplitude of such crustal magnetic field can reach values of 200 nT at altitudes greater than 400 km. The solar wind interaction with Mars appears more complicated than with Venus due to the effect of the crustal magnetic fields.

Space Science Reviews **111:** 223–231, 2004.

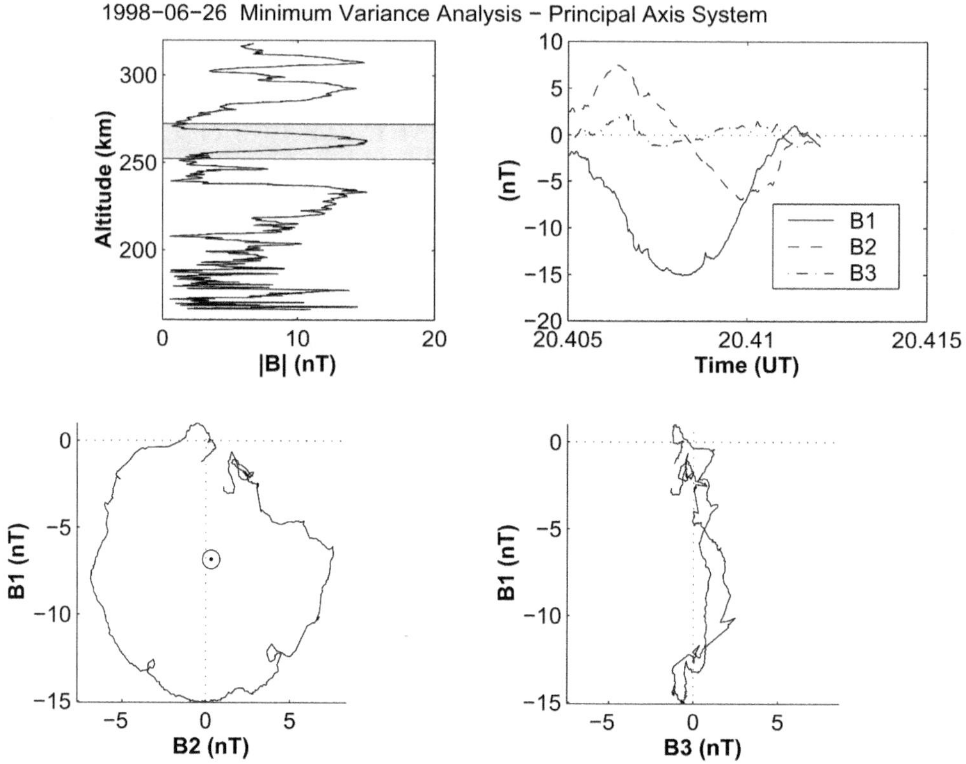

Figure 1. The upper left panel shows the magnetic field amplitude observed as a function of altitude. The shaded spike, above 250 km, corresponds to the flux rope studied. The magnetic field components in the principal axis system of this flux ropes are shown in the upper right panel. The lower panel shows the hodograms in the principal axis system coordinates of the flux ropes selected.

In spite of this complicated effect, flux ropes are observed in the Mars ionosphere. A magnetic flux rope is a cylindrical body of magnetized plasma with magnetic topology characterized by relatively strong axial fields near the center and weaker, more azimuthal fields towards the outer edge. The first case of a flux rope identified at Mars was reported by Cloutier *et al.* (1999). Flux ropes are also observed in the Venus ionosphere, in the geomagnetic tail, and also on the Sun in many different forms (Priest, 1990). For this study we use exclusively the data available during Mars Global Surveyor elliptical orbits. On 5% of these orbits, we observed magnetic flux ropes. We still don't know the mechanism responsible for the flux ropes formation at Mars and Venus. One of the possible mechanisms of formation could be the destruction of the magnetic field pushed down by high dynamical pressure of the solar wind according to Krymskii and Breus (1988). The goal of this paper is to examine the global characteristics of Martian flux ropes.

2. Flux Ropes Identification

The magnetic character of the flux ropes at Mars appears chaotic with no common direction or amplitude. The magnetic field amplitude consists of a series of individual spikes in the Martian atmosphere as seen in the upper left corner of Figure 1. In fact, these spikes show some order: they can be expressed in 2 dimensions. To order the vector data, we use the minimum variance technique also called principal axis technique (Elphic and Russell, 1983a). For such a study, we use the magnetic data with their highest resolution: 8 to 32 samples per second. We use the principal axis technique for the shaded rope in the upper left corner. The upper right corner shows the magnetic field components of this flux rope in the principal axis system. The coordinates B1, B2 and B3 refer to the maximum, intermediate and minimum variance directions, respectively. The features of these components are typical of flux ropes in the principal axis system (Russell, 1990). The lower panels of Figure 1 show the hodograms of the magnetic field of the flux rope in two orthogonal planes. The lower left figure shows the plane of maximum variance: the width of the hodogram shows how twisted is the flux rope. The lower right figure shows the plane of minimum variance: the curvature of this hodogram indicates that the spacecraft has crossed the rope close to its center axis (Russell and Elphic, 1979). Using the principal axis technique, we have found 104 flux ropes in 46 different orbits over 981 orbits studied during MGS elliptical trajectories. The flux ropes are more often observed in the Venus atmosphere: 70% of the orbits of the Pioneer Venus Orbiter (Elphic and Russell, 1983b). In the next section, we examine the global characteristics of the flux ropes identified, and we compare our results with those for Venus.

3. Flux Ropes Properties

Figure 2 shows the properties of the flux ropes found at Mars. The upper left corner gives their altitude occurrence in logarithmic axis. The mean value of the flux ropes altitude is 246 km, but half of them occur below 200 km. In the case of Venus, Elphic and Russell (1983b) found that for low and high solar zenith angle regions, the relative fractional occurrence of flux ropes increases with decreasing altitude. We also note that we find a flux rope at an altitude of 1200 km, much higher than the ionopause. Flux ropes were also found at such an altitude in the case of Venus by Elphic and Russell (1983a). Further, we find very few flux ropes at altitude lower than 160 km.

The upper right corner gives the distribution of the flux ropes as a function of the solar zenith angle. Most of them are identified near the terminator plane; the mean value of the solar zenith angle is 80 deg with a standard deviation of 17 deg. We also note that only 10% of the flux ropes are found for solar zenith angles below 45 deg while 32% of the orbits taken into account in this study have their periapsis

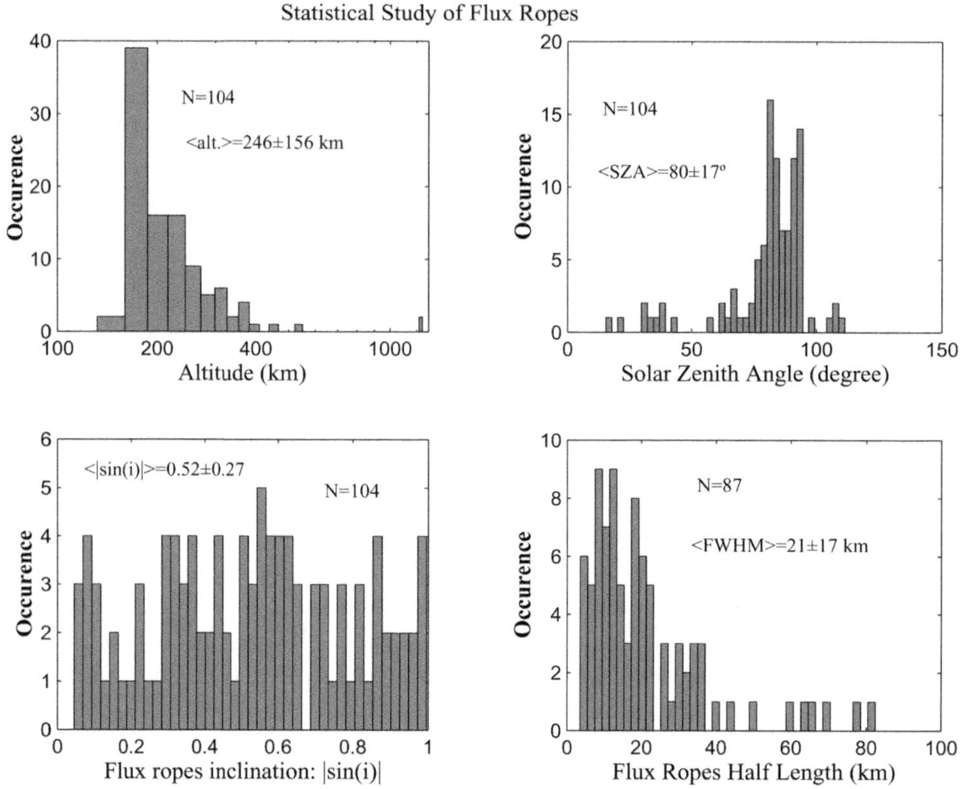

Figure 2. Occurrence of flux ropes as a function of their altitude, solar zenith angle, inclination and cross-section.

for solar zenith angles below 45 deg. Thus, this apparent depletion of flux ropes, close to the subsolar point, do not appear to be due to an orbit artefact.

The large-scale field cases are observed most often in the subsolar regions. Globally, the range of the magnetic field, over three minutes periods, for orbits with solar zenith angles below 30 deg and altitude below 400 km, vary between 4 and 180 nT. While near the terminator, for orbits with solar zenith angle between 60 and 90 deg and altitude below 400 km, the range of the magnetic field vary between 4 and 90 nT. No flux ropes are found in the southern hemisphere of Mars, and few are found near the subsolar point where the large-scale fields are observed. The magnetization of the atmosphere seems to preclude the formation of flux ropes like in the case of Venus (Luhmann *et al.*, 1990).

The lower left corner shows the distribution of the flux ropes inclination. The angle i corresponds to the inclination of the flux ropes to the horizontal axis. It can be obtained from the dot product of the maximum field vector and the local vertical. Small values of $|\sin(i)|$ correspond to horizontal flux ropes. The inclination of the flux ropes appears quite random, while at Venus the inclination is more horizontal near the terminator region and random near the subsolar region (Elphic and Russell,

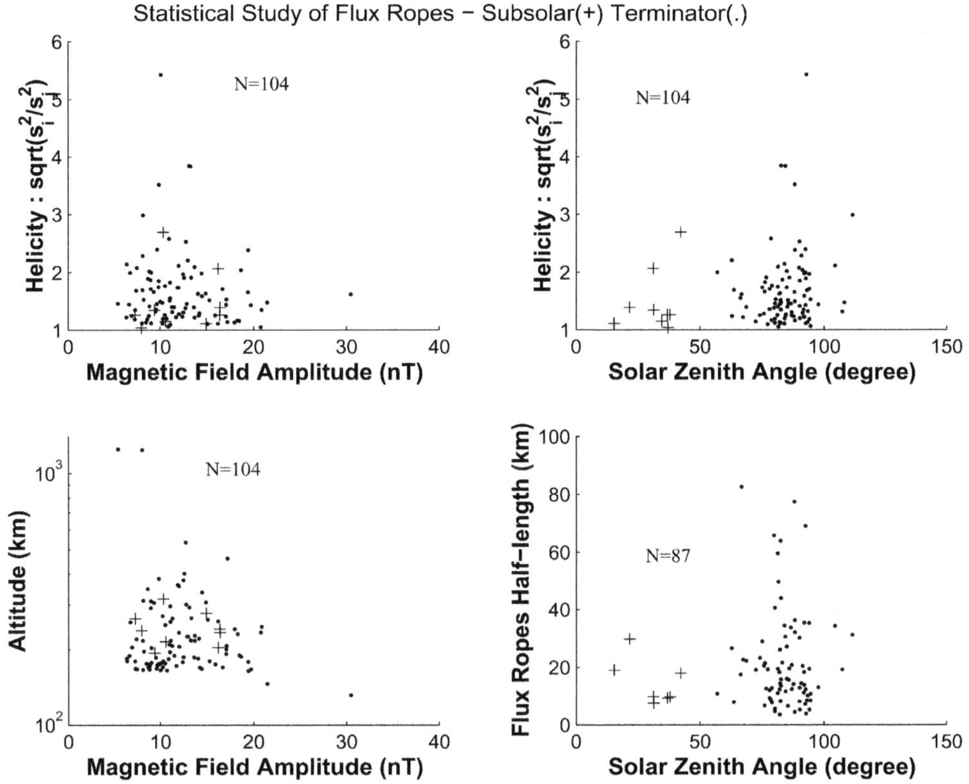

Figure 3. Flux ropes magnetic field amplitude as a function of helicity (upper left) and altitude (lower left). And flux ropes solar zenith angle as a function of helicity (upper right) and half-length (lower right).

1983b). In our case, most of the flux ropes identified occurs for SZA $> 45°$, and we find no preferences for horizontal or vertical oriented flux ropes: the mean value of $\mid \sin(i) \mid$ is 0.52 with a standard deviation of 0.27.

The lower right corner shows the distribution of the flux ropes half-length. Most of them (70%) have a full width at half maximum (FWHM) less than 20 km although some flux ropes are found with FWHM greater than 80 km. These results are similar to those of Venus: the FWHM is between 7.5 to 16 km near the terminator plane and between 6 to 12 km near the subsolar point. Russell (1990) has also found that the flux rope cross-sections decrease with decreasing altitude, suggesting an increased compression of the ropes with decreasing altitude. In the case of Mars, we find that the mean value of the FWHM is nearly equal for altitude greater and lower than 200 km. So, it seems that the ropes are not compressed in the Martian ionosphere.

Figure 3 shows others characteristics of the Martian flux ropes. The upper left corner plots the root variance ratio $\sqrt{(\sigma_i^2/\sigma_j^2)}$ (helicity index) as a function of the magnetic field amplitude of the flux ropes. σ_i and σ_j represent the maximum and intermediate variance for the flux ropes obtained when we use the minimum variance technique to order the vector data. This root variance ratio provides a measure of overall strength of the axial fields (which contributes to σ_i^2) in comparison to the overall strength of the azimuthal fields (which contributes to σ_j^2). When this ratio is close to unity, we find axial and azimuthal fields comparable, meaning greater helicity. The 'plus' signs indicate flux ropes observed for solar zenith angles smaller than 45 deg. For flux ropes with weak axial magnetic field ($\leq 15\,nT$) we find that their helicity is quite variable although we find highly twisted flux ropes for all the magnetic field amplitude ranges.

The upper right corner shows the variation of the solar zenith angle with the root variance ratio. We find that as the flux ropes are closer to the subsolar point, they are less twisted. We also note two gaps in the flux ropes identification for solar zenith angle (SZA) between 40° to 70° and for SZA lower than 20°. This is not due to an artifact of the spacecraft coverage; we have nearly the same number of magnetic field vectors measurement for altitudes below 400 km and for every 10 deg in range of SZA from 0° to 70°. In the case of Venus, Elphic and Russell (1983b), using the data from Luhmann *et al.* (1980), obtained the relative frequency of occurrence of flux ropes orbits as a function of solar zenith angle. Their results also showed a smaller number of flux ropes identified for SZA between 40 to 60 deg and for SZA lower than 20 deg. The large-scale fields more often observed near the subsolar plane explain why we find more flux ropes near the terminator plane. However, in the case of Mars, the magnetic field amplitude is stronger near the subsolar point than for the region with SZA between 40° to 70°. Actually, no reason seems to explain the apparent depletion of flux ropes, for SZA between 40 to 70 deg, which appears at Mars and Venus. This depletion could be related to the mechanisms responsible for the flux ropes formation or evolution.

The lower left corner gives the variation of maximum flux rope field with altitude. The strongest flux rope fields are mainly found at low altitudes < 300 km. However the mean value of the magnetic field amplitude is the same for flux ropes higher and lower than 200 km ($\langle |\mathbf{B}| \rangle >= 12.2$ nT). The standard deviation of the magnetic field amplitude is slightly greater for flux ropes below 200 km. The flux rope strength is more important at Venus: at high altitudes it is about 20 nT increasing to 70 nT at low altitudes (Russell, 1990) but the variation of maximum flux rope field with altitude is similar.

The lower right corner plots the flux ropes half-length as a function of the solar zenith angle. The mean value of the flux rope FWHM, for SZA $> 45°$, is 22 km with a standard deviation of 18 km. We note that the flux ropes half-length are highly variable near the terminator while for SZA, under 45 deg, the mean value of the flux rope FWHM is 17 km with a standard deviation of 8 km. We do not have enough statistics for SZA $< 45°$ to make any assertion about the flux ropes half-length.

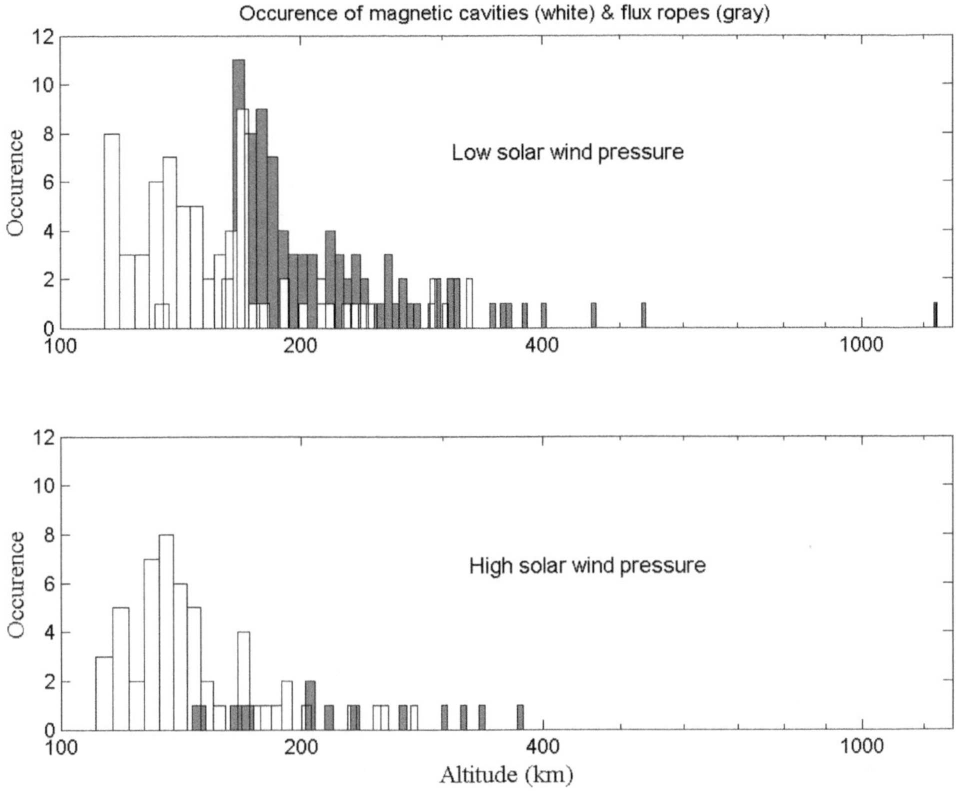

Figure 4. Occurence of magnetic cavities in white and flux ropes in gray as a function of altitude for low solar wind dynamic pressure (upper panel) and high solar wind dynamic pressure (lower panel).

However, we observed a lot of orbits without crustal magnetic fields or flux ropes measured near periapsis. These orbits look like magnetic cavities: the amplitude of the magnetic field is low and no flux ropes are observed. We distinguish the cases where the relative solar wind pressure is high or low. MGS can not measure the solar wind dynamic pressure, so a basic assumption has been made in order to have a rough estimate of the solar wind pressure. As the magnetic pile-up boundary (MPB) appears to be the outer edge of the magnetic barrier, which is the obstacle to the solar wind, it seems reasonable to assume that the magnetic pressure at the MPB and the value of the solar wind dynamic pressure vary in a similar way. As we know the value of the magnetic pressure, we are able to distinguish between high solar wind pressure regime and low solar wind pressure regime (Crider *et al.*, 2001).

As this method enables to have a relative estimate of the solar wind pressure from an orbit to another, we distinguish the cases where the estimated solar wind pressure is low or high. Figure 4 shows the occurence of these magnetic cavities (white) and the occurence of the flux ropes (gray) as a function of their altitude respective for cases of low and high solar wind dynamic pressure. First, the most striking difference between the two panels is the small number of flux ropes observed for high solar wind dynamic pressure: only 10% of all the flux ropes identified. Like in the case of Venus, when the solar wind dynamic pressure increases, the ionosphere begins to become magnetized and the flux rope formation process appears to be suppressed (Russell, 1990). Secondly, most of the flux ropes and magnetic cavities are observed at low altitude, less than 200 km. Further, the magnetic cavities are mainly found at lower altitudes than the flux ropes.

4. Summary

In this paper we have presented some results of the characteristics of the flux ropes observed in the Martian atmosphere. The details of the formation and evolution of the flux ropes is still a subject of investigation. The general properties of these flux ropes, for Venus and Mars, are as follows:

(1) Flux ropes are not observed in regions where the magnetic field is strong.

(2) The magnetic structures are stationary on time relatively to the speed of the spacecraft.

(3) The width of the flux ropes is of the order of a few tens kilometers.

This paper aims to show that we find flux ropes in the Martian atmosphere, which have the same global characteristics of those of Venus although we find less flux ropes overall at Mars, 5% of the orbits, than at Venus, 70% of the orbits.

In the case of Mars, no flux ropes are observed in the southern hemisphere. Most of them occur close to the terminator at high solar zenith angle and at high latitude. As the Martian ionosphere is quite often magnetized by the strong crustal field, this crustal magnetic field seems to inhibit the flux ropes formation in the southern hemisphere. However, we do observed magnetic cavities in the southern hemisphere. Thus we should be able to observed flux ropes in this hemisphere above non-magnetized crust when the solar wind dynamic pressure is weak.

We also find that flux ropes with weak axial magnetic field have a high variable helicity while, close to the subsolar point, they have greater helicity. The cross section of the flux ropes is in the order of few tens of kilometers with mean magnetic field amplitude of 12 nT and random inclination.

We have identified a smaller number of flux ropes for SZA between 40 to 60 deg and for SZA lower than 20 deg, like in the case of Venus. Statistically, the region with SZA range between 40° to 60° is less magnetized than near the subsolar point for Mars. So, others factors seem to suppress the flux rope formation process like the solar wind dynamic pressure. When the solar wind dynamic pressure is high, a large-scale horizontal magnetic field is observed in Mars' ionosphere. As, in these cases of high solar wind dynamic pressure, we only observed magnetic cavities without flux ropes, we deduced that the flux ropes formation process is suppressed by large scale magnetic field. Further, very few flux ropes are observed at altitudes

< 160 km where most of the magnetic cavities are identified. Thus, the flux ropes do not appear to be formed at altitudes < 160 km. The mechanisms of formation and suppression of flux ropes requires an additional investigation that is beyond the scope of the present discussion.

Acknowledgements

This work was performed while D. Vignes held a visitor researcher position sponsored by the Centre National d'Etudes Spatiales (CNES) at NASA Goddard Space Flight Center, Greenbelt, MD, USA.

References

Acuña, M. H. *et al.*: 1999, 'Global Distribution of Crustal Magnetization Discovered by the Mars Global Surveyor MAG/ER Experiment', *Science* **284**, 790–793.

Cloutier, P. A. *et al.*: 1999, 'Venus-Like Interaction of the Solar Wind with Mars', *Geophys. Res. Lett.* **26**, 2685–2688.

Crider, D. H. *et al.*: 2001, 'The Effects of the Solar Wind Pressure on the Martian Magnetic Pileup Boundary', *AGU Fall Meeting*, San Francisco.

Elphic, R. C. and Russell, C. T.: 1983a, 'Magnetic Flux Ropes in the Venus Ionosphere: Observations and Models', *J. Geophys. Res.* **88**, 58–72.

Elphic, R. C. and Russell, C. T.: 1983b, 'Global Characteristics of Magnetic Flux Ropes in the Venus Ionosphere', *J. Geophys. Res.* **88**, 2993–3003.

Krymskii, A. M. and Breus, T. K.: 1988, 'Magnetic Fields in the Venus Ionosphere: General Features', *J. Geophys. Res.* **93**, 8459–8472.

Luhmann, J. G. *et al.*: 1980, 'Observations of Large-Scale Steady Magnetic Fields in the Dayside Venus Ionosphere', *Geophys. Res. Lett.* **7**, 917–920.

Luhmann, J. G. *et al.*: 1990, 'The Solar Wind Interaction With Unmagnetized Planets: A Tutorial', *Geophys. Monograph* **58**, 1–22.

Priest, E. R.: 1990, 'The Equilibrium of Magnetic Flux Ropes (Tutorial Lecture)', *Geophys. Monograph* **58**, 401–411.

Russell, C. T. and Elphic, R. C.: 1979, 'Observation of Magnetic Flux Ropes in the Venus Ionosphere', *Nature* **279**, 616–618.

Russell, C. T.: 1990, 'Magnetic Flux Ropes in the Ionosphere of Venus', *Geophys. Monograph* **58**, 413–423.

UNUSUALLY DISTANT BOW SHOCK ENCOUNTERS AT MARS: ANALYSIS OF MARCH 24, 1989 EVENT

M. I. VERIGIN[1,2], J. SLAVIN[1], A. SZABO[1], G. A. KOTOVA[2], A. P. REMIZOV[2], H. ROSENBAUER[3], S. LIVI[4], K. SZEGÖ[5], M. TÁTRALLYAY[5], K. SCHWINGENSCHUH[6] and T.-L. ZHANG[6]

[1]*NASA Goddard Space Flight Center, Greenbelt, Maryland, U.S.A.*
[2]*Space Research Institute of Russian Academy of Sciences, Moscow, Russia*
[3]*Max-Planck-Institut für Aeronomie, Katlenburg-Lindau, Germany*
[4]*HU Applied Physics Laboratory, Laurel, MD, U.S.A.*
[5]*KFKI Research Institute for Particle and Nuclear Physics, Budapest, Hungary*
[6]*Space Research Institute of Austrian Academy of Sciences, Graz, Austria*

(Accepted in final form 14 March 2003)

Abstract. Detailed analysis of disturbances observed on 24 March, 1989 far upstream of the usual Martian bow shock position was completed with the use of the planetary obstacle and bow shock models relevant for the period of Phobos 2 observations and for low Mach numbers, respectively. It is proven that the system of discontinuities observed in the solar wind between 18:42 and 19:36 UT was the consequence of unusually distant planetary bow shock excursions. The cause was unusually small ρV^2 and M_a values in the solar wind flow.

1. Introduction

Between 21 March 1989 and the loss of the mission on 26 March the Phobos 2 Martian orbiter conducted plasma and magnetic field measurements in a circular orbit close to that of the Phobos moon. About 50% of Phobos 2's ~ 8 hours orbit was spent outside of the planetary bow shock and in the solar wind. A series of weak, quasiperpendicular shocks were observed there by Phobos 2 on 24 March 1989 between 18:42 and 19:36 UT. They were interpreted by Slavin *et al.* (1991, 1993) as unique observations of the Martian bow shock far upstream of its usual position. This interpretation was supported by the unusually small Alfvénic Mach number M_a and ram pressure measured on this day. However a comprehensive analysis has never been conducted due to the absence of reliable models for the Martian magnetopause and bow shock under such extreme solar wind conditions.

The later Phobos 2 orbits were synchronized with the Phobos moon orbit in such a way that Phobos 2 approached it to within about 200 km every time the spacecraft was directly upstream of Mars. These *regular* close approaches of Phobos 2 to Phobos provided attempts to hook up solar wind magnetic field and plasma disturbances (*sometimes* matched in time with above approaches) with the presence of the Phobos moon. Thus Yeroshenko (2000) attributed the event of 18:42–

Space Science Reviews **111:** 233–243, 2004.

19:36 UT on 24 March 1989 and some other events to 'direct interaction of the solar wind with a thin dust disk of Phobos', while Mordovskaya *et al.* (2001) speculated that these events originated from 'the existence of the Phobos magnetic field and magnetosphere'.

After the original interpretation by Slavin *et al.* (1991, 1993), a quantitative model of the Martian magnetopause appropriate for the period of Phobos 2 observations and for different solar wind ram pressures ρV^2 was developed by Verigin *et al.* (1997) as well as a model of planetary bow shocks applicable for low M_a (Verigin *et al.*, 1997a, 1999). Thus, it has become possible to conduct a thorough analysis of the 24 March, 1989 event. Below we present an overview of these boundary models and apply them to the 24 March 1989 events.

2. Motivation of the Models Used

For the physical modeling of the planetary bow shock, information is required regarding the position and shape of the obstacle to supersonic solar wind flow. Verigin *et al.* (1997) developed a model of the Martian magnetopause for the period of maximum solar activity. This model quantitatively describes:

–the observed relation between the solar wind ram pressure ρV^2 and the magnetopause position in the magnetotail;

–the observed relation between ρV^2 and the flaring angle;

–the position of a few magnetopause crossings observed by Phobos 2 above the day side of the planet.

The term 'magnetopause' was used there for the labeling of the Martian magnetotail/magnetosphere boundary revealed in the 1970's after the plasma measurements aboard the Mars 2,3,5 orbiters (see, e.g., Gringauz, 1976). The same boundary was relabeled afterwards by different names including, e.g., 'planetopause' (Trotignon *et al.*, 1996) and 'magnetic pile-up boundary' (Vignes *et al.*, 2000). Dubinin *et al.* (1996) compared positions of different plasma features bounding the Martian magnetotail as found with the ASPERA ion spectrometer data – 'ion composition boundary', 'termination of proton fluxes', peak of electron density', and demonstrated that their positions and variations with ρV^2 generally coincide with these features of magnetopause boundary deduced from TAUS spectrometer data by Verigin *et al.* (1993). After MGS measurements Crider *et al.* (2004) additionally demonstrated that this boundary (magnetic pile-up boundary in their terms) acts as an obstacle to the solar wind flow around Mars.

Shape of the obstacle was determined by Verigin *et al.* (1997) from the pressure balance equation:

$$k\rho V^2 \sin^2 \alpha + p = p_{\text{in}}, \tag{1}$$

where $k = 0.88$, p is the solar wind thermal and magnetic field pressure, and p_{in} is the total pressure at the inner side of the magnetopause. Dependence of the height of the magnetopause subsolar point on the solar wind ram pressure for two versions of the magnetopause model is shown in Figure 1 (detailed tables of planetocentric distance to the magnetopause as function of ρV^2 and solar zenith angle are presented in the paper by Verigin *et al.*, 1997). The same plot can also be interpreted as the approximate $p_{\text{in}}(h)$ height profile (second horizontal axis, shifted by k factor). The term p_{in} was considered in the original model as the sum of magnetic and ionospheric pressures: $p_{\text{in}}(h) = 4f^2M^2/(8\pi(R_m + h)^6) + p_0 \exp(-(r - r_0)/H)$, where $f = 1.22$, R_m is the Martian radius, and M, p_0, r_0, H are constants selected to fit the observations. It is the total inner pressure profile, not the physical sense of its constituents, that is important for the determination of the magnetopause shape and thus for the application of the magnetopause model to the bow shock modeling.

It would be also possible to describe the bending of the curves in Figure 1 either by taking the sum of two exponential pressure terms in the right-hand side of (1) (Verigin *et al.*, 1997) or by assuming a specific multipole profile of the planetary crust magnetic field. In the latter case, due to Chapman-Ferraro currents along the obstacle boundary, $P_{\text{in}}(h) \approx 4f^2B^2_{\text{in}}(h)/8\pi$, and Figure 1 can also be interpreted as the profile of the planetary magnetic field pressure $B^2_{\text{in}}(h)$ or effective profile of planetary magnetic field B_{in} (third and fourth axis, respectively, in Figure 1). This $B_{\text{in}}(h)$ profile though, of course, seems to be simplified reflection of the complicated magnetic field structures found by MGS orbiter at lower altitudes (100–200 km) (Acuña *et al.*, 1999; Connerney *et al.*, 1999), nevertheless is in line ith specific magnetic field intensities $B \sim 100\gamma$ observed at the MGS circular orbit ($h \sim 400$ km), where low altitude, small scale, intensive crustal magnetic field irregularities become smooth and large-scale ones (Ness *et al.*, 2000).

It was found that crustal magnetic field irregularities influence the Martian equatorial magnetopause position downstream of planetary terminator mainly (Verigin *et al.*, 2001). Thus we can expect that the subsolar bow shock position will be established by the large scale position (r_0) and curvature (R_0) of the subsolar part of the obstacle formed by some effective $B_{\text{in}}(h)$ profile. Those profiles presented in Figure 1 describe all the magnetopause positions observed by the Phobos 2 orbiter (Verigin *et al.*, 1997) and are reasonably corresponding to MGS magnetic field measurements. So, for any solar wind ram pressure we will use the Martian magnetopause stand off distance r_0 and its nose curvature radius R_0 deduced from the model by Verigin *et al.* (1997).

Once r_0 and R_0, the geometric parameters of the obstacle are defined, the 3-d shape of the bow shock can be determined for any given set of sonic M_s, Alfvénic M_a Mach numbers, and ϑ_{bv} angle between the solar wind velocity and interplanetary magnetic field vectors. The analytical model by Verigin *et al.* (1997a, 1999) provides a description of the Martian bow shock position observed by Phobos 2 orbiter and explain the very weak dependence of its terminator position on the ρV^2. A refined version of this bow shock model was used for the analysis of the Earth's bow shock based upon Interball/Magion 4 and Wind observations (Verigin *et al.*, 2001a, b). Use of the model permitted to reveal fine effects of the bow shock

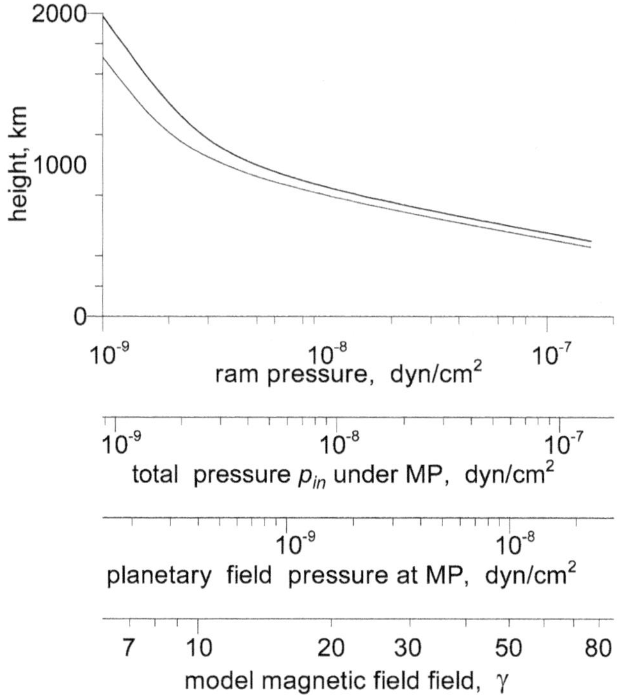

Figure 1. Dependence of the height of the magnetopause subsolar point on the solar wind ram pressure (axis 1), and height profiles of the total pressure at the inner side of the magnetopause (axis 2), planetary magnetic field pressure at MP (axis 3), and model planetary magnetic field (axis 4).

approaching Earth when M_a is decreasing (for field-aligned flow of the solar wind) and asymmetry of the terrestrial bow shock in the terminator plane. Unique cases of the distant subsolar bow shock excursions far from its usual position were well described by the above model as well (see also paper by Faifield *et al.*, 2001). Thus we may reasonably of apply the Verigin *et al.* (2001a, b) model for the study of the Martian bow shock behavior under low M_a conditions.

3. Observations and Modeling

Phobos 2 most frequently crossed the bow shock and exited into the solar wind above the dawn side of Mars in the range of solar zenith angles of 95°–105° (Verigin *et al.*, 1993). But before the unique event of 24 March 1989 Phobos 2 left the magnetosheath a after triple bow shock crossing at 17:51, 17:54, and 18:04 UT in the range of solar zenith angles of 73°–63° (see Figure 2 by Slavin *et al.* (1993) paper), i.e., bow shock was already unusually far from the planet.

Under normal conditions, no additional bow shock crossings would be observed until the orbiter reached the dusk terminator about 4 hours later. Most frequently the dusk bow shock was observed at 90°–95° zenith angles (Verigin *et al.*, 1993).

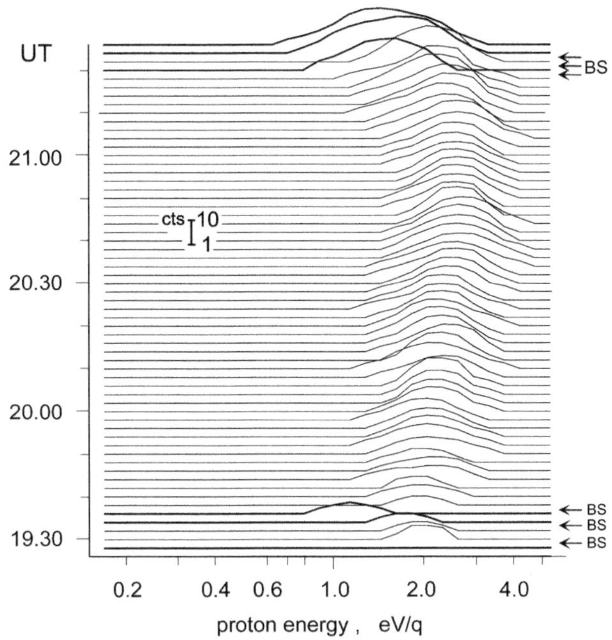

Figure 2. Proton spectra measured by TAUS ion spectrometer from start of its measurements till the dusk bow shock crossing. Spectra measured in the magnetosheath are drawn by thick lines while the moments of the bow shock crossings are marked by arrows.

But on 24 March a series of shock crossings were found in the magnetic field data at a much smaller zenith angles as the orbiter moved outward into the upstream solar wind. According to the Slavin *et al.* (1993) analysis, Phobos 2 again entered the magnetosheath region for about an hour after the triple bow shock crossing at 18:42, 18:45, and 18:54 UT, and left the magnetosheath region for about an hour after the triple bow shock crossing at 19:29, 19:33, and 19:36 UT.

The TAUS plasma measurements started at 19:25:45 UT and covered the period of the last triple shock crossing and subsequent period only. Figure 2 presents proton spectra measured by the TAUS ion spectrometer till the triple dusk bow shock crossing at 21:19, 21:21, and 21:23 UT smoothed over three adjacent energy windows with a running average. These crossings are easily recognized by a specific decrease of the proton velocity (decrease of energy of spectra maximum) and by proton thermalization (increase of spectra width) while entering the magnetosheath.

Similar changes of proton spectra took place at 19:33 and 19:36 UT in line with the original Slavin *et al.* (1993) interpretation suggesting that these discontinuities are bow shocks. The counts in the first two spectra measured by the TAUS instrument (Figure 2) were below the sensitivity threshold thus do not provide direct support, but are not in contradiction to the interpretation that the discontinuity of 19:29 UT was the bow shock also.

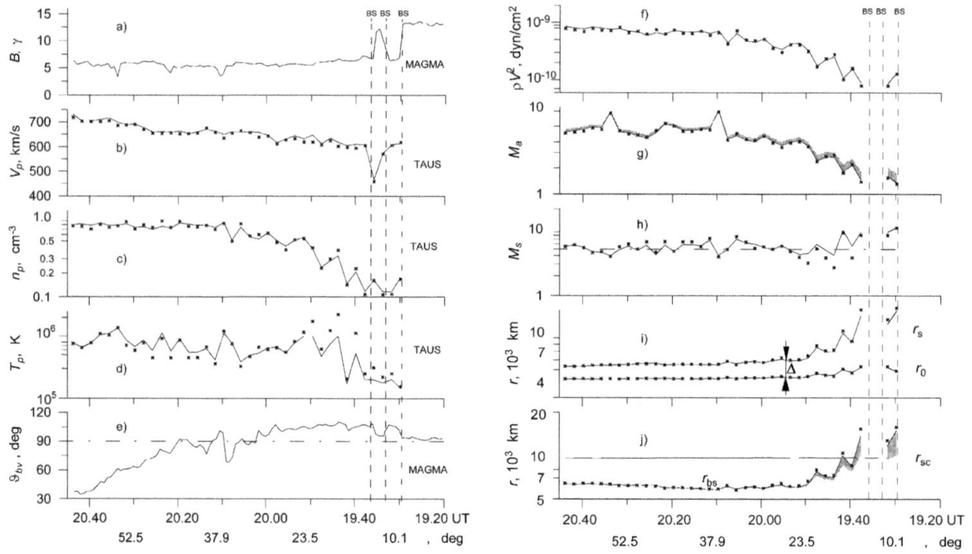

Figure 3. Magnetic field intensity (a) and angle between the magnetic field and X_{ase} directions (e) as measuredby MAGMA magnetometer. Solar wind proton velocity (b), density (c), and temperature (d) were defined overTAUS proton spectra (Figure 2) both by moment calculations (broken lines) and by Maxwellian distribution fitting (asterisks). Solar wind ram pressure (f), Alfvénic (g), and sonic (h) Mach numbers, evaluated over above mentioned measurements, permitted to model nose positions of the obstacle r_0 and bow shock r_s (i) as well as the planetocentric distance to the bow shock r_{bs} in the direction of Phobos2 orbiter (j). Second horizontal axis marks orbiter zenith angle in the reference frame aberrated due to Martian orbital velocity.

After 19:36 UT and before the first crossing of the planetary dusk bow shock at 21:19 UT the magnetic field and plasma instruments did not detect any discontinuity. So the measurements in this period, as well as during the short interval 19:29–19:33 UT, can be considered as being conducted in the solar wind and can be used for model predictions of the Martian bow shock and magnetopause positions.

The essential feature of the solar wind observations between 21:19 UT and 19:36 UT was the systematic decrease of solar wind fluxes (decrease of spectra height in Figure 2) with decrease of time that finally lead to the proton fluxes to drop below the TAUS instrument sensitivity threshold. Figure 3(a–e) presents the results of the MAGMA magnetic field measurements as well as the proton parameters evaluated over the TAUS spectrometer measurements.

These measurements were used for the evaluation of the solar wind ram pressure, Alfvénic and sonic Mach numbers in a manner similar to Verigin *et al.* (1997):

$$\rho V^2 \approx n_p m_p V^2 \left(1 + \frac{n_\alpha}{n_p}\frac{m_\alpha}{m_p}\right), \qquad M_a^2 = \frac{4\pi\rho V^2}{B^2},$$
$$p \approx n_p k T_p \left(1 + \frac{n_\alpha}{n_p}\frac{T_\alpha}{T_p} + \left(1 + \frac{2n_\alpha}{n_p}\right)\frac{T_e}{T_p}\right), \qquad M_s^2 = \frac{\rho V^2}{\gamma p}, \tag{2}$$

where γ is the specific heats ratio, m_p and m_α are the proton and α particle masses, respectively. The ratios of $n_\alpha/n_p \approx 0.047$, $T_\alpha/T_p \approx 4.9$, and $T_e/T_p \approx 1.9$ were adopted for the average solar wind properties according to Feldman *et al.* (1977). Thus calculated ρV^2, M_a, and M_s are presented in Figure 3(f–h), respectively. The use of ρV^2 provides the possibility to model the obstacle stand off distance r_0 (Figure 3(i)) as well as the curvature radius of its nose R_0 with the use of the Martian magnetopause model by Verigin *et al.* (1997).

Then the bow shock nose position r_s (Figure 3(i)) can be deduced from the following relations (Verigin *et al.*, 2001b):

$$r_s = r_0 + R_0 b^{-2/5}(b\varepsilon'/(1.87 + 0.86/(b\varepsilon')^{3/5}))^{2/3}, \tag{3}$$

where $\varepsilon' = \varepsilon(1-\varepsilon)$, $b = 1 - \cos^2\vartheta_{bv}/\varepsilon M_a^2$, and ε is the real root of a following cubic equation (e.g., Zhuang and Russell, 1981):

$$\begin{aligned}\varepsilon^3 - \left(\frac{\gamma-1}{\gamma+1} + \frac{\gamma + (\gamma+2)\cos^2\vartheta_{bv}}{(\gamma+1)M_a^2} + \frac{2}{(\gamma+1)M_s^2}\right)\varepsilon^2 - \\ - \frac{\cos^2\vartheta_{bv}}{(\gamma+1)M_a^4}\left(\gamma - 1 + \frac{2\cos^2\vartheta_{bv}}{M_s^2}\right) + \\ + \frac{1}{(\gamma+1)M_a^2}\left(\gamma(1+\cos^2\vartheta_{bv}) - 2 + \cos^2\vartheta_{bv}\left(\frac{\gamma+1}{M_a^2} + \frac{4}{M_s^2}\right)\right)\varepsilon = 0.\end{aligned} \tag{4}$$

Finally with the use of a quasi-hyperbolic bow shock surface (Verigin *et al.*, 2001b) it is also possible to model the time variation of the planetocentric distance to the bow shock r_{bs} in the direction of the Phobos 2 orbiter (Figure 3(j)).

4. Discussions

The key feature of the event 24 March 1989 was a very low solar wind number density ~ 0.1 cm^3 (Figure 3(c)) that happened when Phobos 2 conducted measurement at small zenith angles of 10–20 deg. Small solar wind number density results in small solar wind ram pressure ($\sim 10^{-9}$ dyn cm^{-2}, Figure 3(d)) that, in turn, results in some expansion of the magnetopause stand off distance r_0 (Figure 3(i)).

Simple expansion of r_0 would be insufficient to explain of the motion of the subsolar bow shock $\sim$9600 km from the planet. The small solar wind number density additionally results in a small Alfvénic Mach number in the solar wind (Figure 3(g)) that, in turn, leads to increased distance $\Delta = r_s - r_0$ between the bow shock and obstacle (Figure 3(i)) due to the quasiperpendicular appearance of the shock (Slavin *et al.*, 1993). Thus the modeled distance to the shock in the direction of the orbiter r_{bs} received a real possibility to achieve the planetocentric Phobos 2 distance r_{sc} (Figure 3(j)).

Figure 4 depicts the orbital position of the Phobos 2 at t_1 = 19:41:45 UT, t_2 = 19:47:34 UT, and t_3 = 20:19:39 UT. Bow shock S_1 was very far from Mars

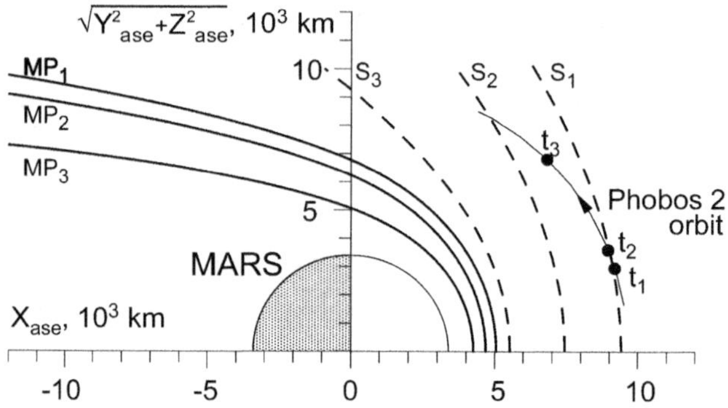

Figure 4. Modeled position of the Martian bow shock $S_{1,2,3}$ and magnetopause $MP_{1,2,3}$ for three positions of Phobos 2 $t_{1,2,3}$ along its orbit. Note distant excursion ofthe BS from the planet at moment t_1 and larger variation of the BS nose position compared to those one of the obstacle.

and just downstream of the orbiter in the moment t_1, moved closer to the planet at t_2, and reached its usual position S_3 by t_3, where it was crossed by Phobos 2 again in the vicinity of the planetary terminator (see top three BS crossings in Figure 2).

It is the small M_a numbers together with the quasiperpendicular appearance of the subsolar BS that was responsible for the distant excursion of the Martian bow shock from the planet and the larger range of r_s variations compared to r_0 variations (Figure 4). Quasiparallel subsolar shock would have a reciprocal tendency to move *to* the obstacle with decreasing M_a (Spreiter and Rizzi, 1974; Verigin *et al.*, 2001a, b).

Some overestimation of the model distance to the bow shock r_{bs} can be explained by the uncertainty of the n_p estimations over the proton spectra measurements close to the instrumental sensitivity threshold. A relatively small increase of M_a by 0.4 (shaded region in Figure 3(g)) leads to a better agreement of the modeled and observed (r_{sc}) positions of the bow shock (shaded region in Figure 3(j)).

So, taking into account the verification of the models used by comparison with Martian magnetopause (Verigin *et al.*, 1997) and bow shock observations (Verigin *et al.*, 1997a, 1999), and by comparison with distant Earth's bow shock observations (Fairfield *et al.*, 2001; Verigin *et al.*, 2001a), we can consider it to be proven that the events observed by Phobos 2 on 24 March 1989 between 18:42 and 19:36 UT are the consequence of the unusually distant planetary bow shock excursion from the planet initiated by unusually small ρV^2 and M_a values that occurred in the solar wind.

Generally, after 21 March 1989 and before the loss of the Phobos 2 mission, there were 12 intervals of about four hours duration when magnetic field and plasma were measured in the region assumed to be solar wind. During each of these intervals the orbiter approached the Phobos moon in about the same manner but only in 4 cases were some disturbances in the measured magnetic field (Yeroshenko, 2000): 22 March at 20:16–22:08 UT, 23 March at 3:48–6:16 UT,

23 March at 11:32–13:16 UT, and the event of 24 March studied in details above. This 25-percentage rate of disturbances must be warning (though not conclusive) against interpreting disturbances to be Phobos moon related. At least, before searching the Phobos related phenomena, all other phenomena related to well know processes in the plasma must be ‘subtracted’ from consideration. The present analysis of the 24 March event left no room for it to be associated with the Phobos moon. What can the remaining 3 events be associated with?

Yeroshenko (2000) noticed that the bigger part of the latest period of Phobos 2 measurements was influenced by an interplanetary recurrent disturbance. The start of the first event of 22 March was characterized by a sharp simultaneous increase of B, n_p, V_p, and T_p. Thus the strong interplanetary shock (V_p was increased from $\sim$450 km s^{-1} to $\sim$600 km s^{-1}) passed through the position of orbiter. Solar wind velocity remained to be about 600 km s^{-1}, and n_p remained to be very large ($>$ 10 cm^{-3}) during the entire period 20:16–22:08 UT marked by Yeroshenko (2000) as this event. In case Phobos 2 observed the solar wind interaction with the dust disk of Phobos (Yeroshenko, 2000) we would expect a solar wind deceleration, whether Phobos 2 crossed the magnetopause of Phobos (see Figure 5 by Mordovskaya *et al.*, 2001) we would expect a drop in n_p. Neither of this phenomena took place, so, both hypothesis failed describing the first event associated with interplanetary shock observation.

Almost the same arguments can be applied against the association of the second (03:48–06:16 UT) and the third (11:32–13:16 UT) events of 23 March with solar wind – dust disk or with solar wind – Phobos magnetosphere interactions. Really, the solar wind velocity rapidly increased from 640 km s^{-1} to 700 km s^{-1} at the start of the third event and continue grow till $\sim$750 km s^{-1} to the end of it thus excluding solar wind – dust disk interaction. Whether Phobos’ magnetopause was absent in the analysis of this event in Figure 6 by Mordovskaya *et al.* (2000), nevertheless we would expect the solar wind deceleration upstream of the Phobos’ magnetic field region (e.g., decelerated magnetosheath’s flow) that was not observed during the third event thus excluding its interpretation as solar wind – Phobos magnetosphere interaction. The second event was analyzed in details neither by Yeroshenko (2000) nor by Mordovskaya *et al.* (2001). The solar wind velocity was very large with several variations in the range from $\sim$650 km s^{-1} to $\sim$700 km s^{-1}, proton density generally decreased from $\sim$4 cm^{-3} to $\sim$1.5 cm^{-2}, proton temperature generally decreased from $\sim$ 4 $\times$ 10^5 K to $\sim$ 2 $\times$ 10^5 K during this event. These values are quite usual for periods of interplanetary recurrent disturbances.

5. Conclusions

Detailed analysis of disturbances observed on 24 March, 1989 far upstream of the usual Martian bow shock position was completed with the use of the planetary obstacle and bow shock models relevant for the period of Phobos 2 observations and for low Mach numbers, respectively.

The results strongly support the hypothesis that the system of discontinuities observed in the solar wind between 18:42 and 19:36 UT was the consequence of unusually distant planetary bow shock excursions initiated by extremely small upstream ρV^2 and M_a values.

Acknowledgement

This work was performed while one of the authors (MV) held a NRC Associateship Award at NASA/GSFC.

References

Acuña, M. H., Connerney, J. E. P., Ness, N. F., Lin, R. P., Mitchell, D., Carlson, C. W., McFadden, J., Anderson, K. A., Rème, H., Mazelle, C., Vignes, D., Wasilewski, P. and Cloutier, P.: 1999, 'Global Distribution of Crustal Magnetization Discovered by the Mars Global Surveyor MAG/ER Experiment', *Science* **284**, 790–793.

Connerney, J. E. P., Acuña, M. H., Wasilewski, P. J., Ness, N. F., Rème, H., Mazelle, C., Vignes, D., Lin, R. P., Mitchell, D. L. and Cloutier, P. A.: 1999, 'Magnetic Lineations in the Ancient Crust of Mars', *Science* **284**, 794–797.

Crider, D. H, Brain, D. A., Acuña, M. H., Vignes, D., Mazelle, C. and Bertucci, C.: 2004, 'Mars Global Surveyor Observations of Solar Wind Magnetic Field Draping around Mars', *Space Sci. Rev.* **111** (1-2).

Feldman, W. C., Asbridge, J. R., Bame, S. J. and Gosling, J. T.: 1977, *Plasma and magnetic fields from the Sun*, in *The Solar Wind Output and Its Variations*, in O.R. White (ed.), Colorado Assoc. University Press, pp. 351–382.

Gringauz, K. I.: 1976, 'Interaction of Solar Wind with Mars as Seen by Charged Particle Traps on Mars 2, 3 and 5 Satellites', *Rev. Geophys. Space Phys.* **14**(3), 391–402.

Dubinin, E., Sauer, K., Lundin, R., Norberg, O., Trotignon, J. G., Schwingenschuh, K., Delva, M. and Riedler, W.: 1996, *J. Geophys. Res.* **101**(A12), 27 061–27 065.

Fairfield, D. H., Cairns, I. H., Desch, M. D., Szabo, A., Lazarus, A. J. and Aellig, M. R.: 2001, 'The Location of Low Mach Number Bow Shocks at Earth', *J. Geophys. Res.* **106**(A11), 25 361 –25 376.

Mordovskaya, V. G., Oraevsky, V. N., Styashkin, V. A. and Rustenbach, J.: 2001, 'Experimental Evidence of the Phobos Magnetic Field', *JETP Lett.* **74**(6), 293–297.

Ness, N. F., Acuña, M. H., Connerney, J. E. P., Kliore, A. J., Breus, T. K., Krymskii, A. M., Cloutier, P. and Bauer, S.J.: 2000, 'Effects of Magnetic Anomalies Discovered at Mars on the Structure of the Martian Ionosphere and Solar Wind Interacion as Follows from Radio Occultation Experiments', *J. Geophys. Res.* **105**(A7), 15 991–16 004.

Slavin, J. A., Schwingenschuh, K., Riedler, W. and Yeroshenko, Ye.: 1991, The Solar Wind Interaction with Mars: Mariner 4, Mars 2, Mars 3, Mars 5 and Phobos 2 Observations of Bow Shock Position and Shape', *J. Geophys. Res.* **96**, 11 235–11 241.

Slavin, J., Verigin, M., Gringauz, K., Kotova, G., Stahara, S., Spreiter, J., Riedler, W., Schwingenschuh, K., Rosenbauer, H. and Livi, S.: 1993, 'The Solar Wind Interaction with Mars: Phobos-2 Bow Shock Observations on 24 March 1989', *Plasma Environments of Non-Magnetic*

Planets, COSPAR Coloquia Series, Vol. 4, in T. I. Gombosi (ed.), Pergamon Press, New York, pp. 279–283.

Spreiter, J. R. and Rizzi, A. W.: 1974, 'Aligned Magnetohydrodynamic Solution for Solar Wind Flow Past the Earth's Magnetosphere', *Acta Astronaut.* **1**, 15–55.

Trotignon, J. G., Dubinin, E., Grard, R., Barabash, S. and Lundin, R.: 1996, 'Martian Planetopause as Seen by the Plasma Wave System Onboard Phobos 24, *J. Geophys. Res.* **101**(A11), 24 965–25 977.

Verigin, M. I., Gringauz, K. I., Kotova, G. A., Remizov, A. P., Shutte, N. M. Richter, A., Riedler, W., Schwingenschuh, K., Szego, K., Apathy, I. and Tatrallyay, M.: 1993, 'The dependence of the Martian Magnetopause and Bow Shock on Solar Wind Ram Pressure According to Phobos 2 TAUS Ion Spectrometer Measurements', *J. Gephys. Res.* **98**, 1303–1309.

Verigin M., Kotova, G., Shutte, N., Remizov, A., Szegö, K., Tátrallyay, M., Apáthy, I., Rosenbauer, H., Livi, S., Richter, A. K., Schwingenschuh, K., Zhang, T.-L., Slavin, J. and Lemaire, J.: 1997, 'Quantitative Model of the Martian Magnetopause Shape and Its Variation with the Solar Wind Ram Pressure Based on Phobos 2 Observations', *J. Geophys. Res.* **102**(A2), 2147–2155.

Verigin, M. I., Kotova, G. A., Remizov, A. P., Shutte, N. M., Schwingenschuh, K., Riedler, W., Delva, M., Rosenbauer, H., Szego, K., Tatrallyay, M. and Styashkin, V.: 1997a, 'Studies of the Martian Bow Shock Response to the Variation of the Magnetosphere Dimensions According to TAUS and MAGMA Meaurements Aboard the PHOBOS 2 Orbiter', *Adv. Space Res.* **20**(2), 155–158.

Verigin, M. I., Kotova, G. A., Remizov, A. P., Styazhkin, V. A., Shutte, N. M., Zhang, T.-L., Riedler, W., Rosenbauer, H., Szego, K., Tatrallyay, M. and Schwingenschuh, K.: 1999, 'Shape and Location of Planetary Bow Shocks', *Cosmic Research* **37**(1), 34–39.

Verigin, M. I., Kotova, G. A., Remizov, A. P., Szegö, K., Tátrallyay, M., Slavin, J., Rosenbauer, H., Livi, S., Riedle, W., Schwingenschuh, K. and Zhang, T.-L.: 2001, 'Evidence of the Influence of Equatorial Martian Crustal Magnetization on the Position of the Planetary Magnetotail Boundary by Phobos 2 data', *Adv. Space Res.* **28**(6), 885–889.

Verigin, M. I., Kotova, G. A., Slavin, J., Szabo, A., Kessel, M., Safrankova, J., Nemecek, Z., Gombosi, T. I., Kabin, K., Shugaev, F. and Kalinchenko, A.: 2001a, 'Analysis of the 3-d Shape of the Terrestrial Bow Shock by Interball/Magion 4 Observations', *Adv. Space Res.* **28**(6), 857–862.

Verigin, M., Kotova, G., Szabo, A., Slavin, J., Gombosi, T., Kabin, K., Shugaev, F. and Kalinchenko, A.: 2001b, 'Wind Observations of the Terrestrial Bow Shock: 3-D Shape and Motion', *Earth Planets and Space* **53**(10), 1001–1009.

Vignes, D., Mazelle, C., Rme, H., Acũna, M. H., Connerney, J. E. P., Lin, R. P., Mitchell, D. L., Cloutier, P., Crider, D. H. and Ness, N. F.: 2000, 'The Solar Wind Interaction with Mars: Locations and Shapes of the Bow Shock and the Magnetic Pile-up Boundary from the Observations of the MAG/ER Experiment Onboard Mars Global Surveyor', *Geophys. Res. Lett.* **27**(1), 49–52.

Yeroshenko Ye. G.: 2000, 'Interaction of the Solar Wind and Martian Magnetosphere with the Dust Disc of Phobos', *Cosmic Res.* **38**(2), 119–130.

Zhuang, H. C. and Russell, C. T.: 1981, 'An Analytic Treatment of the Structure of the Bow Shock and Magnetosheath', *J. Geophys. Res.* **86**, 2191–2205.

MARS EXPRESS AND MARSIS

ERLING NIELSEN
Max-Planck-Institut für Aeronome, 37191 Katlenburg-Lindau, Germany

(Accepted in final form 14 march 2003)

Abstract. The Mars Express mission to be launched in 2003 will provide high resolution measurements of the Martian atmosphere and ionosphere. The neutral density, temperature, and composition will be measured, and MARSIS, a low frequency radar experiment, will probe the Martian plasma environment for electron densities in the range from 100 to 3×10^5 electrons/cm^{-3}. The radar will for the first time make measurements of the ionosphere in the sub-solar and midnight regions which were not accessible to previous missions. The radar will provide global coverage and an increase in spatial resolution of vertical electron density profiles, which can be used to improve understanding relationships between the neutral atmosphere and ionosphere, and the interactions of the solar wind with the planet. The radar may be able to detect effects associated with magnetic merging, plasma clouds, and plasma streams, which have been predicted in the dayside/terminator ionosphere. Measurements on the nightside can be used to search for proposed fine structures ('holes') in the electron density. Hydrodynamic waves are predicted to be excited by the solar wind/ionosphere interaction and their influence on the electron densities can be observed with the radar. Neutral density and temperature profiles are required to understand the details of the vertical variations of the electron density. This includes testing for effects induced on the electron density profiles by dust storms and by precipitations of solar wind ions in cusp-like magnetic regions. The search for trace elements by optical spectrometers may reveal magnesium ions in predicted meteoric layer.

1. Introduction

The environment of a planet is strongly influenced by the radiation from the sun: both the radiation which ionize and heat its atmosphere, and the radiation that has the form of the solar wind. When the solar wind encounters a planet, physical processes occur, which provide the forces, that are required to deflect the solar wind around the planetary obstacle. These processes involve the planets atmosphere and ionosphere, its intrinsic magnetic field, and/or possibly a magnetic field induced in the planet's ionosphere. The Earth with its N_2/O_2 atmosphere is an example of such a system for a planet with a strong intrinsic magnetic field. Mars with its low density CO_2 dominated and occasionally dust filled atmosphere, and its weak and spatially sporadic intrinsic magnetic field, which allows a direct interaction between the solar wind and the ionosphere, is a very different obstacle to the solar wind than is the Earth. The Mars Express mission is being prepared to explore these relationships and their effects on both the dayside and nightside ionosphere.

Space Science Reviews **111:** 245–262, 2004.

TABLE I

Mars Express instrumentation.

HRSC	High-resolution sterio camera
OMEGA	IR spectrometer with low spectral and high spatial resolution
PFS	IR spectrometer with high spectral and low spatial resolution
MARSIS	Low frequency radar
ASPERA	Energetic neutral atoms imager
SPICAM	UV and IR atmospheric spectrometer
MaRS	Mars Radio Science Experiment
Beagle-2	Geochemistry and Exobiology Lander

2. Mars Express

Mars Express is an ESA mission to be launched in the summer of 2003, into a near polar orbit (inclination 86.6°) with perigee and apogee at altitudes of 250 and 10142 km, respectively. The orbital period is 6.75 hours. The spacecraft will arrive at Mars during northern spring. The nominal mission duration is one Martian year, but the spacecraft will have consumables enough for two years, which is the time required to obtain global coverage of the planet.

Mars Express has contributions from Russia, Europe, and USA. It was conceived to replace the ill-fated Russian Mars-96 mission. The scientific objectives of the mission are therefore in part the same as for the Mars-96 mission. The mission objectives are directed towards the solid planet, the atmosphere, and the ionosphere:

- interaction of atmosphere/ionosphere with the interplanetary medium;
- global atmospheric circulation and high resolution mapping of the neutral air pressure, temperature and composition;
- global high resolution photography ($\sim$ 10 m) for geological studies;
- subsurface structures down to a few kilometers;
- global high-resolution mineralogical mapping of surface ($\sim$ 100 m).

In Table I is a list of experiments on Mars Express.

In the following are short descriptions of the experiments that make measurements related to Martian neutral atmosphere and ionized environment.

PFS (Planetary Fourier Spectrometer): It is well known that the Martian atmosphere is about 100 times less dense than the Earth' at ground level, and that it is composed almost entirely of CO_2 ($\sim$98%). The atmospheric temperature range from 300 K to 140 K. The atmosphere contains dust, in large amounts during dust storms. - The instrument will measure absorption spectra in the wavelength intervals 1.2 to 5 μm, and from 5 to 45 μm of sunlight reflected from the planet's surface. The instrument is nadir pointing and functions best near perigee but can make measurements from an altitude of up to $\sim$ 4000 km. In the fundamental mode the instrument will measure the absorption at 1.5 μm (absorption by CO_2) and derive the global vertical density/temperature profiles, and possibly the global circulation pattern. Since CO_2 is the major constituent of the atmosphere these results will reflect the neutral atmosphere behaviour. The instrument will also be operated in a spectral mode in search of trace constituents of the atmosphere, which are thought to be mainly: dust, H_2O, CO, CH_4, and HCHO). The goal is to determine how the composition varies in place and time, and to determine the dust content (the kind, and how much of it there is) and how it is transported.

SPICAM: A major purpose of this instrument is to investigate why the Martian atmosphere is so strongly oxidizing. The reason is thought to involve atmospheric ozone content as well as solar UV light. The instrument consists of two sensors, one for UV light (118–320 nm), and the other for IR light (1–1.7 micron). The experiment will operate in 3 different modes:

- in the nadir pointing mode the absorption spectrum of sun light reflected by the surface will be measured at 1.38 μm (water) and at 250 nm (ozone) to yield the column ($\sim$ 10 km^2) densities of water and ozone over the total surface of the planet at different seasons. From these measurements the density of the radical OH molecule can be estimated. OH is thought to be the sought for radical oxidizing agent;
- in occultation mode pointing to a star or the sun the instrument measures the vertical distribution of ozone and CO_2, and since CO_2 is the major constituent the measurements are used to derive the vertical density and temperature profile of the neutral atmosphere;
- Limb mode; or emission mode: the instrument is pointed across the limb (with no star or sun in the field of view) and measures the emission intensities in the interval 200 to 240 nm (CO_2+ and CO+). This provides the line-of-sight densities of these ions, and eventually yield the global distribution or image of these ions.

ASPERA (Analyser of Space Plasmas and Energetic Atoms): Mars no longer has a magnetic field and the charged particles that make up the solar wind are free to interact unhindered with the outermost atmospheric gases. Atoms of gas that are ionized (electrically charged) by this interaction, or by the action of sunlight, get picked up by the solar wind and swept out to space.

Mars no longer has a magnetic field and the charged particles that make up the solar wind are free to interact unhindered with the outermost atmospheric gases. Atoms of gas that are ionized (electrically charged) by this interaction, or by the action of sunlight, get picked up by the solar wind and swept out to space. Although the loss is of ionized gas, ASPERA will explore the answers to these questions mainly by detecting energetic neutral atoms (ENA; energy range 0.1–10 keV). An ion formed by interaction with the solar wind can become a neutral atom again by stripping an electron off an atom of neutral background gas. As the ion was energetic, the resulting neutral atom is too. By detecting the ENAs and their direction of travel, ASPERA will build up a global image of the region in the upper atmosphere that is interacting with the solar wind, or plasma. The neutral particle imager, one of the four sensors on ASPERA, will be responsible for generating this image.

ASPERA will measure ions and electrons when Mars Express passes close to the red planet in the region where the solar wind and Martian atmosphere interact. An ion mass analyser (IMA) will measure the flux and mass of ions coming from any direction in the energy range 0.01–40 keV).

An electron spectrometer (ELS) will measure electron fluxes from any direction in the energy range of a few to 20 000 eV.

MaRS (Mars Radio Science experiment): There are two radio downlinks on Mars Express, one in the X-band (8.4 GHz) the other in the S-band (2.3 GHz), will be used as radio occultation experiment making use of the fact that changes to the radio signal's frequency or phase and amplitude can say much about the physical mechanisms that brought the changes about. The technique will be used to probe the neutral atmosphere and ionosphere.

3. MARSIS

MARSIS stands for 'Mars Advanced Radar for Subsurface and Ionosphere Sounding' (Picardi *et al.*, 1999). This low frequency radar will be used for the dual purpose of sounding the outer layers of Mars for ice deposits and water, and for sounding the top-side ionosphere for the purpose of studying the interaction of Mars with the solar wind. Since this interaction on Mars takes place directly between the solar wind and the ionosphere, the top-side ionosphere is of particular interest. MARSIS includes a further experiment which measures the local electron density at the spacecraft. This experiment is named PSM (= Passive Spectrum Measurements.

Ionospheric sounding with radio waves is a well known technique, that relies on the fact that an electromagnetic wave can not propagate in a plasma with a plasma frequency larger than the signal frequency. The basic approach is to measure the time delay to the ionospheric reflection point as a function of frequency. The reflection point is reached when the radio wave frequency equals the local ionosphere plasma frequency. The power received back from a reflecting target is,

$$P_r = \frac{P_t G A s_T}{(4\pi)^2 h^4}, \tag{1}$$

$$P_r = \frac{P_t G^2 s_T l^2}{64\pi^3 h^4}, \tag{2}$$

$$P_r = \frac{P_t A^2 s_T}{4\pi h^4 l^2}, \tag{3}$$

where P_t = transmitted power; G = antenna gain; l = signal wavelength; $A = (lG)/4\pi$; s_T = radar cross section = $(\pi h^2 R^2)/(1 + h/a)^2$; h = distance to target; a = radius of target; R = reflectivity of target = 1 for ionosphere.

Reflections from the ionosphere give a received power of

$$P_r = \frac{P_t A^2 a^2}{4h^2(a+h)^2 l^2}. \tag{4}$$

This equation can be used to calibrate the received power, or the antenna gain, G.

3.1. MARSIS AS A TOP-SIDE IONOSONDE

The ionosphere is a dispersive medium with a refractive index, n, which in a collisionless ionosphere is given by

$$n^2 = 1 - X = 1 - \left(\frac{f_p}{f}\right)^2, \tag{5}$$

where

$$f_p^2 = \frac{n_e e^2}{m e_o 4\pi^2}. \tag{6}$$

Inserting parameter values we find the plasma frequency

$$f_p = 9 \times 10^3 \sqrt{n_e(\text{el cm}^3)} \quad [\text{Hz}]. \tag{7}$$

The radar frequency band is from 0.1 to 5.4 MHz corresponding to electron densities between 1.25×10^2 to 3.6×10^5 el cm^{-3}.

In an ionosphere with collisions the refractive index is

$$n^2 = 1 - \frac{X}{1 - iZ} = \frac{\frac{f_p^2}{f}}{1 - i\frac{v}{2\pi f}}. \tag{8}$$

Thus, collisions are included by replacing ne with

$$n_e \rightarrow \frac{n_e}{1 + \left(\frac{v}{f}\right)^2}, \tag{9}$$

i.e., collisions bring the refractive index closer to '1' and thus makes the medium less dispersive. For the altitudes of interest in the Martian ionosphere the collision frequency is much less than the radio wave frequency (see next section), and Equation (5) is therefore relevant for the radar experiment except maybe for the lowest frequencies.

On the spacecraft passing across the top of the ionosphere MARSIS determines the time delay between transmission of a pulse and the reception of the signal reflected from the regions in the ionosphere where the signal frequency equals the plasma frequency. However, the time delay is a function of the refractive index; the wave velocity in the ionosphere is given by c/n, where c is the velocity of light in vacuum. The apparent range obtained from the measured time delay must therefore be corrected for this effect to yield the actual range (Zhang *et al.*, 1990a).

Sounding from a spacecraft can give direct information about the electron density variations down to the altitude of maximum density. However, at higher frequencies the signal will penetrate the ionosphere, be reflected from the ground, penetrate the ionosphere again, and what remains of the signal can be detected at the spacecraft. Since this signal has traversed the ionosphere twice, it contains information also about the bottom side ionosphere.

If the local plasma frequency, f_p, at the spacecraft is within the range of transmitted frequencies, then no signal can be transmitted below the plasma frequency. At the plasma frequency the local plasma will be excited and emit a strong plasma line. If the plasma is magnetized with a field of magnitude B, then plasma lines will be excited at the gyro-frequency, f_c, and for combinations of f_p and f_c ($f_p \pm f_c$, $2f_c$, $3f_c$, etc.). The gyro-frequency is given by 28B [Hz], where the magnetic field strength is in nano Teslar (nT). For a possible magnetic field of order 1000 nT the gyro-frequency will be near 30 kHz, well beyond the frequency resolution of the radar. Furthermore, to complicate matters, a wave propagating in a magnetic field splits into two waves, an Ordinary wave and an Extraordinary wave.

3.2. Radio Wave Absorption

The radio wave absorption depends on the radio wave frequency, the momentum-transfer collision frequency of the thermal electrons, and of the electron density. The momentum-transfer electron collision frequency is governed by the main molecular component in the Martian atmosphere, which is CO_2. For the relative cool Martian ionosphere, with plasma temperatures of the order 1000 K, the thermal mean electron energy (kT) is about 0.1 eV. Laboratory measurements of in a gas of CO_2 molecules show the momentum-transfer collision frequency of such electrons to be close to 10^{-7} per molecule s^{-1} (Hake and Phelps, 1967; Sen and Wyller, 1960). Multiplying the collision frequency by the CO_2 densities then yield the altitude variation of the momentum-transfer collision frequency (see, for example, Nielsen, 1998). It is noted that the momentum-transfer collision frequency in the Martian ionosphere is much larger (close to a factor 10 larger) than in the Earth's atmosphere. The reason is that the collision frequency with the N_2 molecules, which is the dominating molecule in the Earth's atmosphere, is much lower at 2×10^{-9} per molecule s^{-1}. Strong absorption of radio waves is a striking aspect of the Martian ionosphere. During the daytime sounding of the Martian subsurface will be limited owing to the strong radio wave absorption in the 'normal' ionosphere.

TABLE II
MARSIS top-side ionosonde parameters.

Frequency range	0.1–5.4 MHz
Unmodulated CW pulses	
TX power	100 W peak
Pulse width	91.4 μm
Range resolution	13.71 km
PRF	125 Hz
Pulse interval	8 ms
Max. radar altitude	1200 km
Frequency step size	10.937 kHz
Number of frequency points	485
An ionogram per	3.9 s

If the proposed secondary meteoric layer (see below) is real then the radio wave absorption at 5 MHz can be very large in these layers (maybe more than 100 dB).

3.3. Technical Parameters

The radar will sound in the frequency range 0.1 to 5.4 MHz, covering an electron density range from 100 to 3×10^5 el cm^{-3}. The sounding will be done with unmodulated CW pulses (using a TX peak power of ~ 100 W). The pulse length is ~ 100 μs, and the range resolution is ~ 15 km. The pulse repetition frequency (PRF) is 125 Hz, which corresponds to a time interval of length 8 ms; this is the time it takes the signal to travel a distance of 2400 km in vacuum. Thus a spacecraft altitude of more than 1200 km would require interleaving of TX and RX operations. Measurements will be made in ~ 10 kHz steps across the frequency range, yielding a number of steps of ~ 531. The time resolution of the ionograms will be the number of frequency steps times 8 ms, or ~ 4.2 s. For a spacecraft horizontal speed of 5 km s^{-1} this corresponds to a horizontal spatial resolution of ~ 20 km. Owing to demands of the technical realization the actual radar parameters are slightly different from the nominal ones; the actual parameters are listed in Table II.

The MARSIS instrument can be considered composed of three subsystems: antenna -, analog radio frequency -, and digital electronics subsystem. The primary antenna is a 40 m tip-to-tip dipole which is used for both transmission and reception. The secondary antenna is a 6 m monopole used for reception only. The radio frequency part consists of a power amplifier (transmitter), a transmit/receive switch connected to the dipole antenna, and two receivers to collect echoes from each antenna. The receiver outputs are digitized and send to the digital electronics unit. This unit has the task of operating the on-board processing of the signals from the two separate channels corresponding to each of the antennas. It also contains the sounder frequency generator, as well as the logic for the instrument and interfaces to the spacecraft.

MARSIS includes an experiment which can measure the local electron density and temperature. These experiments provides the electron density at the top of the region traversed by the radar signals. This measurement can be used to improve the conversion of the ionogram to an electron density profile. The reason is that the radar signal (its propagation velocity) will be influenced by all the plasma between the spacecraft and the point of reflection. Electron density 'anchor points', at the spacecraft and at the point of reflection, are therefore essential.

PSM (Passive Spectrum Measurements): This experiment will measure the local electron density and temperature at the spacecraft. It is known that a weak thermally excited emission line exists at the electron plasma frequency in a plasma, or at the upper hybrid frequency in the case of a magnetized plasma. Provided the antenna length, L, is long compared to the Debey length, D, in the plasma ($D = \sqrt{T[\mathrm{K}]/n[\mathrm{el\ cm}^{-3}]}$) [cm], the local plasma frequency, and hence the local electron density, can be determined directly from the frequency of the thermal plasma line. The condition $L \gg D$ is nearly always satisfied in the Martian ionosphere. The detailed shape of the plasma line has also been exploited to give the electron temperature.

The main disadvantage of this technique is that the thermal plasma line may be very weak (a few mV m^{-1}), and measurements therefore requires very low instrument noise levels to achieve good results. On Mars Express we expect to have low noise levels, below the cosmic noise level, over most of the frequency range.

4. Areas of MARSIS Contributions

Topside soundings have been made at the Earth, but not at Mars, or at any other planet.

Note that earlier measurements of Martian electron density profiles were made using the radio occultation technique. Since Mars is an exterior planet measurements could only be made in a ± 45 deg interval around the day/night terminator region. We have currently no observations of the ionosphere near the sup-solar region or near the midnight region on Mars. The MARSIS radar will be able (depending on actual orbit) to make measurements at all solar zenith angles. The radio occultation technique must assume spatial homogeneity of the ionosphere across a fairly large azimuthal range ($\sim 150°$). This is actually unlikely to be a good assumption around the terminator. Thus, the (horizontal) spatial resolution of 20 km for Marsis is a major improvement on the resolution of available data.

The key parameters in the physical processes that causes deflection of the solar wind around Mars, are the properties of the atmosphere, the ionosphere, and the magnetic field. Experiments on Mars Express will observe the atmosphere, and

MARSIS will provide global coverage and an increase in spatial resolution of vertical electron density profiles. From these observations the plasma scale height can be determined as a function of altitude. The scale height could be strongly influenced by an intrinsic magnetic field and by the solar wind/ionosphere interaction. These measurements can be used to improve understanding relationships between the neutral atmosphere and ionosphere, and the interaction of the solar wind with the planet.

The radar may be able to detect effects associated with possible magnetic merging, plasma clouds, and plasma streams, which may occur in the dayside/terminator ionosphere. Measurements on the nightside may be used to search for proposed fine structures ('holes') in the electron density.

4.1. Mars' intrinsic magnetic field

MARSIS measurements will be of a spatial scale comparable to the scale available for the Martian intrinsic magnetic field (Acuña *et al.*, 2001; Acuña *et al.*, 1998, 1999). The 'global' equivalent magnetic dipole moment of Mars is currently at a value of 2×10^{21} [G cm^{-3}]. However, the Mars Global Surveyor mission has demonstrated that the Martian magnetic field is not a dipole, but has the characteristics of many spatially separated crustal regions where the main field component is vertical (pointing upward or downward), such that in effect the Martian magnetic field may be described as being composed of several magnetic loops distributed across the Martian surface. These crustal fields are especially pronounced in the southern hemisphere.

A map of the vertical (radial) component of magnetic fields reveal that there are magnetic band structures in the southern hemisphere around 180 deg longitude, with alternating polarities as one moves from south to north. The vertical fields are approaching 1500 nT at the spacecraft altitude of about 110 km. For comparison, the magnetic field at the Earth's surface is about 30 000 nT, or a third of a Gauss.

Mars' crustal magnetic fields are a mystery, because they are so intense. Furthermore, they are arrayed in east-west bands of alternating polarity, expending for over 1000 km north-south like a bar-code on the planet's surface (Connerray *et al.*, 1999).

4.2. Solar wind interaction with local crustal magnetic fields

Regions of localized crustal magnetic fields have been found to be a common feature of the southern hemisphere of Mars. The interaction of localized crustal magnetic fields with the solar wind may be similar to processes occurring on the Moon. One of the surprises of the Apollo program was the detection of crustal magnetic fields on the Moon, with magnitudes of ~ 1000 times less than the fields observed at Mars. Analysis of Electron Reflectometer measurements at the Moon lead Lin *et al.* (1998) to propose that a small magnetosphere can form around a region of local magnetic fields centered on the Moon surface. It appears that a shock wave is formed and the solar wind and its magnetic field deflected around this local magnetic object. The intrinsic magnetic field in itself may be sufficient to deflect the solar wind. Alternatively, the thermal pressure of the ionosphere and/or an induced magnetic field may also contribute to the deflecting forces. The detailed interaction remains to be determined. The possibility that small magnetospheres with convection, field aligned currents, and ionosphere currents are parts of the Mars/solar wind interaction scenario must be investigated.

The field lines of the intrinsic magnetic field can be pictured as half cylinders lined up side by side on the surface, oriented east-west. The place where the half-cylinders touch are areas of strong vertical magnetic field (cusp-like regions), where ionized hydrogen and helium of the solar wind are able to funnel down to low altitudes. The tops of the cylinders are areas of strong horizontal magnetic fields, which acts as an umbrella to shield the underlying atmosphere from the solar wind. Precipitation of solar wind ions in the cusp-like regions may add enough energy to the atmosphere to have a significant influence on atmosphere/ionosphere densities and temperatures.

Ness *et al.* (2000) reexamined old radio occultation measurements for correlation's with these magnetic fields. They found that in regions of pre-dominantly vertical fields the ionosphere tended to extend towards higher altitudes than in regions where a horizontal field is dominating. It confirmed that in regions of vertical fields the ionosphere plasma reach diffusive equilibrium above the photo equilibrium region; examples of a scale height of 65 km were identified. If there is an induced horizontal magnetic field component in such a region, there would be no significant curbing of vertical plasma motion, if its magnitude is small compared to the intrinsic vertical component. In the available radio occultation profiles there are only a few cases which can be related to the known vertical field regions; the magnetic field pattern is small scale, and the associated ionosphere pattern is therefore difficult to measure using the radio occultation method. MARSIS will provide measurements with the ionospheric pattern of a spatial scale that compare better to the spatial scale of the crustal magnetic field.

More recently observations of low energy electrons on board Mars Global Surveyor lend further new evidence to support the view that vertical magnetic fields allow diffusive equilibrium to be established at altitudes above the photo equilibrium region, so that the ionosphere extends to higher altitudes than in regions of horizontal magnetic field (Mitchell *et al.*, 2000, 2001). The MGS/'Electron Reflectometer' yield measurements that allow to determine if the spacecraft is inside the ionosphere or outside the ionosphere in the solar wind. The data from the instrument were used to determine when the spacecraft orbiting 400 km above the surface, was inside the planet's ionosphere or outside in the solar wind. This is possible because the energy spectrum of ionosphere electrons is distinctly different from that of the solar wind electrons. The probability for being in the ionosphere was determined as a function of longitude and latitude. Comparing this plot with plots of Martian vertical magnetic fields, reveal that there is a high probability to be in the ionosphere when the spacecraft is inside regions of large vertical magnetic field. Over areas of horizontal magnetic fields the ionosphere rarely reached as high as the spacecraft orbit at 400 km. This is essentially the same result as reported in the previous paragraph, that the ionosphere extends to higher altitudes when diffusive equilibrium can be established above ~ 180 km, when free vertical transport is possible.

Outside the crustal magnetic anomalies in the southern hemisphere it appears that the solar wind interaction leads to an induced , mainly horizontal, magnetic field in the ionosphere (see below), which slow vertical plasma motion and prevent diffusive equilibrium to be established; as a result the plasma (electron) density in the high ionosphere is depleted, and the scale height reduced relative to what it would be in an assumed field free ionosphere, or in an ionosphere with a vertical magnetic field (Krymskii *et al.*, 1995; Hantsch and Bauer, 1990; Bauer and Hantsch, 1989; Hanson *et al.*, 1977; Kliore *et al.*, 1972).

Inside the crustal magnetic anomalies where the intrinsic magnetic field is mainly horizontal the average scale height is close to that typical of the region outside the crustal anomalies.

Inside the crustal magnetic anomalies, where the intrinsic magnetic field is mainly vertical, i.e., in regions of cusp-like magnetic structures, the solar wind enters the cusp, and heat the atmosphere, which then inflates. This inflation causes the homopause to increase its altitude, as is the case during heating associated with dust storms, and the altitude of the electron density maximum therefore may also increase (see below).

4.3. The Solar Wind Interaction with the Ionosphere

The planet Venus is different from Mars therein that it has no intrinsic magnetic field; a further difference between the two planets is that Mars is much smaller, and thus has a much weaker (thinner) atmosphere, and therefore an ionosphere with corresponding lower densities, and lower thermal pressures. The interaction of the solar wind with Mars is of interest (not only because it has an intrinsic magnetic field) but also because this planet has a low thermal pressure, so that forces additional to the thermal pressure and pressure from the (spatially sporadic) intrinsic fields, may be required to deflect the solar wind (Luhman *et al.*, 1987; Breus and Krymskii, 1987; Luhman, 1986; Phillips *et al.*, 1984; Slavin and Holzer, 1982; Schunk, 1980; Brace *et al.*, 1980; Johnson and Hanson, 1979).

If the solar wind interaction with the planet occurs at an altitude where the plasma is collisionless, i.e., a plasma with low ohmic losses, the movement of ions and electrons across the direction of the magnetic field in the interaction region is severely reduced. The plasma moves freely along the magnetic field lines, but the plasma is tied to and moving with the magnetic field lines in the directions perpendicular to the direction of the field lines. The free motion of the plasma along field lines, leads to a removal of (ionospheric) plasma on the interplanetary side of the magnetic pile up region. This results in the formation of an 'ionopause', where the electron density at the top of the ionosphere is rapidly decreasing with altitude. On the other hand if collisions between the constituents of the plasma as well as with neutrals occurs in the interaction region, it is possible for the magnetic field to diffuse through the plasma. In this latter case the solar wind interaction with the planet causes an induced magnetic field to appear in the ionosphere, and this magnetic field will effect the composition, the vertical density variations, as well as the flow patterns of the plasma.

Theoretical models have difficulties reproducing the observed vertical electron density profiles.

Theoretical results. Calculations of a multispecies magnetohydrodynamic theoretical model of the dayside ionosphere of Mars were carried out by Shinagawa and Cravens (1988, 1989). Quantitative altitude variations of the major ion species of Mars were calculated in a one-dimensional (vertical) model for a variety of boundary conditions. A main conclusion is that a considerable horizontal transport away from the sub solar region has to be assumed in order to account for the observed low ion (electron) densities in the high altitude ionosphere.

The authors first considered a totally shielded ionosphere, in photo chemical and diffusive equilibrium, and with no magnetic field. They found the densities below 200 km to be in reasonable agreement with observations, but the model produces too much O^+ and O_2^+ at higher altitudes. The model reproduces correctly that O_2^+ is dominating at all altitudes, but the scale height is too large compared to observations.

Adding an intrinsic magnetic field ($\sim$ 50 nT) the authors found nearly no change from the former case for altitude below 200 km; the reason is that photochemistry is the most important process at those altitudes. However, at higher altitudes the densities of O_2^+ and especially O^+ are still too high. O^+ is the dominating ion at high altitudes contrary to observations.

Instead of an intrinsic magnetic field the authors then considered an induced magnetic field applied on the top boundary of the ionosphere. Magnetic flux is moving downward with the plasma motion, reaching a peak velocity of of about -12 m s^{-1} at 170 km, after which the flux is quickly dissipated in the magnetic diffusion region below 170 km owing to collisions. The magnetic field magnitude at the top boundary is 57 nT. Above 300 km the ion densities are considerably decreased from the previous case (intrinsic field), as a result of the downward plasma motion. However, the O+ densities are still too large.

In order to reduce the ion densities at high altitudes the authors now introduce an ion loss term into the former model with an induced magnetic field. They assume the horizontal magnetic pressures varies as $\cos^2$(zenith angel), for a Newtonian variation of the solar wind pressure leading to horizontal winds and consequently to ion losses. Ion densities are then in much improved agreement with observations. However, the O^+ densities are still dominating at high altitudes, in conflict with observations. The altitude profile of the Viking observed electron densities is in good agreement with the calculated profile.

This model could not reproduce the observed distribution of ions with altitude, especially not reproduce the dominance of O_2^+ at high altitudes. To obtain a good fit to observations it was necessary to introduce a horizontal loss term (associated with a horizontal divergence of the horizontal component of the magnetic field). Actually, horizontal flows of ions has been observed on Mars with speeds up to 300 m s^{-1} (Stewart and Hanson, 1982).

Krymskii *et al.* (1995) treated the magnetic field somewhat differently from Shinagawa and Cravens (1989) and found better agreement with the observed vertical variation of ions with O_2^+ being the dominant ion at all altitudes. These authors also had to introduce a horizontal velocity to reproduce the electron density versus altitude observed on the Viking Landers. Krymskii *et al.* (1995) also concluded that Mars must have an intrinsic magnetic field. A key point in their argument is the observation of a small plasma scale height also at large solar zenith angles, in a region where the ionosphere thermal pressure is sufficient to hold off the solar wing without requiring an induced magnetic field. That the scale height under these conditions nevertheless is small is taken as evidence of a (horizontal) intrinsic magnetic field which impeed vertical transport.

4.4. Hydrodynamic pulsations

At the Earth the solar wind interacts primarily with the geomagnetic field, which shields the atmosphere and ionosphere from direct contact with the solar wind. At Mars the solar wind interacts directly with the ionosphere. Recent theoretical investigations (Wang and Nielsen, 2001, 2002) indicate that the direct interaction between the ionosphere and the solar wind may excite hydrodynamic waves in the ionospheric plasma. The driving force for these waves is the fluctuating solar wind pressure applied to the top of the ionosphere. Some of these waves propagate downward in the ionosphere and give rise to significant fluctuations in the vertical ionospheric density profiles (up to 20% of the background density).

4.5. Secondary electron density peaks

Radar signals with frequencies exceeding the maximum plasma frequency in the ionosphere will pass through the whole ionosphere, be reflected from the planet's surface and pass through the ionosphere again before being detected at the spacecraft. Thus these high frequencies signals contain effects from the whole ionosphere, also that part located below the electron density maximum. The primary electron density peak in the Martian ionosphere is governed by photo ionization processes, and is located typically at an altitude of 125 km. Below this altitude secondary density maxima has been partly observed and partly proposed.

4.6. Soft Solar X-rays

In Kliore *et al.* (1972), in Figure 29, an electron density profile is shown with a distinct secondary peak near ~105 to 110 km altitude, below the main density peak. It is also noticed that there generally appears to be a second maximum with a peak near 110 km, below the altitude of the primary peak electron density. It has been suggested that this secondary lower altitude maximum is the result of precipitation of soft solar X-rays (Fox *et al.*, 1995). The authors suggest that this peak is produced by soft X-ray ionization (1.8–20 nm). The peak ion density is composed mainly of O_2^+ and NO^+ ions, where NO^+ is due to partial O_2^+ loss via reactions with NO and N.

4.7. Meteoric Magnesium Layer

Pesnell and Grebowshy (2000) calculated the amount of neutral Mg resulting from ablation of the meteoroids as they enter the Martian atmosphere, calculated the result of diffusing Mg vertically through the atmosphere, and obtained the profile of the Mg^+ ion distribution resulting from photochemical equilibrium considering a large set of ion-neutral reactions. The authors predicted that meteoric influx into the Martian atmosphere will lead to formation of a magnesium layer near an altitude of 80 km with a maximum density within an order of magnitude of 104 ion cm^{-3} in the sub-solar region. Since the intensity of the solar radiation is a controlling factor, the density maximum is dependent on solar zenith angle. This is a theoretical prediction so far with little supporting experimental evidence.

It has already been pointed out that the Martian is likely to be a strong absorber of radio waves. The meteoric layer is at so low altitudes that the momentum transfer collision frequency is so large that for electron densities as mentioned above, the radio wave absorption at 5 MHz may be or the order of 100 dB, i.e., if such a layer exists the radar signals will be effectively extinguished in the layer (Witasse *et al.*, 2001).

4.8. The 'Dusty' Atmosphere

MARSIS measures the altitude and magnitude of the electron density maximum. The maximum is located in the photo ionization region, is essentially controlled by the neutral air density in that region. The Martian neutral atmosphere is subject to turbulent mixing up to an altitude of 125 to 150 km, called the homopause (Izakov, 2001). Below the homopause the ratio of neutral components is constant owing to turbulent mixing, and above this altitude the concentration of each component varies with altitude according to its scale height determined by the molecular mass of the component (Murphy *et al.*, 1990; Stewart and Hanson, 1982). The homopause is defined as the altitude where the molecular and turbulent diffusion coefficients are equal. During dusts storms it was found that the neutral temperature increased in the upper atmosphere (15 to 20 deg, or as in a recent dust storm in

August/September 2001, up to 45 deg), owing to increased solar energy absorption by the dust, while the temperature near the surface, being blanketed by the dust, decreased. It has been proposed that the heating of the atmosphere leads to an expansion of the atmospheric column, and the decreased temperature gradient to an increase of the turbulent diffusion coefficient. The two effects combined may cause an increase in altitude of the homopause. The expansion of the atmosphere and the displacement of the homopause to a higher altitude is associated with an increase of the neutral air density, which affects the altitude distribution of energy precipitated by solar radiation and charged particles. The increase in neutral air density during dust storms causes the solar induced ionization rate profiles, or ion/electron production rate profiles, to be shifted to higher altitudes. This leads to an increase of the altitude where the solar induced (photoionization) electron density is maximum (Stephen *et al.*, 2001; Bougher *et al.*, 2001; Zank *et al.*, 1990b; McElroy *et al.*, 1977).

Direct interaction between the solar wind and the atmosphere/ionosphere is associated with a significant mass loss to the solar wind (mass loading). Ions are picked-up by the solar wind and accelerated. However, some accelerated ions are subsequently neutralized; some of these neutral atoms are lost to the planet, but others will precipitate into the atmosphere influencing the ionization altitude profile.

4.9. Nightside Ionosphere

For solar zenith angles larger than 90° the night side electron density peaks at an altitude between ~ 140 and 180 km, has a peak density between about ~ 2 and $\sim 30 \times 103$ el cm^{-1} near the terminator and the density decreases towards very low values near 125-deg zenith angle (the limit of current observations). The density layer has a width of roughly 20 km.

Mars' nightside ionosphere is thought to have its main source in transport of plasma from the dayside. As long as the altitude region of transport (~ 250 km) is below the ionopause, or rather below the altitude of the solar wind, then a nightside ionosphere can be generated by transport. Thus, if the ionopause is at a low altitude, say owing to strong pressure of the solar wind, then the nightside ionosphere will disappear. Alternatively, electron precipitation sources (the solar wind) could contribute to the nightside peak density (Zang *et al.*, 1990c). Note that meteoric ions are long lived, and can survive transport from the dayside into the nightside. If that also is the case at Mars, and if meteoric layers exists there, then high radio wave absorption amy also be expected during martian night.

5. Summary

Mars Express is designed to contribute to evaluating:

(1) the global vertical electron density profiles,

(2) the global profiles of neutral air densities and temperature,

(3) the global atmospheric circulation,

(4) the global water and ozone column distribution, and

(5) to search for and identify trace elements and determine how the neutral air composition varies in place and time

The Observations can be used to address many of the issues treated in this report. The increase in detail of vertical electron density profiles and their variations with longitude, latitude, and solar zenith angle, can together with the spatial variations of the crustal magnetic field be used to better understand the interaction of the solar wind with the planet. MARSIS can for the first time measure electron density profiles in the regions located at silar zenith angles less than 45 deg or more than 135 deg from the sub-solar region. Measurements on the nightside will be used to search for proposed fine structures ('holes') in the electron density. Hydrodynamic waves are predicted to be excited by the solar wind/ionosphere interaction and their influence on the electron densities can be observed by MARSIS. The observed neutral density and temperature profiles can be used to understand the details of the vertical variations of the electron density. This includes testing for effects induced on the electron density profiles by dust storms and by precipitations of solar wind ions in cusp-like magnetic regions. The search for trace elements may reveal magnesium ions in the predicted meteoric layer.

References

Acuña, M. H., Connerney, J. E. P., Wasilewski, P., Lin, R. P., Anderson, K. A., Carlson, C. W., McFadden, J., Curtis, D. W., Mitchell, D., Reme, H., Mazelle, C., Sauvaud, J. A., D'Uston, C., Cros, A., Medale, J. L., Bauer, S. J., Cloutier, P. A., Mayhew, M., Winterhalter, D., and Ness, N. F.: 1998, 'Magnetic Field and Plasma, Observations at Mars: Initial Rersults of the Mars Global Observer Mission', *Science* **279**, 1676.

Acuna, M. H., Connerney, J. E. P., Ness, N. F., Lin, R. P., Mitchell, D., Carlson, C. W., McFadden, J., Anderson, K. A., Reme, H., Mazelle, C., Vignes, D., Wasilewski, P., and Cloutier, P. A.: 1999, 'Global Distribution of Crustal Magnetization Discovered by Mars Global Observer', *Science* **284**, 790.

Acuna, M. H., Connerney, J. E. P., Wasilewski, P., Lin, R. P., Mitchell, D., Anderson, K. A., Carlson, C. W., McFadden, J., Reme, H., Mazelle, C., Vignes, D., Bauer, S. J., Cloutier, P. A., and Ness, N. F.: 2001, 'Magnetic Field of Mars: Summary of Results from the Aerobraking and Mapping Orbits', *J. Geophys. Res.* **106**, 23404.

Bauer, S. J. and Hantsch, M. H.: 1989, 'Solar Cycle Variation of the Upper Atmosphere of Mars', *Geophys. Res. Lett.* **16**, 373.

Brace, L. H., Theis, R. F., Hoegy, W. R., Wolfe, J. H., Mihalov, J. D., Russel, C. T., Elphic, R. C., and Nagy, A. F.: 1980, 'The Dynamic Behaviour of the Venus Ionosphere in Response to Solar Wind Interactions', *J. Geophys. Res.* **85**, 7663.

Bougher, S. W., Engel, S., Hinson, D. P., and Forbes, J. M.: 2001, 'Mars Global Surveyor Radio Science Electron Density Profiles: Neutral Atmosphere Implications', *Geophys. Res. Lett.* **28**, 3091.

Breus, T. K. and Krymskii, A. M.: 1987, 'Some Problems of the Solar Wind Interaction with Venus', *Planetary Space Sci.* **35**, 1213.

Connerney, J. E. P., Acuña, M. H., Wasilewski, P., Ness, N. F., Reme, H., Mazelle, C., Vignes, D., Lin, R. P., Mitchell, D., and Cloutier, P. A.: 1999, 'Magnetic Lineations in the Ancient Crust of Mars', *Science* **284**, 794.

Fox, J. L., Zhou, P., and Bougher, S. W.: 1996, 'The Martian Thermosphere/Ionosphere at High and Low Solar Activities', *Adv. Space Res.* **17**(11), 203–218.

Hake, R. D. and Phelps, A. V.: 1967, 'Momentum-Transfer and Inelastic-Collision Cross Section for Electrons in O_2, CO, and CO_2', *Physical Rev.* **158**, 70.

Hanson, W. B. and Mantas, G. P.: 1988, 'Viking Electron Temperature Measurements: Evidence for a Magnetic Field in the Martian Ionosphere', *J. Geophys. Res.* **93**, 7538.

Hanson, W. B., Sanatani, S. and Zucarro, D. R.: 1977, 'The Martian Ionosphere as Observed by the Viking Retarding Potential Analyzers', *J. Geophys. Res.* **82**, 4317.

Hantsch, M. H. and Bauer, S. J.: 1990, 'Solar Control of the Mars Ionosphere', *Planetary Space Sci.* **38**, 539.

Izakov, M. N.: 2001, 'Turbulence and Anomalous Heat Fluxes in the Atmospheres of Mars and Venus', *Planetary Space Sci.* **49**, 47.

Johnson, F. S. and Hanson, W. B.: 1979, 'A New concept for the Daytime Magnetosphere of Venus', *Geophys. Res. Lett.* **6**, 581.

Kliore, A. J., Cain, D. L., Fjeldbo, G., Seidel, B. L., Sykes, M. J., and Rasool, S. I.: 1972, 'The Atmosphere of Mars from Mariner 9 Radio Occultation Measurements', *Icarus* **17**, 484.

Krymskii, A. M., Breus, T. and Nielsen, E.: 1995, On Possible Observational Evidence in Electron Density Profiles of a Magnetic Field in the Martian Ionosphere', *J. Geophys. Res.* **100**, 3721.

Lin, R. P., Mitchell, D. L., Curtis, D. W., Anderson, K. A., Carlson, C. W., McFadden, J., Acuna, M. H., Hood, L. L. and Binder, A.: 1998, 'Lunar Surface Magnetic Fields and Their Interaction with the Solar Wind: Results from Lunar Prospector', *Science* **281**, 1480.

Luhman, J. G.: 1986, 'The Solar Wind Interaction with Venus', *Space Sci. Rev.* **44**, 241.

Luhman, J. G., Russell, C. T., Scarf, F. L., Brace, L. H. and Knudsen, W. C.: 1987, 'Characteristics of the Marslike Limit of the Venus-Solar Wind Interaction', *J. Geophys. Res.* **92**, 8545.

McElroy, M. B., Kong, T. Y. and Yung, V. L.: 1977, 'Photochemistry and Evolution of Mars' Atmosphere: A Viking Perspective', *J. Geophys. Res.* **82**, 4379.

Mitchell, D. L., Lin, R. P., Reme, H., Crider, D. H., Cloutier, P. A., Connerney, J. E. P., Acuña, M. H. and Ness, N. F.: 2000, 'Oxygen Auger Electrons Observed on Mars', *Geophys. Res. Lett.* **27**(13), 1871.

Mitchell, D. C., Lin, R. P., Mazelle, C., Reme, H., Cloutier, P. A., Connerney, J. E. P., Acuña, M. H. and Ness, N. F.: 2001, 'Probing Mars's Crustal Magnetic Field and Ionosphere with MGS Electron Reflectometer', *J. Geophys. Res.* **106**(E10), 23419.

Murphy, J. R., Toon, O. B., Haberle, R. M. and Pollack, J. B.: 1990, 'Numerical Simulations of the Decay of Martian Global Dust Storms', *J. Geophys. Res.* **95**, 14629.

Ness, N. F., Acuna, M. H., Connerney, J. E. P., Kliore, A. J., Breus, T. K., Krymskii, A. M., Cloutier, P. and Bauer, S. J.: 2000, 'Effects of Magnetic Anomalies Discovered at Mars on the Structure of the Martian Ionosphere and Solar Wind Interaction as Follows from Radio Occultation Experiments', *J. Geophys. Res.* **105**(A7), 15991.

Nielsen, E.: 1998, Mars Riometer System. *Proc. 32nd ESLAB Symp. Remote Sensing Methodology for Earth Observations and Planetary Exploration*, ESA SP-423, ESTEC, Noordwijk, The Netherlands.

Pesnell, W. D. and Grebowsky, J.: 2000, 'Meteoric Magnesium Ions in the Martian Atmosphere', *J. Geophys. Res.* **105**, 1695.

Phillips, J. L., Luhmann, J. G. and Russell, C. T.: 1984, 'Growth and Maintenance of Large-Scale Magnetic Fields in the Dayside Venus Ionosphere', *J. Geophys. Res.* **89**, 10676.

Picardi, G., Sorge, S., Seu, R., Fedele, G., Frederico, C. and Orosei, R.: 1999, *MARSIS: Models and System Analysis*, INFO-COM, University of Rome, Technical Report, MRS-001/005/99.

Schunk, R. W. and Nagy, A. F.: 1980, 'Ionospheres of the Terrestrial Planets', *Rev. Geophys. and Space Phys.* **18**(4), 813.

Sen, H. K. and Wyller, A.: 1960, 'On the Generalization of the Appleton-Hartree Magnetoionic Formulas', *J. Geophys. Res.* **65**, 3931.

Shinagawa, H. and Cravens, T. E.: 1988, 'A One-Dimensional Multispecies Magnetohydromagnetic Model of the Dayside Ionosphere of Venus', *J. Geophys. Res.* **93**, 11263.

Shinagawa, H. and Cravens, T. E.: 1989, 'A One-Dimensional Multispec Magnetohydrodynamical Model of the Dayside Ionosphere of Mars,' *J. Geophys. Res.* **94**, 6506.

Slavin, J. A. and Holzer, R. E.: 1982, 'The Solar Wind Interaction with Mars Revisited', *J. Geophys. Res.* **87**, 10285.

Stewart, A. J. and Hanson, W. B.: 1982, 'Mars Upper Atmosphere: Mean and Variations', *Adv. Space Res.* **2**, 87.

Wang, J.-S. and Nielsen, E.: 2002, 'Possible Hydrodynamic Waves in the Topside Ionospheres of Mars and Venus', *J. Geophys. Res.*, in press.

Wang, J.-S. and Nielsen, E.: 2001, 'Numerical Simulation of One-Dimensional Hydrodynamic Waves in Mars Topside Ionosphere', *J. Geophys. Res.*, submitted.

Witasse, O., Nouvel, J.-F., Lebreton, J.-P. and Kofman, W.: 2001, 'HF Radio Wave Atteniuation Due to a Meteoric Layer in the Atmosphere of Mars', *Geophys. Res. Lett.* **28**, 3039.

Zhang, M. H. G., Bauer, S. J. and Luhmann, J. G.: 1990a, *Ionogram Analysis for a Mars Topside Sounder*, Report of Space Research Institute of the Austrian Academy of Sciences, University of Graz.

Zhang, M. H. G., Luhmann, J. G., Kliore, A. J. and Kim, J.: 1990b, 'A Post-Pioneer Venus Reassessment of the Martian Dayside Ionosphere as Observed by Radio Occultation Methods', *J. Geophys. Res.* **95**, 14829.

Zhang, M. H. G., Luhmann, J. G. and Kliore, A. J.: 1990c, 'An Observational Study of the Nightside Ionosphere of Mars and Venus with Radio Occultation Methods', *J. Geophys. Res.* **95**, 17095.

LIST OF PARTICIPANTS

Mario Acuña	Goddard Space Flight Center/NASA, Greenbelt, U.S.A.
Ian Axford	Max Planck Institut, Lindau, Germany
Klaus Baumgaertel	Max Planck Institut, Lindau, Germany
Siegfried Bauer	University of Graz, Graz, Austria
Cesar Bertucci	Centre d'Etude Spatiale des Rayonnements/CNRS, Toulouse, France
David Brain	University of Colorado, Boulder, U.S.A.
Stephen Brecht	Bay Area Res Corp, Orinda, CA 94563, U.S.A.
Tamara Breus	Space Research Institute, Moscow, Russia
Jack Connerney	Goddard Space Flight Center/NASA, Greenbelt, U.S.A.
Tom Cravens	University of Kansas, Lawrence, U.S.A.
Dana Crider	Goddard Space Flight Center/NASA, Greenbelt, U.S.A.
Magda Delva	Österr. Akademie der Wissenschaften, Graz, Austria
Eduard Dubinin	Max Planck Institut fur Aeronomie, Katlenburg-Lindau, Germany
Mats Holmstrom	Swedish Institute of Space Physics, Kiruna, Sweden
Galina A. Kotova	Space Research Institute, Moscow, Russia
Esa Kallio	Finnish Meteorological Institute, Geophysical Research, Helsinki, Finland
Alexander M. Krymskii	Rostov State University, Rostov-on-Don, Russia
Robert Lin	University of California, Berkeley, U.S.A.
Kiyoshi Maezawa	Institute of Space and Astronautical Science, Tokyo, Japan
Christian Mazelle	Centre d'Etude Spatiale des Rayonnements/CNRS, Toulouse, France
David Mitchell	University of California, Berkeley, U.S.A.
Diedrich Möhlmann	Deutsches Zentrum für Luft- und Raumfahrt, Cologne, Germany
Andy Nagy	University of Michigan, Ann Arbor, U.S.A.
Norman Ness	Bartol Research Institute, Newark, U.S.A.
Erling Nielsen	Max Planck Institut, Lindau, Germany
Marit Oieroset	University of California, Berkeley, U.S.A.
Henri Reme	Centre d'Etude Spatiale des Rayonnements/CNRS, Toulouse, France
Konrad Sauer	Max Planck Institut, Lindau, Germany
Gerald Schubert	University of California, Los Angeles, U.S.A.
Jim Slavin	Goddard Space Flight Center/NASA, Greenbelt, U.S.A.
Tilman Spohn	Westfälische Wilhelms-Universität, Münster, Germany
Jean-Gabriel Trotignon	LPCE/CNRS, Orleans, France
Misha I. Verigin	Russian Academy of Sciences, Moscow, Russia
Didier Vignes	Goddard Space Flight Center/NASA, Greenbelt, U.S.A.
Daniel Winterhalter	Jet Propulsion Laboratory/Caltech, Pasadena, U.S.A.
Alexander Zakharov	Space Research Institute, Moscow, Russia

Space Science Series of ISSI

1. R. von Steiger, R. Lallement and M.A. Lee (eds.): *The Heliosphere in the Local Insterstellar Medium.* 1996 ISBN 0-7923-4320-4
2. B. Hultqvist and M. Øieroset (eds.): *Transport Across the Boundaries of the Magnetosphere.* 1997 ISBN 0-7923-4788-9
3. L.A. Fisk, J.R. Jokipii, G.M. Simnett, R. von Steiger and K.-P. Wenzel (eds.): *Cosmic Rays in the Heliospehre.* 1998 ISBN 0-7923-5069-3
4. N. Prantzos, M. Tosi and R. von Steiger (eds.): *Primordial Nuclei and Their Galactic Evolution.* 1998 ISBN 0-7923-5114-2
5. C. Fröhlich, M.C.E. Huber, S.K. Solanki and R. von Steiger (eds.): *Solar Composition and its Evolution – From Core to Corona.* 1998 ISBN 0-7923-5496-6
6. B. Hultqvist, M. Øieroset, Goetz Paschmann and R. Treumann (eds.): *Magnetospheric Plasma Sources and Losses.* 1999 ISBN 0-7923-5846-5
7. A. Balogh, J.T. Gosling, J.R. Jokipii, R. Kallenbach and H. Kunow (eds.): *Co-rotating Interaction Regions.* 1999 ISBN 0-7923-6080-X
8. K. Altwegg, P. Ehrenfreund, J. Geiss and W. Huebner (eds.): *Composition and Origin of Cometary Materials.* 1999 ISBN 0-7923-6154-7
9. W. Benz, R. Kallenbach and G.W. Lugmair (eds.): *From Dust to Terrestrial Planets.* 2000 ISBN 0-7923-6467-8
10. J.W. Bieber, E. Eroshenko, P. Evenson, E.O. Flückiger and R. Kallenbach (eds.): *Cosmic Rays and Earth.* 2000 ISBN 0-7923-6712-X
11. E. Friis-Christensen, C. Fröhlich, J.D. Haigh, M. Schüssler and R. von Steiger (eds.): *Solar Variability and Climate.* 2000 ISBN 0-7923-6741-3
12. R. Kallenbach, J. Geiss and W.K. Hartmann (eds.): *Chronology and Evolution of Mars.* 2001 ISBN -7923-7051-1
13. R. Diehl, E. Parizot, R. Kallenbach and R. von Steiger (eds.): *The Astrophysics of Galactic Cosmic Rays.* 2001 ISBN 0-7923-7051-1
14. Ph. Jetzer, K. Pretzl and R. von Steiger (eds.): *Matter in the Universe.* 2001 ISBN 1-4020-0666-7
15. G. Paschmann, S. Haaland and R. Treumann (eds.): *Auroral Plasma Physics.* 2002 ISBN 1-4020-0963-1
16. R. Kallenbach, T. Encrenaz, J. Geiss, K. Mauersberger, T.C. Owen and F. Robert (eds.): *Solar System History from Isotopic Signatures of Volatile Elements.* 2003 ISBN 1-4020-1177-6
17. G. Beutler, M.R. Drinkwater, R. Rummel and R. Von Steiger (eds.): *Earth Gravity Field from Space – from Sensors to Earth Sciences.* 2003 ISBN 1-4020-1408-2
18. D. Winterhalter, M. Acuña and A. Zakharov (eds.): *"Mars" Magnetism and its Interaction with the Solar Wind.* 2004 ISBN 1-4020-2048-1

Kluwer Academic Publishers – Dordrecht / Boston / London

GPSR Compliance

The European Union's (EU) General Product Safety Regulation (GPSR) is a set of rules that requires consumer products to be safe and our obligations to ensure this.

If you have any concerns about our products, you can contact us on ProductSafety@springernature.com

In case Publisher is established outside the EU, the EU authorized representative is:

Springer Nature Customer Service Center GmbH
Europaplatz 3
69115 Heidelberg, Germany

Batch number: 10823959

Printed by Printforce, the Netherlands